Advances in Reservoir Simulation

Advances in Reservoir Simulation

Guest Editors

Haifeng Zhao
Yang Xia

MDPI

Basel • Beijing • Wuhan • Barcelona • Belgrade • Novi Sad • Cluj • Manchester

Guest Editors
Haifeng Zhao
School of Petroleum Engineering
China University of Petroleum
Beijing
China

Yang Xia
School of Petroleum Engineering
China University of Petroleum
Beijing
China

Editorial Office
MDPI AG
Grosspeteranlage 5
4052 Basel, Switzerland

This is a reprint of the Special Issue, published open access by the journal *Energies* (ISSN 1996-1073), freely accessible at: https://www.mdpi.com/journal/energies/special_issues/Reservoir_Simulation.

For citation purposes, cite each article independently as indicated on the article page online and as indicated below:

Lastname, A.A.; Lastname, B.B. Article Title. *Journal Name* **Year**, *Volume Number*, Page Range.

ISBN 978-3-7258-3816-5 (Hbk)
ISBN 978-3-7258-3815-8 (PDF)
https://doi.org/10.3390/books978-3-7258-3815-8

Contents

About the Editors

Haifeng Zhao

School of Petroleum Engineering, China University of Petroleum, Beijing 102249, China. Interests: tight oil and gas and shale gas volume fracturing technology; coalbed methane drilling and fracturing technology; carbonate rock acid fracturing technology; wellbore stability mechanics and chemistry; oil and gas well sand production prediction and prevention.

Yang Xia

School of Petroleum Engineering, China University of Petroleum, Beijing 102249, China. Interests: tight oil and gas reservoir simulations; shale gas flow; production prediction of unconventional reservoirs; fluid–solid coupling of unconventional reservoirs; fracturing technology of unconventional reservoirs.

Preface

This volume advances numerical simulation, data assimilation, and multi-physics coupling to address reservoir heterogeneity, fracture dynamics, and operational safety. Ensemble-based algorithms like ES-MDA reduce reservoir uncertainties by 15–20% through 4D seismic and production data integration, as demonstrated by 8% recovery gains in Middle East carbonates. Fracture studies combine hydro-mechanical models with true triaxial experiments on hydrate-bearing sediments, guiding discrete fracture network (DFN)-optimized completions that boost shale gas productivity by 40% in China's Sichuan Basin. To enhance computational efficiency, proxy models such as INSIM-FT accelerate interwell connectivity analysis by 90%, while physics-informed neural networks (PINNs) embed flow equations into AI architectures for rapid production forecasting. Machine learning (XGBoost) resolves permeability heterogeneity with a 92% accuracy in carbonates, and well-test analyses quantify fracture-driven permeability contrasts exceeding 1,000x.

Geomechanical integrity frameworks address salt cavern storage risks, where coupled creep-thermal stress modeling reduced cement sheath failure risks by 60% and poro-elastic simulators optimize fracture networks under depletion. Cross-disciplinary trends fuse data-driven and physics-based models, enabling digital twins for real-time field management. Challenges remain in proppant transport modeling and sparse 4D data utilization. Emerging priorities include quantum computing for fracture simulation and adapting reservoir engineering to carbon sequestration, positioning the field as critical for sustainable energy transitions.

Haifeng Zhao and Yang Xia

Guest Editors

energies

MDPI

Article

Four-Dimensional History Matching Using ES-MDA and Flow-Based Distance-to-Front Measurement

Eduardo Barrela [1,*], Philippe Berthet [1,*], Mario Trani [1], Olivier Thual [2] and Corentin Lapeyre [2]

[1] TotalEnergies S.E.—Centre Scientifique & Technique Jean Féger, Av. Larribau, 64000 Pau, France
[2] Centre Européen de Recherche et de Formation Avancée en Calcul Scientifique, 42 Av. Gaspard Coriolis, 31100 Toulouse, France
* Correspondence: eduardo-jose.airoso-barrela@totalenergies.com (E.B.); philippe.berthet@totalenergies.com (P.B.)

Abstract: The use of 4D seismic data in history matching has been a topic of great interest in the hydrocarbon industry as it can provide important information regarding changes in subsurfaces caused by fluid substitution and other factors where well data is not available. However, the high dimensionality and uncertainty associated with seismic data make its integration into the history-matching process a challenging task. Methods for adequate data reduction have been proposed in the past, but most address 4D information mismatch from a purely mathematical or image distance-based standpoint. In this study, we propose a quantitative and flow-based approach for integrating 4D seismic data into the history-matching process. By introducing a novel distance parametrization technique for measuring front mismatch information using streamlines, we address the problem from a flow-based standpoint; at the same time, we maintain the amount of necessary front data at a reduced and manageable amount. The proposed method is tested, and its results are compared on a synthetic case against another traditional method based on the Hausdorff distance. The effectiveness of the method is also demonstrated on a semi-synthetic model based on a real-case scenario, where the standard Hausdorff methodology could not be applied due to high data dimensionality.

Keywords: four-dimensional seismic; history matching; ensemble smoother with multiple data assimilation; distance-to-front; streamlines

Citation: Barrela, E.; Berthet, P.; Trani, M.; Thual, O.; Lapeyre, C. Four-Dimensional History Matching Using ES-MDA and Flow-Based Distance-to-Front Measurement. *Energies* **2023**, *16*, 7984. https://doi.org/10.3390/en16247984

Academic Editors: Haifeng Zhao and Yang Xia

Received: 2 October 2023
Revised: 20 November 2023
Accepted: 25 November 2023
Published: 9 December 2023

1. Introduction

In this paper, we aim to explore and demonstrate the effectiveness of front re-parametrization methods in the context of history matching for hydrocarbon reservoirs. Specifically, we focus on a novel distance-to-front method using streamlines obtained from full physics flow simulation aimed at enhancing the accuracy and efficiency of history-matching processes in reservoir management. History matching is an important stage in the development of every hydrocarbon field, playing a crucial role in both the modeling and simulation phases. During the process of history matching, key properties of the reservoir model are calibrated to match past production data. This not only allows inference of reservoir properties from production data but also, ultimately, allows obtaining predictive model(s) that reduce uncertainty in forecasting future production. The end goal is to use predictive reservoir models to aid the development of future production strategies and support management decisions. The reliability of a reservoir model can be estimated when it is able to reproduce all available data as accurately as possible. Currently, many oil and gas fields have long historical data that can help constrain the history matching procedure, as well as seismic data acquired during their production life.

The incorporation of time-lapse seismic data (4D seismic data) into the history-matching workflow has been a topic of great interest over the years. Four-dimensional seismic data can provide information in areas of the reservoir where no data are available; more specifically, the data can be used as a tool to monitor changes in the subsurface originating from fluid substitution due to production [1–5] or changes in other subsurface properties such as temperature or pressure [6]. These insights are vital for enhancing the accuracy of history matching by aiding in more precisely calibrating reservoir models to reflect actual conditions. Thus, the integration of 4D seismic data plays a pivotal role in overcoming the limitations of conventional data, offering a more comprehensive understanding of the reservoir and ensuring more reliable and effective history-matching outcomes.

From an industry standpoint, 4D seismic data have been primarily used as a qualitative constraint on the reservoir model [7–12] as the understanding of the reservoir evolves and revisions are made in a multidisciplinary framework, encompassing the domains of geology, seismic, and reservoir engineering [13]. However, while it is possible to identify the changes occurring on the subsurface, like saturation and pressure variations (e.g., [14,15]), using 4D data in a qualitative way, we have no information on the magnitude of the change or with what certainty.

On top of that, seismic interpretation can be subject to variations in its interpretation, even when facing easy-to-interpret seismic data. In a study by Rankey and Mitchell [16], the authors focus on the subjectivity of seismic data interpretation by presenting the same seismic data to different interpreters. The interesting conclusion was that although the data were unanimously considered easy to interpret, overconfidence in the interpreters led to variations in the interpretation, which in turn reflected considerable differences in volumetrics estimation. Apart from that, other uncertainties arising from non-repeatability effects [17], noise, and imaging are present in 4D seismic data. On the plus side, this can provide a great opportunity for the quantitative integration of 4D seismic data in history-matching workflows as it allows access to uncertainty quantification over model parameter estimates and fluid production forecasts analysis (e.g., [18–21]). Advancements over the past decades, leading to the proposal and development of a range of stochastic seismic inversion techniques, have provided ways of generating an ensemble of alternative heterogeneous impedance representations that agree with the 3D seismic volume, accounting for the non-uniqueness of the inversion process. Nonetheless, the full integration of quantitative 4D seismic data interpretation into the history-matching procedure is far from straightforward, remaining a challenge to be addressed as well as a topic of great interest. Several examples of quantitative approaches for integrating 4D seismic data exist in the literature (e.g., [10,22–31]). Problems identified with the adoption of such techniques on an industrial level are related to the practical feasibility and the inexistence of a fully integrated software solution that can easily handle the integration of both production and seismic data in a computer-aided history matching loop. However, the main concern is related to the computational feasibility of incorporating the large amount of data associated with seismic acquisition into existing workflows [32]. In particular, aspects related to the high nonlinearity of the problem at hand include the limitation on the number of degrees of freedom associated with the amount of data to be assimilated, questions on how to address the contribution in the assimilation procedure of the different types of data in consideration, how to elect relevant parameters for matching both seismic data and fluid flow production while staying within the boundaries of a plausible geological and physical model, over-conservative prior assumptions, and errors arising from the forward modelling of seismic data and attributes [33]. Therefore, due to the complexity and computational cost associated with the modeling of seismic attributes, presented workflows often fall into the categories of either being able to provide only unique solutions to the problem or requiring a significant reduction of the uncertainty space [34]. For this reason, alternative methods for adding seismic data information to the history-matching procedure have been explored. Recently, Rollmann et al. [35] presented a method using a convolutional neural network trained to fit observed seismic history. However, results were only shown in a synthetic case, and the overhead cost of gathering the necessary amount of training data (large amounts of models that need to be classified) as well the time spent in appropriate architecture development (which can be very case-specific) and computational costs associated with the training of the network are still big disadvantages. Furthermore, the integration of such methods into the 4D history-matching workflow carries additional challenges (especially under real or realistic case scenarios), remaining a topic for future research.

In an attempt to address the problem related to the varying nature of data within the 4D seismic history matching procedure, Tillier et al. [36] proposed a method based on the local modified Hausdorff distance for measuring the dissimilarity between observed and simulated seismic attributes. This idea was later expanded on by Abadpour et al. [37], where coupled with an Ensemble Kalman Filter (EnKF) [38] workflow, the same Hausdorff distance metric was used to compute the distance between observed and saturation fronts, synthesized by binary image-based data reduction. The method showed promising results in a synthetic case; however, unless some other precursory data reduction steps are considered, the direct application of this method, particularly within the scope of ensemble-based workflows, becomes difficult to achieve in large-scale models as it implies computing the inverse of a gain matrix that could be at least the square of the number of cells in the model. There are also other issues related to binary image-based approaches, such as the double-penalty effect (i.e., when a feature is predicted where it should not be and is not predicted where it should), which, together with the amount and extension of data, add complexity to the minimization procedure even if the mismatch can be easily quantifiable.

Other binary image-based methods have also been proposed over the past decade (e.g., [39–43]). Similarly, such methods mainly focus on the conversion of hardening and softening signals of 4D seismic data to a binary image. However, while providing an effective and straightforward way of integrating 4D seismic information by reducing the level of information in a continuous 4D signal into discrete states (0 and 1), the main drawback is still related to the amount of data resulting from the computation of the difference between observed and measured responses.

Observed seismic amplitude fronts can be compared to saturation fronts when under cases where pressure variation or compaction effects [1,34] or the effect of variation in porosity or net-to-gross ratio can be neglected [33]. These can capture the main information related to the drained area of a reservoir under production. Like binary image-based methods, they can be looked at as a solution for applying data reduction to the problem at hand. This type of front re-parametrization reduces the amount of seismic data to be used by representing swept regions through a saturation front, which can capture the most significant 4D seismic information. A reduction in the nonlinearity of the problem is also achieved since front positions are closely related to uncertain petrophysical parameters of the model (before or after the front location). Finally, full seismic inversion procedures are avoided, and the method remains an option even when facing low-quality seismic data sets. All of these qualities place such methods as good candidates for application on history-matching workflows with the potential for increasing the performance of history-matching workflows.

Kretz [44] proposed a history-matching workflow to match front positions based on streamlined simulation. In their proposal, model permeabilities along the streamlines were modified in order to match observed and simulated saturation front positions derived from 4D seismic data. The discrepancy between front positions was provided by the difference in time-of-flight measured from the streamlines. The method showed great promise and served as a starting point for research on other front parametrization-based methods used in 4D history-matching workflows (e.g., [34,39,45,46]). However, the main drawback with this method relates to the properties being updated only along cells intersected by the path of the streamlines. This leads to models that lose their geologic consistency, no longer honoring geostatistical assumptions, and could also lead to overfitting of the matched production data. Finally, the application of the proposed method to realistic 3D cases was not discussed, and results were presented only in simple 2D synthetic examples.

Trani et al. [34] proposed the re-parameterization of saturations extracted from 4D seismic data in terms of front arrival times. The main disadvantage of this method was the need to run the fluid flow simulations beyond the update time at which the seismic data are available (i.e., until all cells of the observed front location have been flooded). Later, Trani [47,48] concluded that an ensemble of complete predictions with extended simulation times could be replaced by an approximation of late arrival time by an arbitrarily large value. In an attempt to address the problem with extended simulation times and based on the work by Tillier et al. [36], Leeuwenburgh and Arts [45] and Zhang and Leeuwenburgh [46,47] proposed re-parameterizating front time-of-arrivals into distance-to-fronts. In the work presented by Leeuwenburgh and Arts [45], the authors assumed that uniform velocities in a monotonically expanding 3D front could be converted to distances calculated using a fast-marching algorithm [49] for the solution of Eikonal equations to cartesian grids [50]. These computed distances were then used as innovations in an ensemble history-matching framework using the EnKF. Later, Zhang and Leeuwenburgh [46,47] proposed an improvement over the method presented by Leeuwenburgh and Arts [45], where the use of the fast-marching method was extended for applicability on corner point grids, thereby improving the overall accuracy of the method. They presented their method in a simple 2D synthetic case and applied it to the Norne field using the ensemble smoother with multiple data assimilation (ES-MDA) [51], Appendix A.

In this paper, we focus on the simplicity and applicability of front re-parametrization methods and present a flow-based alternative to previous distance-to-front methods. For this, based on the work presented by Kretz [44], we propose a distance-to-front method using streamlines obtained from full physics flow simulation.

In the following section, we introduce the concept of distance-to-front measurement (Section 2.1), then we introduce fluid flow streamlines as a post-processing of flow simulation and propose a method for calculating distances to fronts using streamline information obtained from full-physics flow simulation (Section 2.2). We then present a set of numerical experiments on a 2D synthetic case (Section 3.1) and a realistic 3D case based on a real case scenario (Section 3.2), where the proposed method is applied and the obtained results are discussed. Finally, the main conclusions of the work are presented in Section 4.

2. Methodology

2.1. Distance-to-Front Measurement

Measuring the distance to a front requires front extraction, normally from a seismic attribute that can capture spatial changes in saturation or pressure over the subsurface. These changes can often refer to a timelapse confirmation on features related to the displacement of oil by water and/or gas, dissolution effects, or significant pressure changes. By posing the problem in this way, we assume that the shape or boundary of these features is enough to capture the relevant phenomena as opposed to using the original individual cell amplitudes of 4D seismic data. Added to that, reducing the information to a relevant shape or boundary can often be more informative and reliable while being an advantage in terms of computational cost and efficiency.

Normally, within the feature-based 4D history-matching domain, grid-based geometrical distance measurements are used (e.g., Euclidean, Hamming, fast-marching methods, Chain-Vese, Hausdorff, etc.). However, one should address the choice of measuring subsurface changes from a dynamic perspective, as subsurface changes derive from dynamic mechanisms related to fluid production and subsurface geology.

2.2. Using Streamlines for Distance-to-Front Calculation

In this regard, streamlines can be a solution to link both subsurface geology and fluid production. Streamlines and streamline-based properties are valuable tools for understanding reservoir connectivity and fluid flow patterns for large, heterogeneous models and can be easily obtained through the post-processing of full-physic simulations. In order to define a fluid flow streamline, three key properties are necessary: flow rate, time of flight, and a cell ID pointing to a given cell in the reservoir grid. The grid cell ID is used to extract relevant information from the grid and map solution variables between the streamline and the global numerical grid.

Streamline geometries (used for visualization purposes only) are then obtained through the Pollock method [52,53], starting from a single cell of the model with the calculation of a flow rate calculated for each of the cell faces (assumed to be uniform along the faces). Along with pressure, the total flow rate in and out of each of the faces can then be calculated based on the total Darcy velocity:

$$\vec{v}_t = -\lambda_t \vec{\nabla} P_0 + \left(\sum j \lambda_j \rho_j\right) \vec{g}, \quad (1)$$

where $\vec{v}$ is the flow velocity, λ is the phase mobility, $\vec{\nabla}P$ is the pressure gradient, ρ is the mass density, and $\vec{g}$ is the acceleration due to gravity.

To conform with the orthogonal grid assumptions of the Pollock method (Figure 1), an isoparametric transformation is applied to all of the grid cells of the model onto a unit cube grid [54]. Consequently, for visualization purposes, this representation can be transformed back into the spatial coordinate system (e.g., corner point gridding format). This way, by being able to obtain a set of streamlines branching all cells of the model where flow is occurring from a given source to a sink, we can use streamlines as a flow-based real coordinate system medium to measure the distance between the true location of water-saturated fronts and simulated saturation fronts obtained from the flow simulation of candidate models.

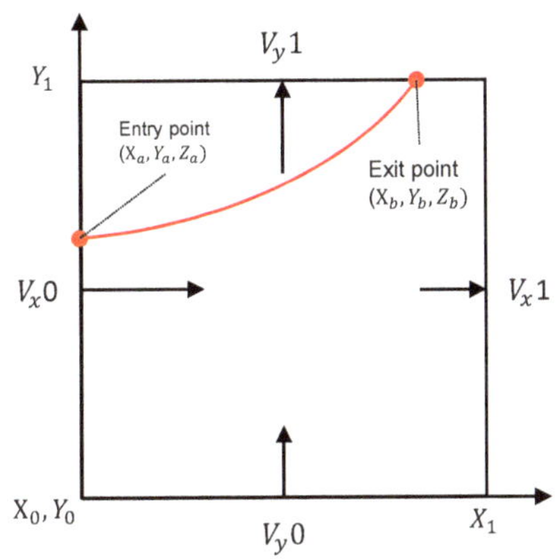

Figure 1. Streamline tracing scheme where a single streamline enters the Y0 face of a single cell and exits the X1 face through the exit point. The curvature along the cell is guided by the calculated in/out flow velocities of all faces of the cell.

As shown in Figure 2a, a simple 2D five-spot reservoir model was used to illustrate the method. Two fronts are presented, one originating from either the inversion or interpretation of 4D seismic data (observed front, in black) and a second one representing the same front on a given candidate model after flow simulation (simulated front, in grey). By computing the fluid-flow streamlines over the simulation period, we obtained a set of streamlines that connects both fronts. This way, it became possible to obtain the distance between both features using flow-based distance measurement supported by streamlines and to use the computed distance vector as an innovation in the history-matching workflow. The same concept is shown for cases where the resulting simulated flow patterns were obtained using an anisotropic permeability field (Figure 2b) and for the example of a more heterogenous geological scenario representing a channelized structure (Figure 2c). From Figure 2c, it also becomes clear that by applying this method, we gained added advantage of capturing the distance between both fronts along the geological representation of our model (grid geometry and petrophysical properties driving fluid flow) as opposed to using a purely geometrical approach where such detail might be lost. A distance value of zero was assigned for all positions where both observed and measured front data coincided; for the remaining unmatched locations, an added physical meaning was obtained by assigning positive distances for locations where the measured front was ahead of the observed front and vice versa.

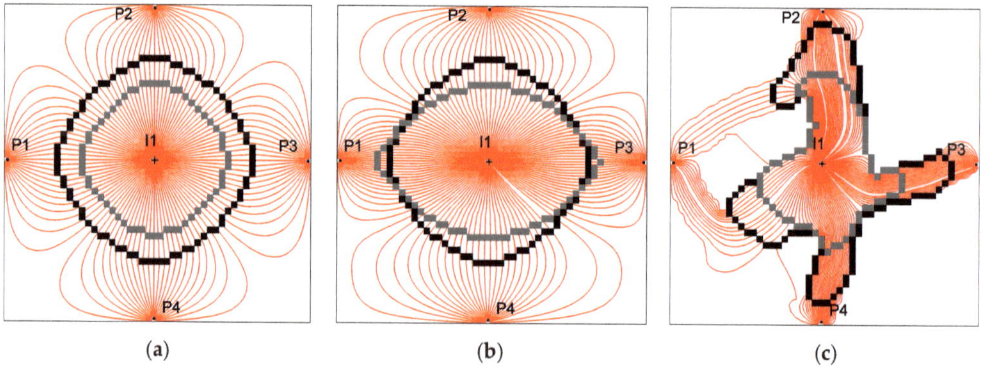

Figure 2. Streamline tracing with observed and predicted waterfronts for different model geologies at the same time step. The observed fronts are indicated in black, simulated fronts in grey and production streamlines in red. (**a**) streamlines obtained in an isotropic model, (**b**) streamlines obtained in an anisotropic model, (**c**) streamlines obtained in an model representing a channelized structure.

Under cases where geology is uncertain (e.g., multiple different geological scenarios), the same distance metric can be used. For this case, we considered an uncertain geological scenario where we took two different channelized structures, honoring the well data but with different orientations.

Figure 3 illustrates the application of the proposed method, where we evaluated the 4D seismic distance between both model propositions on a given monitor date.

To obtain the waterfronts at a given time t_f (Figure 3b,e), we calculated the difference between the water saturation at t_f (Figure 3a,d) with the initial water saturation at t_0. A threshold was applied to binarize the information into drained/undrained regions, and finally, we extracted the contour of the drained region (a perimeter in 2D and a surface in 3D) (Figure 3c). Since different candidate models have different simulated front shapes (e.g., in 2D having smaller or larger perimeters over a different number of model cells), distances must be computed from the location of the observed front, which should be sourced from seismic 4D interpretation and expertise related to the mechanisms driving the production of a given reservoir. This way, the observed front location, comprising a set of cells to which the distances are to be computed, is input into the algorithm. The final innovation vector is calculated by gathering all the distance measurements obtained from the set of shortest distance paths provided by computed streamlines connecting the observed front to the simulated front. For illustration purposes, Figure 4 is a spatial representation of the obtained innovation vector on top of the observed front location.

The same calculation can be carried out for different models, with the result being a vector of innovations with the size of the number of cells on the observed front. By being able to obtain sets of equally sized innovation vectors for a variety of models, the proposed method can be easily integrated into any ensemble-based assimilation approach. Algorithm 1 summarizes the use of the innovation vector in the context of an ensemble-based history-matching workflow.

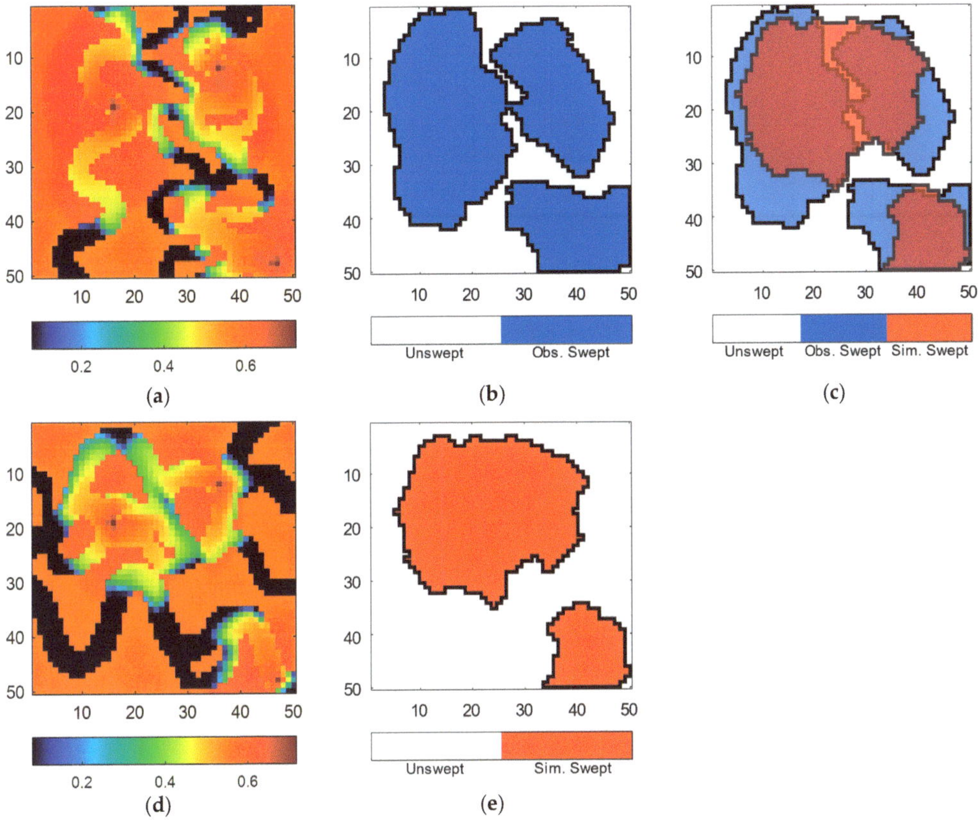

Figure 3. Example of front extraction (**c**) for a true model (**a**) and a candidate model (**d**), given the binarized ΔS_W obtained from both solutions after thresholding (**b**,**e**). The colormap for (**a**,**d**) represents S_W on the monitor date (dimensionless).

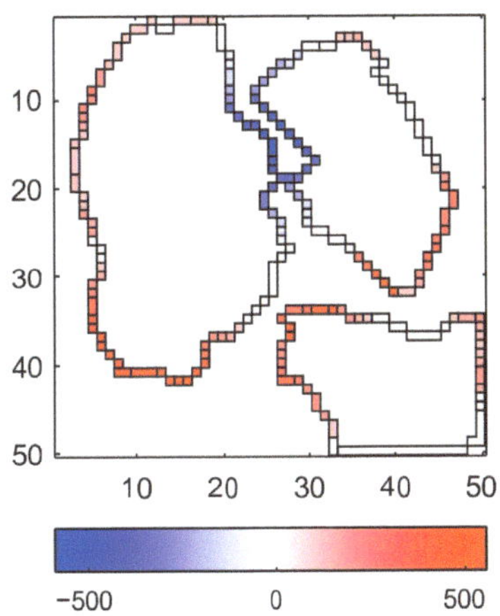

Figure 4. Distance between observed and simulated fronts streamlines represented over all cells of the observed saturation front (i and j indexes are grid cells, and colorbar units are m).

Algorithm 1: Four-dimensional distance-to-front using streamlines	
FOR EACH ENSEMBLE UPDATE STEP	
1.	Input array(s) of cells representing observed front(s) in the model grid.
2.	Full physics flow simulation of all candidate members of the ensemble.
3.	Calculate the difference between saturation data from the date of the seismic survey(s) to t_0.
4.	Binarize the output of step 3 into flooded/non-flooded regions according to a threshold.
5.	Calculate the contour of the flooded region at the date of seismic survey(s).
6.	Matching locations for the output of step 5 and step 1 are assigned a distance of 0.
7.	Post-processing of streamlines coming from step 2.
8.	Extract the shortest distance given by the streamlines (step 7) connecting the observed front (step 1) with the simulated front (step 5).
9.	Merge array(s) of distances computed at step 6 and step 8.
CONTINUE TO THE NEXT UPDATE STEP	

3. Results and Discussion

3.1. Synthetic 3D Case

In this section, we resort to a simple 3D channel reservoir to showcase the performance of the proposed method for 4D seismic history matching using streamlines for distance-to-front measurement. The reservoir comprises an anticlinal trap with three facies, containing a North-West to South-East trending leaking normal fault with a throw towards North-East. It has an extension of 5900 × 3800 m with a thickness of 52 m and is discretized into a 60 × 39 × 5 grid with 11,682 active cells in total. The used model attempted to represent the flow of two immiscible phases (oil and water) with a connate water saturation of 0.15, residual oil saturation of 0.15, and initial formation pressure of 400 bar. Out of the three facies in the model, two were permeable, and one acted as a horizontal barrier in the middle layer of the reservoir. The model had two injectors and two producing wells located in the best-quality facies. The true model was randomly sampled from an ensemble of realizations generated through truncated Gaussian simulation (TGS) [55,56] and conditioned to the information at well locations. The average permeability was 650 mD and 150 mD in the best and background facies, respectively. The average porosity in the best facies was 0.3, and in the background facies, it was 0.18. Field production spanned a total of 20 years, with production occurring through the intervention of two producing wells located on each side of the normal fault and an injector well located in the south.

The considered uncertain parameters in the reservoir were facies location, populated by the respective spatial distributions of porosity, the permeability and net-to-gross ratio (NTG) of each of the facies, and an ensemble of 100 realizations used to sufficiently avoid sampling errors and rank deficiency in the updated procedure.

True front positions for dates were obtained by running a flow simulation on a model outside of the ensemble until, respectively, 8 and 16 years after the initial production date. The resulting saturation maps were then processed in order to extract waterfront positions through the binarization of the saturation differences between monitor dates and initial condition, with a threshold (a threshold of 0.02 was used for this case) to obtain a swept region and interpretation of the front position over the binarized swept region (Figure 5). The Bayesian formalism on data assimilation problems normally requires the likelihood function to be responsible for assigning the weighting of data mismatch terms [33]. Regardless, there are examples in the literature of 4D history-matching attempts where observation errors were selected according to what the authors believed to be an "acceptable" result in terms of match quality [47,57–59]. In our case, we considered the interpreted front position to have an uncorrelated error with a standard deviation of 150 m, which consisted of approximately the size of three grid blocks along the XY direction. The resulting front positions for all seismic monitor dates, along with the historical production data, were used in an assisted history-matching scheme using ES-MDA with five iterations and 100 candidate models per iteration to update the underlying uncertain parameters towards matching both production and seismic data. For all candidate models, we resorted to a full physic reservoir simulator to run fluid flow simulation and obtain saturation data on monitor dates as well as average bottom-hole-pressure (WBPC3), flow rates (oil and water), and water cut. A set of three experiments were run to show the capabilities of the proposed method. We considered a scenario where only production data were used for history matching (NO4D), a scenario where the Hausdorff distance [37] was used to measure waterfront position mismatches (4DHDF), and finally, our proposed method, using streamlines for distance-to-front measurement (4DSLN).

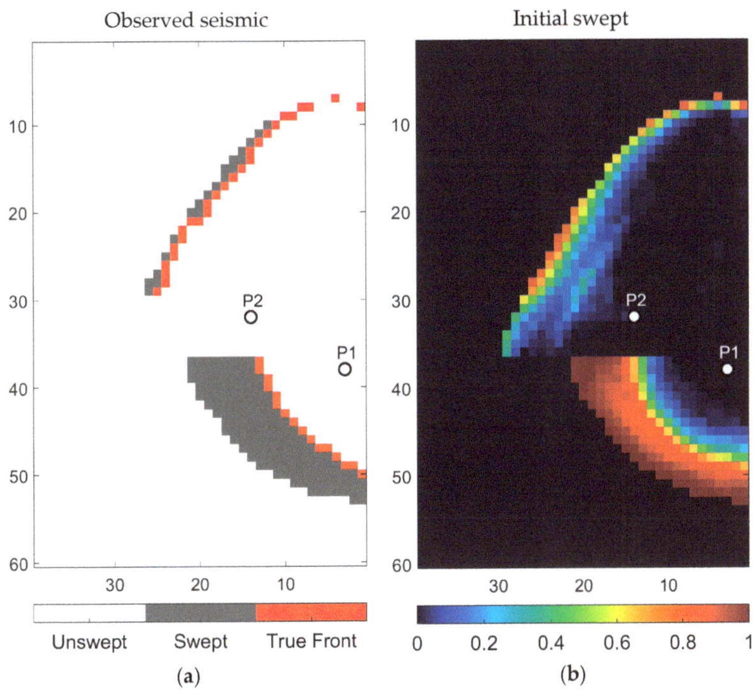

Figure 5. Observed seismic information for the ground truth model (**a**) and initial ensemble grid block average of swept area (**b**), where the spread of initial candidate solutions can be observed. Colormap represents block average binarized ΔS_W measured over the initial ensemble (dimensionless).

Figure 5a shows the observed seismic data in a binarized form (initial minus monitor date saturation) for one of the provided seismic monitors of the experiment and also the true front that we intended to match. Figure 5b shows the distribution of the simulated swept regions provided by the same binarized saturation information but averaged over all grid blocks for all members of the ensemble. We observed that there was a significant spread of the simulated swept regions (blue to red variation, Figure 5b), especially to the north of the reservoir where the model was less swept at this stage of the production schedule.

The synthesized results of the three ran experiments are presented in Figure 6, where we can observe the match of the swept regions at the end of the update procedure. From the obtained results, we can see that the absence of additional seismic information of the NO4D method provided a poor match concerning the observed seismic data (Figure 6b). The initial and final spreads of the swept regions also did not vary substantially, although we arrived at a fairly good match concerning the production data (Figure 7a). This reinforces the idea that seismic data integration is crucial for history matching as, while the solutions provided might be considered moderately accurate in terms of matching well production data, they are, in fact, inaccurate in terms of matching real waterfront movement along the reservoir as the reservoir is being produced, greatly reducing any forecasting capabilities of the model. For the case of the experiment using 4DHDF, the final solution was a close match when compared to the observed seismic information (Figure 6c). This was also accompanied by a very good match in production data (Figure 7b). However, for cases where the grid cell count is higher, the applicability of the method is debatable. In fact, we were able to run the experiment on this simplistic and small 3D model, but the method became more unfeasible with increased model sizes at operational levels (often in the order of millions of cells). Under such conditions, calculating the ensemble update can rapidly become computationally intractable as it requires the inversion of a square matrix at least the size of the grid. Regarding the proposed method (4DSLN), we observed that the final match of the swept regions was nearly identical to the one obtained by the 4DHDF method (Figure 6d), with the same being observed on the match of production data (Figure 7c).

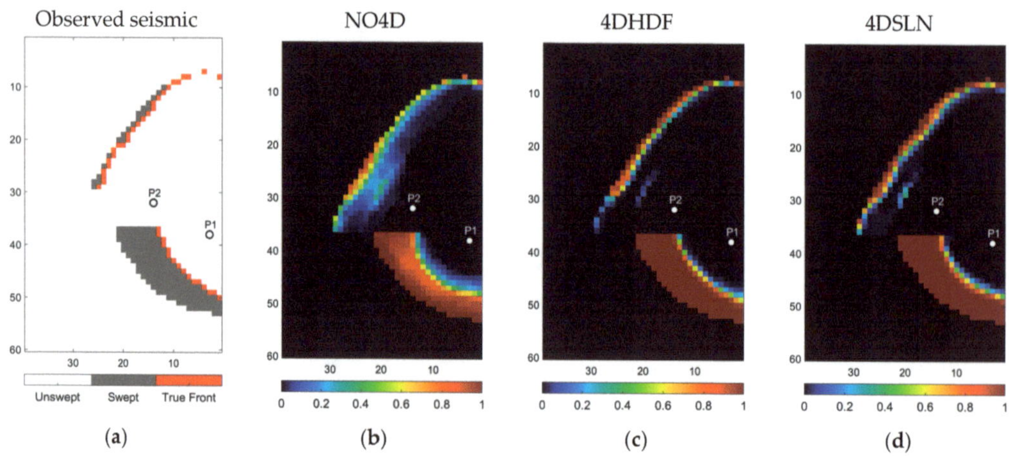

Figure 6. Results obtained from running the three experiments in terms of the final swept area. Observed seismic information for the ground truth model (**a**) and final ensemble grid block average of swept areas for NO4D (**b**), 4DHDF (**c**), and 4DSLN (**d**). Colormap represents block average binarized ΔS_W measured over the final ensemble (dimensionless).

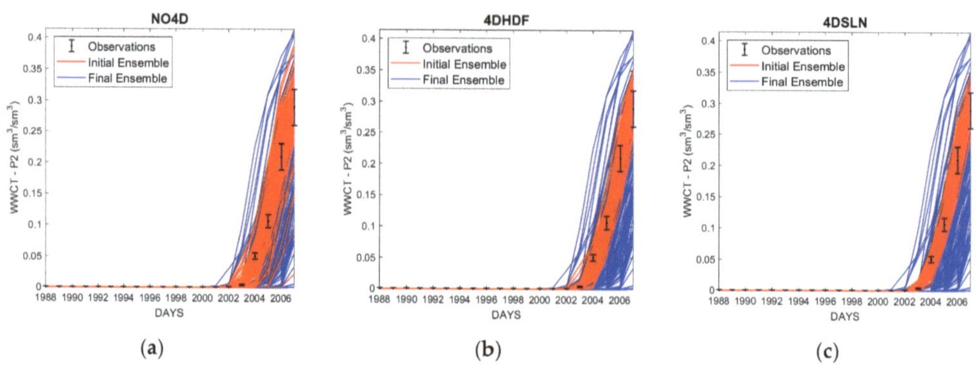

Figure 7. Evolution of production match quality from initial (blue) to final (red) ensembles, along with the available observed production data and associated uncertainty (black error bars). NO4D (**a**), 4DHDF (**b**), and 4DSLN (**c**).

Figure 8 further explores the obtained results by showing the difference (grid block average) obtained between the swept regions of the final ensemble and the ground truth for all three ran experiments. We can again see that not using seismic information renders the final solution far from the truth (Figure 8a), whereas using seismic data provides a much closer match (Figure 8b,c). Moreover, we can see that using the proposed streamlined distance-to-front method provides very similar results when compared to 4DHDF, the advantage of using a substantially reduced amount of data (Figure 9).

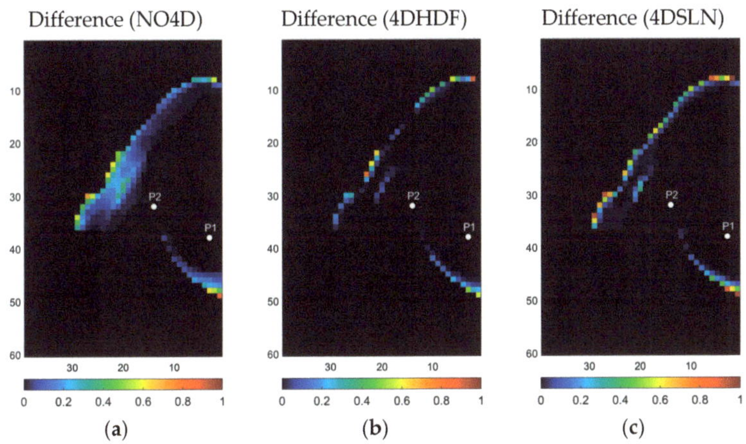

Figure 8. Grid block difference between the obtained swept areas of the final ensemble and the ground truth. NO4D (**a**), 4DHDF (**b**), and 4DSLN (**c**). Units are dimensionless.

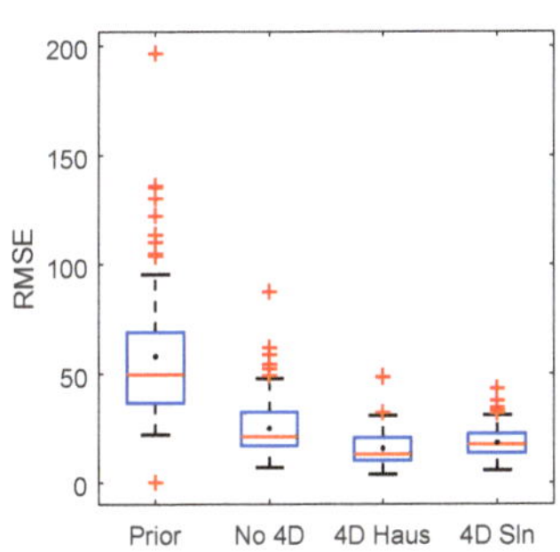

Figure 9. Boxplot for RMSE of all production data mismatches (P1, P2 with WWCT, WWPR, and WBPC3). The blue bounding boxes represent the 25 and 75% quantiles; the whiskers are the extremes, the red line in the box is the median, the black dot is the mean, and the red plus signs are the outliers.

Regarding the production match of the field-producing wells (two producers and one injector), Figure 8 shows the final obtained results in terms of well water cut (WWCT) for producer P2. Well P2 can be taken as a sufficient example as in all cases, the same match quality was observed for all present wells in the model. We can see that in all ran experiments, a reduction of the production ensemble spread was observed. As mentioned previously, and as can be observed in Figure 8, not using seismic data can potentially lead to a final ensemble where production data matching is improved; however, the inclusion of seismic information drastically increases the quality of the final match. In fact, both experiments using seismic data information for the update procedure provided very good matches with similar qualities when coupled with the 4DSLN method, obtaining residually improved results when compared to 4DHDF.

The results for all ran experiments are further demonstrated in Figure 9, depicting the root mean square error (RMSE) defined as follows:

$$\text{RMSE} = \sqrt{\frac{\sum_{i=1}^{n} (\hat{y}_i - y_i)^2}{n}}, \quad (2)$$

where n is the number of observations, and $\hat{y}$ and y are, respectively, the predicted and observed values. Figure 9 shows the minimization of the obtained production data mismatch for all run experiments. Immediately, we can observe that although the incorporation of production data considerably reduced production data mismatch relative to the prior ensemble (NO4D), the added value for the inclusion of seismic data was reflected in the ability to further improve predictions (4DHDF and 4DSLN). We also observed that the final production match quality was similar for both methods using seismic information.

In order to examine the quality of the updates to the facies model, we compared the obtained solutions to the elected ground truth model. In Figure 10, we show the probability map for good facies (shown in dark red) in layer 3 of the reservoir for the initial and final ensembles. Immediately, we can observe the potential for the inclusion of seismic data, as the experiment that used only well production data to constrain the update (NO4D) was unable to capture the spatial distribution patterns of the true model. Furthermore, the convergence towards the optimum solution is hardly visible, as almost no change occurred between the initial and final ensembles. This information, when considering the somewhat satisfactory results for production mismatch shown in Figures 6b and 8a for the same experiment, serves as a good example to stress the importance of using seismic information, illustrating a case of non-uniqueness observed in history matching. In fact, for the NO4D experiment, while many solutions may adequately fit the production data, the lack of seismic information to further constrain the update procedure will render the final ensemble of models less reliable for forecasting future production data. On the other hand, we can observe that 4DHDF and 4DSLN are able to arrive at a final ensemble that better resembles the true model. Using a full grid distance measurement, 4DHDF was able to accurately capture the spatial distribution of the facies locations in the final ensemble. Similarly, 4DSLN was able to arrive at a final ensemble that captured the most relevant spatial distribution patterns for fluid flow prediction, namely the connection of the North-East and South-East regions to the center of the model, with the advantage of using a much lower amount of seismic data for this purpose.

3.2. Realistic 3D Case

For the final example, we applied the proposed method to a realistic 3D case based on a real field, where the application of 4DHDF could not be achieved due to the large amount of data. The reservoir is a turbiditic depositional environment located in offshore Africa. The grid size was 194 × 203 × 48 with cell dimensions of 50 × 50 × 3 m on average over the i, j, and k directions. The field was produced by eight producing wells, and five injector wells provided pressure support. Twelve years of historical oil and water production, as well as bottom-hole pressure, were available. The model had 13 different flow units (Figure 11) populated by a total of three different facies types with different spatial continuity patterns (realizations are obtained using the TGS algorithm), with each facies type having specific petrophysical property distributions and dynamic parametrizations according to the quality of the rock they represented (poor to good sands). The reservoir was compartmentalized by a total of 25 faults with varying transmissibilities.

For our experiment, we considered both geological and engineering uncertain parameters. Facies locations over the different flow units were considered to be uncertain geologic parameters, and a set of connectivity regulating parameters (fault transmissibilities, sedimentological and aquifer connection, and region transmissibilities), pore volume and productivity multipliers, were assumed to be uncertain engineering parameters, 28 in total. Table 1 presents a summary of the uncertain parameters as well as their assumed uncertainty bounds.

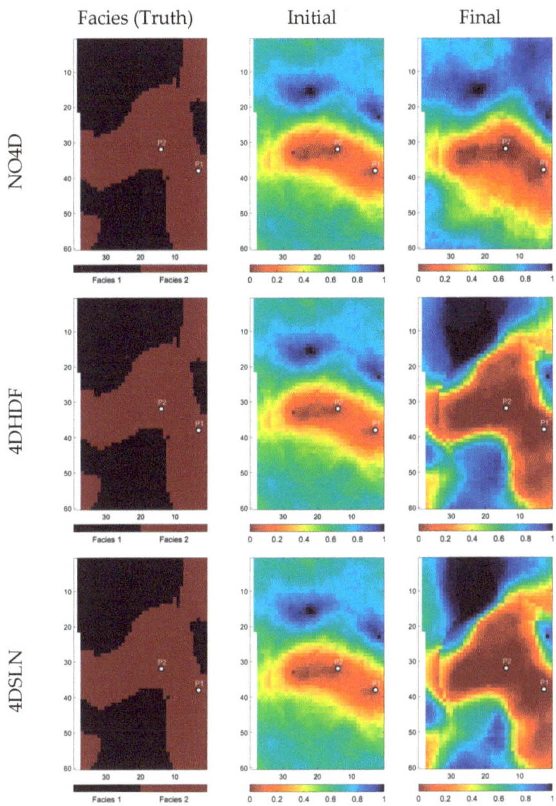

Figure 10. Comparison of match estimates for facies location for all experiments.

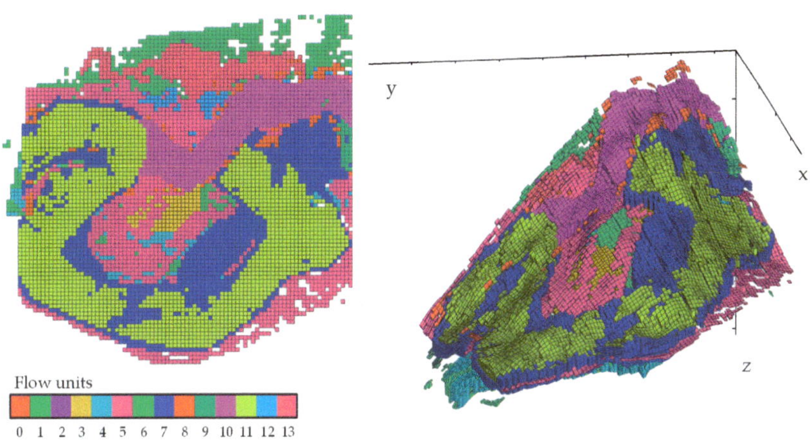

Figure 11. Top view (**left**) and 3D view (**right**) of the flow units for the realistic 3D case. Different colors represent different flow units with different rock properties and spatial continuity patterns, 13 in total.

Table 1. Uncertain parameters and their associated uncertainty bounds.

Parameter Type	Count	Minimum	Maximum
Pore Volume Multipliers	13	0.85	1
Fault Transmissibility Multipliers	10	1×10^{-6}	1
Region Transmissibility Multipliers	3	1×10^{-6}	1
Productivity Index Multipliers	2	0.001	1
Total	**28 Parameters**		

We generated an initial ensemble of 100 equiprobable candidate models using TGS to model facies locations based on variograms estimated from well log data. To sample over the uncertain engineering parameters, we resorted to Latin hypercube sampling based on the available prior knowledge. An extra model, elected as the ground truth, was also generated based on the same geological and engineering prior assumption. The location of the true seismic fronts at both monitor dates was assumed to be obtained from a standard seismic inversion procedure followed by additional processing and interpretation and corresponding to the information obtained after 6.2 and 9.7 years of fluid flow simulation on the ground truth model. The control parameters for the fluid flow simulation were set to operate the producers at historic reservoir volume rates and the injectors at historic fluid rates. The interpreted front locations were assumed to have uncorrelated errors with a standard deviation of 200 m (around four grid blocks).

Figure 12 shows the initial water saturation on a randomly selected model from the initial ensemble and the corresponding streamlines obtained on both monitor dates. The interpreted waterfronts for the ground truth model are also superimposed on the image.

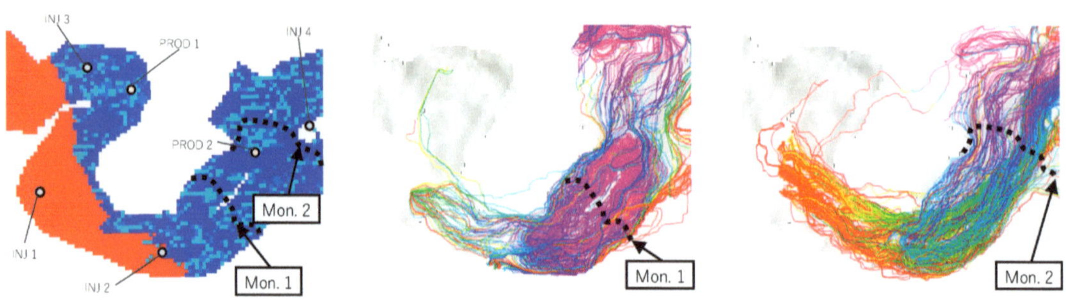

(a) (b) (c)

Figure 12. Top view of the initial S_w (blue—0 to red—1) for a realization of the realistic 3D. Well locations (circles with black borders) and two arbitrarily interpreted fronts (dashed black lines) extracted at two different seismic monitoring dates (**a**). Post-processed fluid flow streamlines (colors represent the time of flight) obtained at the dates of the first and second seismic monitors (**b**,**c**).

In the same randomly selected model from the initial ensemble, using a threshold to binarize the change in saturation at the seismic monitor dates, we obtained the swept regions on each monitor date. Figure 13 shows the swept regions and front locations of the same randomly selected model on both monitor dates against the ones obtained from the ground truth. The mismatch between the observed and simulated fronts was observed, resulting from excessive water sweeping over the model.

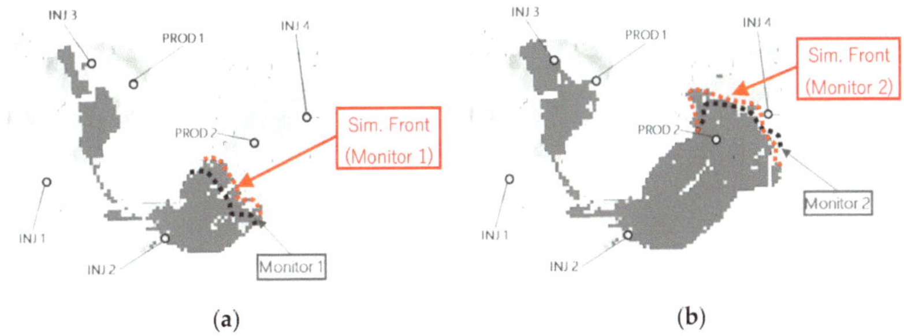

Figure 13. Top view of the front mismatch at monitor date one (**a**) and two (**b**). The property shown (dark gray) is the simulated response in terms of binarized ΔS_w along with the associated simulated front (red) obtained for this realization.

We set up the experiment using an ES-MDA update scheme of the 100 members of the initial ensemble over five iterations. Figure 14 compares, at the first monitor date and on layer 12 of the model, the true front location with a random unmatched realization from the initial ensemble (center) and the best solution found on the final ensemble. We can observe that the final best-matching solution is a clear improvement when compared to the initial guess example. Further, we superimposed the water cut ratio of a producing well (PROD 4) associated with the front arrival at the date of the seismic acquisition. The best matching realization (shown in red) was able to match the water cut ratio perfectly. We can also see that the front arrival time, closely related to the date of the seismic monitor (red dashed line), matches the water breakthrough of the well, meaning we successfully matched the correct time of waterfront arrival and the correct water volume production.

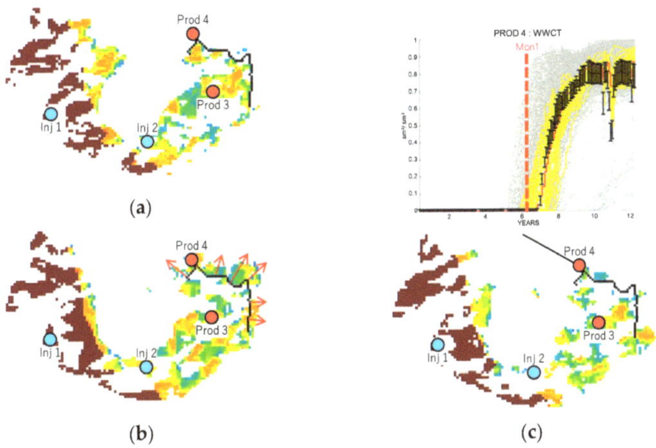

Figure 14. (**a**) Ground truth S_w along with the associated true front (black) for the first monitor date. (**b**) Example of simulated S_w of a randomly selected model from the initial ensemble, highlighting an excessive water sweep going over the front (red arrows). (**c**) Final best-matched model of S_w along with WWCT production curves for PROD4 (c-top). Injectors producers are represented respectively in blue and red circles.

The same match quality was observed for the second monitor date (Figure 15) on a different layer of the reservoir (layer 14).

By comparing the swept regions and front location of the ground truth with the best-obtained solution, we observed a very good match concerning the front location and the production data of an associated well at the true front location (PROD 6).

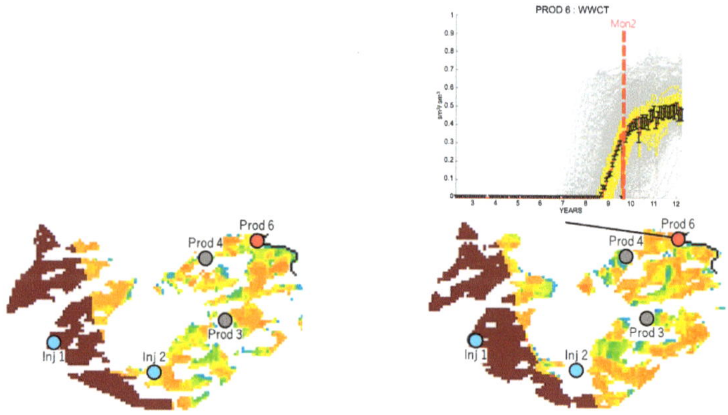

(a) (b)

Figure 15. (**a**) Ground truth S_w along with the associated true front (black) for the second monitor date. (**b**) Final best-matched model of S_w along with WWCT production curves for PROD6 (b-top). Injectors producers are represented respectively in blue and red circles.

A summary of the run, in terms of front location matches, from initial to final ensemble is illustrated in Figure 16. We can see the average and standard deviations of the distances measured to the front at every grid block point where the true front is located (up). We can also observe that before and after history matching, the final obtained front locations much more closely matched the truth (as they are closer to 0) and that, at the same time, the uncertainty over the final ensemble of solutions was substantially reduced.

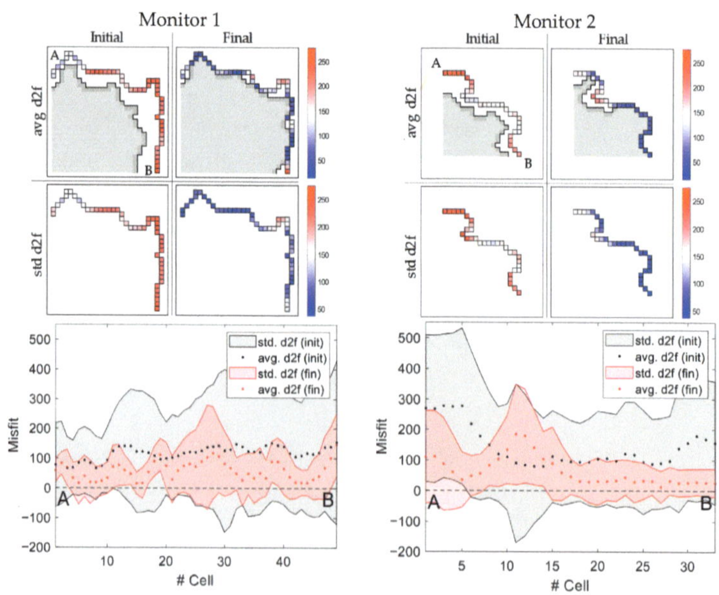

(a) (b)

Figure 16. Summary of the front matches for a given layer of the model at the first (**a**) and second (**b**) monitor dates with the grid block mean and standard deviations of distance-to-front obtained in the initial and final ensembles (**top**). A plot of the most representative initial and final waterfront locations in regard to the mean distances measured over the ensemble is represented in gray shade. In the bottom, the initial and final grid block average (red and black dots) and standard deviation (red and grey area) of distance-to-front obtained for all layers of the model where a front was interpreted.

Regarding the convergence on uncertain parameters over the run, Figure 17 shows the probability of facies locations from the initial to final ensemble, compared to the ground truth. We can immediately observe that, for this particular model, the most predominant facies type is facies 1 (in blue), taking the bulk of the task of governing fluid flow production and reservoir connectivity. The final ensemble shows the convergence towards the true locations of the three facies types, but most importantly, the correct spatial continuity and connectivity patterns that can be observed in the true model.

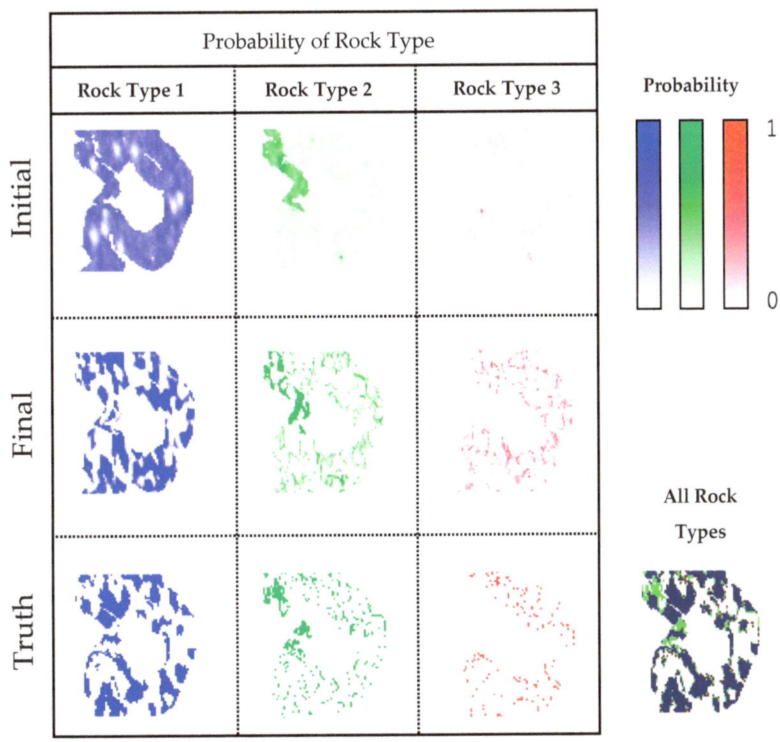

Figure 17. Summary of the probability of rock type obtained from the run experiment. The initial (**top**) and final (**middle**) matches of facies type (blue, green, red) match are compared against the ground truth (**bottom**).

An example of the convergence that can be observed on selected uncertain engineering parameters for the experiment is shown in Figure 18. During the course of the run, we observed the convergence of the parameter distributions towards true values at the same time that the misfit was also being reduced.

The same can be seen for the production plots of water cut and bottom-hole pressure for all producing wells of the model (Figure 19). We can clearly observe the gradual reduction of the mismatch from the initial to the final iteration, with the final ensemble providing good match quality over all wells.

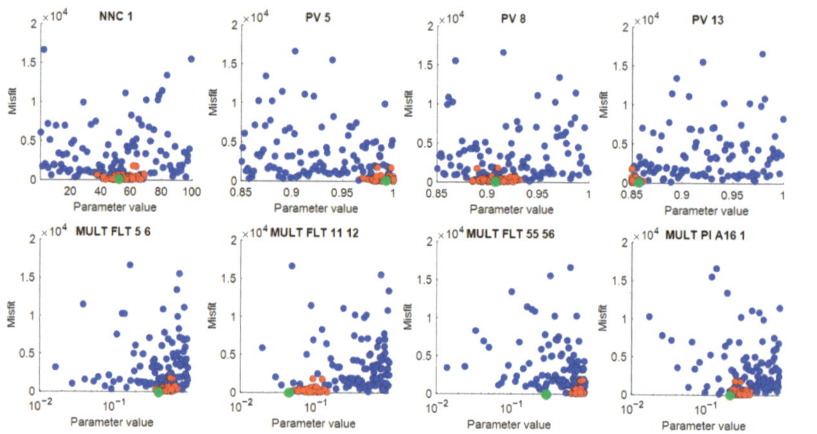

Figure 18. Scatter plot of the parameter value (x-axis) vs. misfit (y-axis) for the convergence of a selected set of parameters (true value in green circle) from the initial (blue) to the final iteration (red).

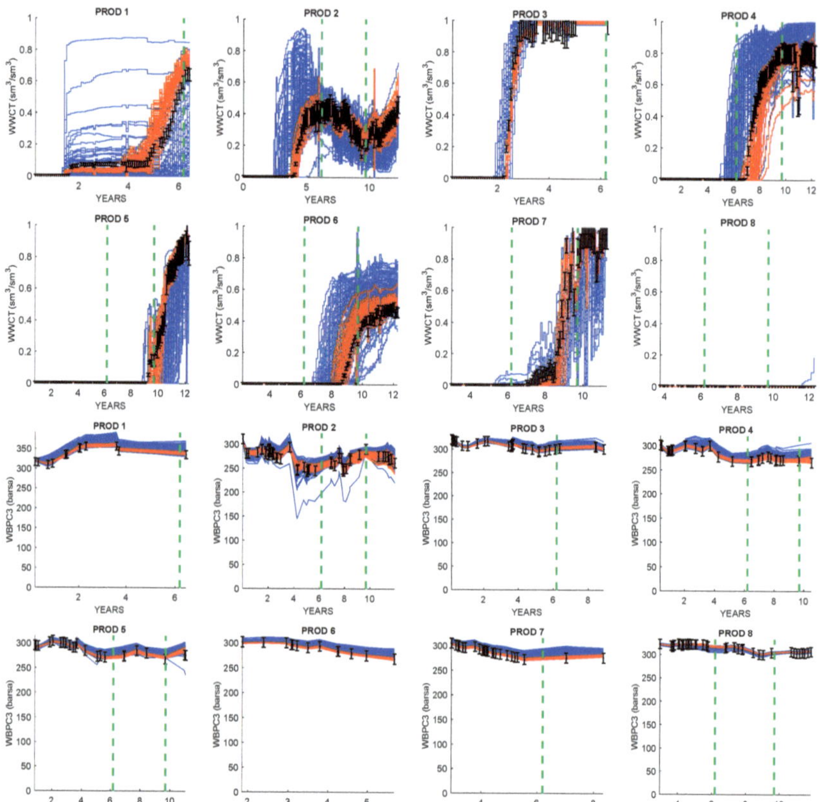

Figure 19. Run evolution in terms of production match for WWCT (first and second rows) and WBPC3 (third and fourth rows) from the initial (blue) to the final iteration (red), along with the observations (black error bars) and acquisition date of used seismic monitors (dashed green lines).

4. Conclusions

A novel parametrization scheme was proposed to quantify the differences between simulated and measured seismic data in terms of distances between fluid front positions measured by streamlines. The methodology was presented with the help of a simplistic 2D case and applied both on a synthetic and realistic 3D case with seismic and production data, highlighting the advantages and good performance of the proposed scheme when compared to other commonly used schemes. Although the presented cases were of synthetic and semi-synthetic nature, the application of the proposed method to real-case scenarios should not inherently present any disadvantages. Naturally, the success in achieving positive outcomes, as with all methods, hinges on the quality of the models and the data at hand.

The obtained results show that the proposed parametrization can achieve similar results when compared to other methods resorting to a reduced amount of data. The streamlined information used for distance measurement can be easily obtained as post-processing of standard full-physics simulation outputs.

Despite the advantages of using streamlines for distance measurement, there could be some potential disadvantages when processing a large number of streamlines, as could be the case with very large models. However, these challenges can be mitigated through parallelization methodologies or, for example, by refining the way streamlines are post-processed. For instance, focusing the processing on the cells where fluid fronts are located can significantly speed up the process. Some limitations in terms of precision may potentially arise when the models, and consequently the simulated fronts, are too far from the ground truth. This is due to the inherent requirement of obtaining streamlines that intersect both fronts in order to accurately compute distance data through them. Despite this, any missing information can be easily complemented by other distance metrics.

The capabilities of the method were not only showcased on a simplistic 3D model with simple reservoir conditions and production schedules but also on a more complex and realistic 3D model, where the Hausdorff distance method becomes computationally intractable, having obtained encouraging results both on fluid front and production data match. The method is expected to be effectively applicable to a range of reservoir types beyond those demonstrated in this study. Furthermore, the parametrization scheme is able to be adaptable and functional across diverse geological settings.

While the application of the innovative parametrization scheme was showcased within the framework of assisted history matching using an ensemble history matching methodology (ES-MDA), its application can be easily extended to any other history-matching workflow or even to different domains of application other than hydrocarbon exploration, e.g., CO_2 monitoring.

Future areas of research may involve advancements in how streamlines are calculated to be fit-for-purpose for the type of methodology presented here. This may include methods relying on the focusing of information on regions pertaining to both the simulated and real fronts. Another avenue of investigation that could prove interesting is implementing localization techniques using information derived from streamlines. This approach could potentially enhance the method's performance by refining the accuracy and efficiency of the parameterization scheme, especially in complex geological settings where the alignment of simulated and real fluid fronts is critical. These developments could lead to more nuanced and effective history matching in reservoir simulation, thus broadening the scope and applicability of the proposed method.

5. Patents

Berthet, P., Trani, M. [60]. A method for obtaining at least one physical property of a subsurface volume of a hydrocarbon reservoir over time (European priority Application, filing date 4 December 2020, publication number EP4009086).

Author Contributions: Conceptualization, E.B., P.B. and M.T.; methodology, E.B., P.B. and M.T.; validation, E.B., P.B. and O.T.; formal analysis, E.B.; investigation, E.B.; resources, P.B.; data curation, E.B.; writing—original draft preparation, E.B.; writing—review and editing, E.B., P.B., M.T., O.T. and C.L.; visualization, E.B.; supervision, P.B., O.T. and C.L.; project administration, P.B.; funding acquisition, P.B. All authors have read and agreed to the published version of the manuscript.

Funding: This research was funded by TotalEnergies S.E. and the ANRT—Association Nationale de la Recherche et de la Technologie, grant number 2020/0558.

Data Availability Statement: Data are contained within the article.

Acknowledgments: The authors acknowledge the support from the "Association Nationale de la Recherche et de la Technologie" for funding this project and colleagues from the TotalEnergies S.E. affiliate for the valuable discussions and for providing the realistic 3D dataset used in this work.

Conflicts of Interest: The authors declare no conflict of interest.

Appendix A

First introduced by Emerick and Reynolds [51,52] as an extension to the standard ES method, the multiple data assimilation (MDA) method was developed with the purpose of enhancing the performance of the ensemble Kalman filter [38] and ensemblesmoother [58] under nonlinear conditions. In this iterative approach, all data are assimilated multiple times by applying an inflation parameter α_k to the covariance matrix of measurement errors. This was proven to be equivalent to the single data assimilation case for linear-Gaussian systems, given that the measurement error covariance matrix is appropriately scaled [51,52].

Therefore, considering the standard ES update equation, the inflation parameter α_k is included as follows:

$$m_j^{k+1} = m_j^k + \hat{C}_{md}^k \left(\hat{C}_{dd}^k + \alpha_k C_d\right)^{-1} \left(d_{obs} - g\left(m_j^k\right) + e_j\right), \tag{A1}$$

where $k+1$ and k are the indexes of the iterative procedure, where $k = 1, 2, \ldots, N_a$ with N_a being the total number of assimilations; $\hat{C}_{md}^f$ is the cross-covariance matrix between the a priori vector of model parameters, m^f, and the predicted data vector, $g(m^f)$; $\hat{C}_{dd}^f$ is the covariance matrix of the predicted data of size $N_d \times N_d$; d_{obs} is the vector of observed data; e_j is the perturbation vector added to the observed data; C_d is the covariance matrix of the observed data error of size $N_d \times N_d$.

To implement ES-MDA, the values of the inflation factor α_k in each iteration need to be defined. The necessary condition for choosing the inflation factor is the following:

$$\sum_{k=1}^{N_a} \frac{1}{\alpha_k} = 1. \tag{A2}$$

A lack of consensus exists on how to choose the values of α_k, with $\alpha_k = N_a$ being commonly used. On the same note, the number of iterations N_a must be set beforehand, with generally accepted values ranging from 4 to 10 iterations. Improved versions of the method have been proposed where α_k and N_a are chosen automatically [59]. For the present work, the authors opted to use $\alpha_k = N_a = 5$.

References

1. Landrø, M. Discrimination between pressure and fluid saturation changes from time-lapse seismic data. *Geophysics* **2001**, *66*, 836–844. [CrossRef]
2. Lumley, D.E. Time-lapse seismic reservoir monitoring. *Geophysics* **2001**, *66*, 50–53. [CrossRef]
3. Thore, P.; Hubans, C. 4D seismic-to-well tying, a key step towards 4D inversion. *Geophysics* **2012**, *77*, R227–R238. [CrossRef]
4. Thore, P.; Blanchard, T.D. 4D propagated layer-based inversion. *Geophysics* **2015**, *80*, R15–R29. [CrossRef]
5. Grana, D.; Mukerji, T. Bayesian inversion of time-lapse seismic data for the estimation of static reservoir properties and dynamic property changes. *Geophys. Prospect.* **2014**, *63*, 637–655. [CrossRef]

6. Maharramov, M.; Biondi, B.L.; Meadows, M.A. Time-lapse inverse theory with applications. *Geophysics* **2016**, *81*, R485–R501. [CrossRef]
7. Kazemi, A.; Stephen, K.D.; Shams, A. Seismic History Matching of Nelson Using Time-Lapse Seismic Data: An Investigation of 4D Signature Normalization. *SPE Reserv. Evaluation Eng.* **2011**, *14*, 621–633. [CrossRef]
8. Maleki, M.; Davolio, A.; Schiozer, D.J. Qualitative time-lapse seismic interpretation of Norne Field to assess challenges of 4D seismic attributes. *Lead. Edge.* **2018**, *37*, 754–762. [CrossRef]
9. Lygren, M.; Husby, O.; Osdal, B.; El Ouair, Y.; Springer, M. History matching using 4D seismic and pressure data on the Norne field. In Proceedings of the 67th EAGE Conference & Exhibition, Madrid, Spain, 13–16 June 2005; European Association of Geoscientists & Engineers: Utrecht, The Netherlands, 2005; p. cp-1. [CrossRef]
10. Roggero, F.; Ding, D.Y.; Berthet, P.; Lerat, O.; Cap, J.; Schreiber, P.-E. Matching of Production History and 4D Seismic Data--Application to the Girassol Field, Offshore Angola. In Proceedings of the SPE Annual Technical Conference and Exhibition, Anaheim, CA, USA, 11–14 November 2007; OnePetro: Richardson, TX, USA, 2007. [CrossRef]
11. Castro, S.A.; Caers, J.; Otterlei, C.; Meisingset, H.; Hoye, T.; Gomel, P.; Zachariassen, E. Incorporating 4D seismic data into reservoir models while honoring production and geologic data: A case study. *Lead. Edge.* **2009**, *28*, 1498–1505. [CrossRef]
12. Le Ravalec, M.; Tillier, E.; Da Veiga, S.; Enchery, G.; Gervais, V. Advanced integrated workflows for incorporating both production and 4d seismic-related data into reservoir models. *Oil Gas Sci. Technol. Rev. d'IFP Energies Nouv.* **2012**, *67*, 207–220. [CrossRef]
13. Roggero, F.; Lerat, O.; Ding, D.; Berthet, P.; Bordenave, C.; Lefeuvre, F.; Perfetti, P. History matching of production and 4D seismic data: Application to the Girassol field, offshore Angola. *Oil Gas Sci. Technol. Rev. d'IFP Energies Nouv.* **2012**, *67*, 237–262. [CrossRef]
14. Byerley, G.; Singer, L.; Rose, P. Resaturated pay: A new infill target type identified through the application and continuous improvement of 4D seismic at the Forties Field. *Lead. Edge.* **2016**, *35*, 831–838. [CrossRef]
15. Calvert, M.A.; Hoover, A.R.; Vagg, L.D.; Ooi, K.C.; Hirsch, K.K. Halfdan 4D workflow and results leading to increased recovery. *Lead. Edge.* **2016**, *35*, 840–848. [CrossRef]
16. Rankey, E.C.; Mitchell, J.C. That's why it's called interpretation: Impact of horizon uncertainty on seismic attribute analysis. *Lead. Edge.* **2003**, *22*, 820–828. [CrossRef]
17. Zhou, W.; Lumley, D. Nonrepeatability effects on time-lapse 4D seismic full-waveform inversion for ocean-bottom node data. *Geophysics* **2021**, *86*, R547–R561. [CrossRef]
18. Sarkar, S.; Gouveia, W.P.; Johnston, D.H. On the inversion of time-lapse seismic data. In Proceedings of the 2003 SEG Annual Meeting, Dallas, TX, USA, 26–31 October 2003; OnePetro: Richardson, TX, USA, 2003. [CrossRef]
19. Buland, A.; El Ouair, Y. Bayesian time-lapse inversion. *Geophysics* **2006**, *71*, R43–R48. [CrossRef]
20. Suman, A.; Fernández-Martínez, J.L.; Mukerji, T. *Joint Inversion of Production and Time-Lapse Seismic Data: Application to Norne Field*; Stanford University: Stanford, CA, USA, 2013. [CrossRef]
21. Alvarez, E.; MacBeth, C.; Brain, J. Quantifying remaining oil saturation using time-lapse seismic amplitude changes at fluid contacts. *Pet. Geosci.* **2016**, *23*, 238–250. [CrossRef]
22. Arenas, E.; van Kruijsdijk, C.; Oldenziel, T. Semi-automatic history matching using the pilot point method including time-lapse seismic data. In Proceedings of the SPE Annual Technical Conference and Exhibition, New Orleans, LA, USA, 30 September–3 October 2001; OnePetro: Richardson, TX, USA, 2001. [CrossRef]
23. Fagervik, K.; Lygren, M.; Valen, T.S.; Hetlelid, A.; Berge, G.; Dahl, G.V.; Sønneland, L.; Lie, H.E.; Magnus, I. A method for performing history matching of reservoir flow models using 4d seismic. In Proceedings of the 2001 SEG Annual Meeting, San Antonio, TX, USA, 9–14 September 2001; OnePetro: Richardson, TX, USA, 2001. [CrossRef]
24. Gosselin, O.; Berg, S.v.D.; Cominelli, A. Integrated history-matching of production and 4D seismic data. In Proceedings of the SPE Annual Technical Conference and Exhibition, New Orleans, LA, USA, 30 September–3 October 2001; OnePetro: Richardson, TX, USA, 2001. [CrossRef]
25. Gosselin, O.; Aanonsen, S.I.; Aavatsmark, I.; Cominelli, A.; Gonard, R.; Kolasinski, M.; Ferdinandi, F.; Kovacic, L.; Neylon, K. History matching using time-lapse seismic (HUTS). In Proceedings of the SPE Annual Technical Conference and Exhibition, Denver, CO, USA, 5 October 2003; OnePetro: Richardson, TX, USA, 2003. [CrossRef]
26. van Ditzhuijzen, R.; Oldenziel, T.; van Kruijsdijk, C. Geological parameterization of a reservoir model for history matching incorporating time-lapse seismic based on a case study of the Statfjord field. In Proceedings of the SPE Annual Technical Conference and Exhibition, New Orleans, LA, USA, 30 September–3 October September 2001; OnePetro: Richardson, TX, USA, 2001. [CrossRef]
27. Dong, Y.; Oliver, D.S. Quantitative Use of 4D Seismic Data for Reservoir Description. *SPE J.* **2005**, *10*, 91–99. [CrossRef]
28. Haverl, M.; Aga, M.; Reiso, E. Integrated Workflow for Quantitative Use of Time-Lapse Seismic Data in History Matching–A North Sea Field Case (SPE94453). In Proceedings of the 67th EAGE Conference & Exhibition, Madrid, Spain, 13–16 June 2005; European Association of Geoscientists & Engineers: Utrecht, The Netherlands, 2005; p. cp-1. [CrossRef]
29. Portella, R.C.M.; Emerick, A.A. Use of Quantitative 4D-Seismic Data in Automatic History Match. In Proceedings of the SPE Latin American and Caribbean Petroleum Engineering Conference, Rio de Janeiro, Brazil, 20 June 2005; OnePetro: Richardson, TX, USA, 2005. [CrossRef]
30. Stephen, K.D.; Soldo, J.; Macbeth, C.; Christie, M.A. Multiple model seismic and production history matching: A case study. *SPE J.* **2006**, *11*, 418–430. [CrossRef]

31. Dadashpour, M.; Kleppe, J.; Landro, M. Porosity and permeability estimation by gradient based history matching using time-lapse seismic data. In Proceedings of the SPE Middle East Oil and Gas Show and Conference, Red Hook, NY, USA, 11–14 March 2007; OnePetro: Richardson, TX, USA, 2007. [CrossRef]
32. Luo, X.; Bhakta, T.; Jakobsen, M.; Nævdal, G. Efficient big data assimilation through sparse representation: A 3D benchmark case study in petroleum engineering. *PLoS ONE* **2018**, *13*, e0198586. [CrossRef]
33. Oliver, D.S.; Fossum, K.; Bhakta, T.; Sandø, I.; Nævdal, G.; Lorentzen, R.J. 4D seismic history matching. *J. Pet. Sci. Eng.* **2021**, *207*, 109119. [CrossRef]
34. Trani, M.; Arts, R.; Leeuwenburgh, O. Seismic history matching of fluid fronts using the ensemble Kalman filter. *SPE J.* **2012**, *18*, 159–171. [CrossRef]
35. Rollmann, K.; Soriano-Vargas, A.; Almeida, F.; Davolio, A.; Schiozer, D.J.; Rocha, A. Convolutional Neural Network Formulation to Compare 4-D Seismic and Reservoir Simulation Models. *IEEE Trans. Syst. Man, Cybern. Syst.* **2021**, *52*, 3052–3065. [CrossRef]
36. Tillier, E.; Le Ravalec, M.; Da Veiga, S. simultaneous inversion of production data and seismic attributes: Application to a synthetic sagd produced field case. *Oil Gas Sci. Technol. Rev. d'IFP Energies Nouv.* **2012**, *67*, 289–301. [CrossRef]
37. Abadpour, A.; Bergey, P.; Piasecki, R. 4D seismic history matching with ensemble Kalman filter-assimilation on Hausdorff distance to saturation front. In Proceedings of the SPE Reservoir Simulation Symposium, Woodlands, TX, USA, 18–20 February 2013; OnePetro: Richardson, TX, USA, 2013. [CrossRef]
38. Evensen, G. The Ensemble Kalman Filter: Theoretical formulation and practical implementation. *Ocean Dyn.* **2003**, *53*, 343–367. [CrossRef]
39. Jin, L.; Alpak, F.O.; Hoek, P.J.v.D.; Pirmez, C.; Fehintola, T.; Tendo, F.; Olaniyan, E.E. A comparison of stochastic data-integration algorithms for the joint history matching of production and time-lapse-seismic data. *SPE Reserv. Eval. Eng.* **2012**, *15*, 498–512. [CrossRef]
40. Jin, L.; Weber, D.; Hoek, P.v.D.; Alpak, F.; Pirmez, C. 4D Seismic history matching using information from the flooded zone. *First Break.* **2012**, *30*, 11. [CrossRef]
41. Obidegwu, D.; Chassagne, R.; MacBeth, C. Seismic assisted history matching using binary maps. *J. Nat. Gas Sci. Eng.* **2017**, *42*, 69–84. [CrossRef]
42. Tillier, E.; Da Veiga, S.; Derfoul, R. Appropriate formulation of the objective function for the history matching of seismic attributes. *Comput. Geosci.* **2012**, *51*, 64–73. [CrossRef]
43. Davolio, A.; Schiozer, D.J. Probabilistic seismic history matching using binary images. *J. Geophys. Eng.* **2017**, *15*, 261–274. [CrossRef]
44. Kretz, V.; Vallès, B.; Sonneland, L. Fluid front history matching using 4D seismic and streamline simulation. In Proceedings of the SPE Annual Technical Conference and Exhibition, Houston, TX, USA, 26–29 September 2004; OnePetro: Richardson, TX, USA, 2004. [CrossRef]
45. Leeuwenburgh, O.; Arts, R. Distance parameterization for efficient seismic history matching with the ensemble Kalman Filter. *Comput. Geosci.* **2014**, *18*, 535–548. [CrossRef]
46. Zhang, Y.; Leeuwenburgh, O. Ensemble-based seismic history matching with distance parameterization for complex grids. In Proceedings of the ECMOR XV-15th European Conference on the Mathematics of Oil Recovery, Amsterdam, The Netherlands, 29 August–1 September 2016; European Association of Geoscientists & Engineers: Utrecht, The Netherlands, 2016; p. cp-494. [CrossRef]
47. Zhang, Y.; Leeuwenburgh, O. Image-oriented distance parameterization for ensemble-based seismic history matching. *Comput. Geosci.* **2017**, *21*, 713–731. [CrossRef]
48. Trani, M.; Moncorgé, A.; Bergey, P.; Chen, Y. Fluid Front History Matching Using an Iterative Ensemble Smoother. In Proceedings of the 77th EAGE Conference and Exhibition 2015, Madrid, Spain, 1–4 June 2015; European Association of Geoscientists & Engineers: Utrecht, The Netherlands, 2015; Volume 2015, pp. 1–5. [CrossRef]
49. Sethian, J.A. A fast marching level set method for monotonically advancing fronts. *Proc. Natl. Acad. Sci. USA* **1996**, *93*, 1591–1595. [CrossRef] [PubMed]
50. Hassouna, M.S.; Farag, A.A. MultiStencils fast marching methods: A highly accurate solution to the eikonal equation on cartesian domains. *IEEE Trans. Pattern Anal. Mach. Intell.* **2007**, *29*, 1563–1574. [CrossRef]
51. Emerick, A.A.; Reynolds, A.C. History matching time-lapse seismic data using the ensemble Kalman filter with multiple data assimilations. *Comput. Geosci.* **2012**, *16*, 639–659. [CrossRef]
52. Emerick, A.A.; Reynolds, A.C. Ensemble smoother with multiple data assimilation. *Comput. Geosci.* **2013**, *55*, 3–15. [CrossRef]
53. Pollock, D.W. Semianalytical computation of path lines for finite-difference models. *Groundwater* **1988**, *26*, 743–750. [CrossRef]
54. Sovold, K.; Rian, D.T.; Sandvik, A. Front Tracking Applied to the Simulation of Water Flooding in a Braided River System. In Proceedings of the SPE Latin America Petroleum Engineering Conference, Rio de Janeiro, Brazil, 14–19 October 1990; OnePetro: Richardson, TX, USA, 1990. [CrossRef]
55. Journel, A.G.; Isaaks, E.H. Conditional indicator simulation: Application to a Saskatchewan uranium deposit. *J. Int. Assoc. Math. Geol.* **1984**, *16*, 685–718. [CrossRef]
56. Matheron, G.; Beucher, H.; de Fouquet, C.; Galli, A.; Guerillot, D.; Ravenne, C. Conditional simulation of the geometry of fluvio-deltaic reservoirs. In Proceedings of the Spe Annual Technical Conference and Exhibition, Dallas, TX, USA, 27–30 September 1987; OnePetro: Richardson, TX, USA, 1987. [CrossRef]

57. Avansi, G.D.; Maschio, C.; Schiozer, D.J. Simultaneous history-matching approach by use of reservoir-characterization and reservoir-simulation studies. *SPE Reserv. Eval. Eng.* **2016**, *19*, 694–712. [CrossRef]
58. Skjervheim, J.-A.; Evensen, G.; Hove, J.; Vabø, J.G. An ensemble smoother for assisted history matching. In Proceedings of the SPE Reservoir Simulation Symposium, The Woodlands, TX, USA, 21–23 February 2011; OnePetro: Richardson, TX, USA, 2011. [CrossRef]
59. Emerick, A.A. Analysis of the performance of ensemble-based assimilation of production and seismic data. *J. Pet. Sci. Eng.* **2016**, *139*, 219–239. [CrossRef]
60. Berthet, P.; Trani, M. A Method for Obtaining at Least One Physical Property of a Subsurface Volume of a Hydrocarbon Reservoir Over Time (European Priority Application, Filing Date 4 December 2020, Publication Number EP4009086). Available online: https://www.sumobrain.com/patents/wipo/Method-obtaining-at-least-one/WO2022117735A1.html (accessed on 1 November 2023).

energies

MDPI

Article

Joint History Matching of Multiple Types of Field Data in a 3D Field-Scale Case Study †

William Chalub Cruz [1,*], Xiaodong Luo [2] and Kurt Rachares Petvipusit [3]

1 University of Stavanger, 4021 Stavanger, Norway
2 Norwegian Research Centre (NORCE), 5008 Bergen, Norway
3 Equinor, 5254 Bergen, Norway
* Correspondence: william.chalubcruz@uis.no

† This paper is modified from the material presented in SPE Norway Subsurface Conference, Bergen, Norway, 27 April 2022.

Abstract: This work presents an ensemble-based workflow to simultaneously assimilate multiple types of field data in a proper and consistent manner. The aim of using multiple field datasets is to improve the reliability of estimated reservoir models and avoid the underestimation of uncertainties. The proposed framework is based on an integrated history matching workflow, in which reservoir models are conditioned simultaneously on production, tracer and 4D seismic data with the help of three advanced techniques: adaptive localization (for better uncertainty quantification), weight adjustment (for balancing the influence of different types of field data), and sparse data representation (for handling big datasets). The integrated workflow is successfully implemented and tested in a 3D benchmark case with a set of comparison studies (with and without tracer data). The findings of this study indicate that joint history matching using production, tracer and 4D seismic data results in better estimated reservoir models and improved forecast performance. Moreover, the integrated workflow is flexible, and can be extended to incorporate more types of field data for further performance improvement. As such, the findings of this study can help to achieve a better understanding of the impacts of multiple datasets on history matching performance, and the proposed integrated workflow could serve as a useful tool for real field case studies in general.

Keywords: joint history matching; iterative ensemble smoother; production, tracer and 4D seismic data

Citation: Cruz, W.C.; Luo, X.; Petvipusit, K.R. Joint History Matching of Multiple Types of Field Data in a 3D Field-Scale Case Study. *Energies* **2022**, *15*, 6372. https://doi.org/10.3390/en15176372

Academic Editors: Haifeng Zhao and Yang Xia

Received: 2 August 2022
Accepted: 28 August 2022
Published: 31 August 2022

Publisher's Note: MDPI stays neutral with regard to jurisdictional claims in published maps and institutional affiliations.

1. Introduction

In the petroleum industry, reservoir simulation plays an important role for the decision-making process of oil field developments, where a common practice is to apply a closed-loop workflow [1] to develop and manage reservoir performance under different scenarios and conditions, as shown in Figure 1. This type of simulation is commonly used to predict and/or optimize future oil, gas, and water productions in a field (referred to as system). However, uncertainties and simplifications are involved in the construction of reservoir models (system models), due to the lack of sufficient information in, e.g., geology, seismic data, well logs and tests. Consequently, reservoir parameters (e.g., permeability, and porosity) are not accurately obtained, which makes reservoir models uncertain for production forecasts, and the corresponding decision-making process unreliable for reservoir management. In order to improve the reliability of production forecasts, one needs to condition reservoir models on sensored field datasets, through a process known as history matching in the literature. In this way, the estimated parameters of reservoir models can generate simulated outputs that are as close as possible to the field datasets.

In a standard history matching process, the field data used to estimate uncertain reservoir model parameters are well production data, which include bottom hole pressure, and oil, gas, and water rates. These kinds of measurements are frequent in time and relatively cheap in terms of acquisition cost, but their usefulness is often somewhat insufficient for the purpose of reducing uncertainties in production forecasts, due to their limited spatial coverage of the reservoir [2,3]. In addition to production data, other types of field data, e.g., inter-well tracer and 4D seismic data, can be used as complementary sources of information to further reduce the uncertainties-hence improve the reliability-of reservoir models, which constitutes the focus of the current study.

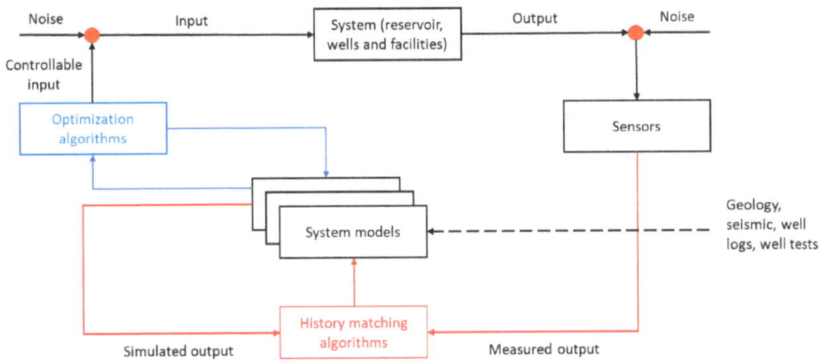

Figure 1. Closed-loop reservoir development and management [1].

Inter-well tracer test (IWTT) has been proven as an efficient technology to obtain information of fluid dynamics, well-to-well communication, and heterogeneities such as fractures and flow barriers (for a review of tracer studies see [4]). In the tracer technology, inert compounds (e.g., radioactive, chemical, or natural tracers) are used to label water or gas from specific wells, and trace fluid movements as the tracer flow moves through the reservoir together with the injection fluid. After the first breakthrough in a producer, IWTT provides a reliable and definite information of a well-to-well communication.

On the other hand, 4D seismic data contain much more information in space, especially about pressure and fluid saturation changes. Therefore, seismic data provide another useful source of information for understanding reservoir behavior, and identifying zones of remaining oils [5]. When used in combination with production and tracer data, they could further improve the qualities of reservoir models through history matching, and thus generate more reliable production forecasts for better decision making.

Despite the appealing features of tracer and 4D seismic data, they are still often qualitatively employed for reservoir monitoring and management in the petroleum industry, but are relatively less used in a quantitative way, due to various reasons arising from both theoretical understandings and practical applications to jointly match both types of field data in a coherent way. Tracer has been incorporated into history matching by a few authors [3,5–10]. Meanwhile, recently, several authors have demonstrated the benefits of incorporating seismic data into a history matching process [2,11–15]. Nevertheless, methodologies for jointly history matching production, tracer, and 4D seismic data have not been previously studied in the literature yet. Thus, there is a potential to include both tracer and 4D seismic data into a history matching workflow to further reduce the uncertainties and improve the reliability of reservoir models.

In the literature, there are different schools of approaches to tackling history matching problems. For instance, history matching can be formulated as a conventional minimization problem, in which an iterative algorithm is used, in combination with certain geostatistical modelling methods [16,17], to find the best matched model(s). Among the iterative history matching algorithms, ensemble-based methods have received immense attention in the petroleum industry, for their convenience in implementation and computational efficiency to handle big models and big data sets when compared to other conventional methods [18]. For instance, the ensemble Kalman filter (EnKF) introduced by [19] was initially extensively used for history matching problems [20,21]. As a sequential algorithm, however, the EnKF requires frequent simulator restarts, which may cause significant additional simulation time and other practical challenges (e.g., more prone to failed reservoir simulations) [22]. To avoid these noticed issues, ensemble smoother (ES) [23] and its iterative versions [24–28] can be adopted instead. In comparison to the EnKF, ES and iterative ES (IES) do not need simulator restarts and have less model variables to update during history matching, and are practically easier to implement than the EnKF for history matching problems. In addition, various numerical results (e.g., [25,26]) indicate that the IES tends to perform better than the EnKF and the ES. As a result, currently, the IES appears to be among the state-of-the-art approaches to history matching problems, and is thus adopted in the current work.

The recent developments of ensemble-based history matching algorithms make the quantitative use of multiple field data sets easier and faster. Despite this technological advancement, to the best of our knowledge, there are no existing studies demonstrating the benefits of simultaneously assimilating production, tracer and 4D seismic data into reservoir models through an ensemble-based history matching algorithm. As a main focus (and contribution) of this study, our primary objective is to demonstrate the benefits of assimilating tracer and 4D seismic data (in addition to production) in an integrated, ensemble-based history matching workflow, through a 3D field-scale case study.

This work is organized as follows: First, we provide details of the integrated ensemble-based history matching workflow, which includes the formulation of an IES algorithm and a correlation-based adaptive localization scheme. Second, we apply the history matching workflow to a 3D field case, the Brugge benchmark [29], and examine the performance of the proposed workflow. Finally, we conclude the work with some technical discussions and our future research plan.

2. Ensemble-Based History Matching Workflow

Figure 2 illustrates the flowchart of the integrated history matching workflow using multiple types of field data. To formulate the history matching problem, we assume there is a forward simulator $\mathbf{g}$ (e.g., a numerical reservoir and/or seismic simulator) which outputs a $N_{\rm d}$-dimensional vector containing the simulated data $\mathbf{d}^{sim}$, given a $N_{\rm m}$-dimensional vector of reservoir model $\mathbf{m}$ as the input:

$$\mathbf{d}^{sim} = \mathbf{g}(\mathbf{m}). \tag{1}$$

Note that, in Step 1 of Figure 2, the forward simulator generates an ensemble of simulated production and tracer data, with an ensemble of input reservoir models, while the forward seismic simulator produces an ensemble of simulated amplitude-versus-angle (AVA) data, with the associated petro-elastic model (PEM) and AVA at Steps 2 and 3, respectively, based on the simulated pressure and saturation profiles from the forward reservoir simulator, and the simulated velocities and density profiles from the PEM, respectively. The formulation of the forward seismic simulator is presented in Appendix A.

Furthermore, we have the observed (field) data $\mathbf{d}^o$ which are obtained through the following noisy observation system:

$$\mathbf{d}^o = \mathbf{g}(\mathbf{m}_{\rm true}) + \delta, \tag{2}$$

where $\mathbf{m}_{\rm true}$ stands for the ground-truth model (unknown in a real reservoir), $\mathbf{g}(\mathbf{m}_{\rm true})$ for the true output of the forward model, and δ for a $N_{\rm d}$-dimensional vector of contamination noise that may be present in the course of field data acquisition, which are assumed to follow a multivariate Gaussian distribution with zero mean and covariance (referred to as measurement error-covariance matrix) represented by $\mathbf{C}_{\rm d}$, i.e., $\delta \sim N(\mathbf{0}, \mathbf{C}_{\rm d})$.

With the above observation system, in Step 4 of Figure 2, a history matching algorithm is then adopted to find one or multiple reservoir models $\mathbf{m}$ whose simulated output $\mathbf{g}(\mathbf{m})$ matches the observed data $\mathbf{d}^o$ reasonably well. In practice, when the size of the observed data is much smaller than of the reservoir model parameters (e.g., $N_{\rm d} \ll N_{\rm m}$), history matching is an under-determined and high-dimensional inverse problem, which makes the history matching problem challenging, in the sense that in principle there could exist an infinite number of reservoir models matching the observed data equally well (non-uniqueness), due to the large degree of freedom (DOF) from the model side [30].

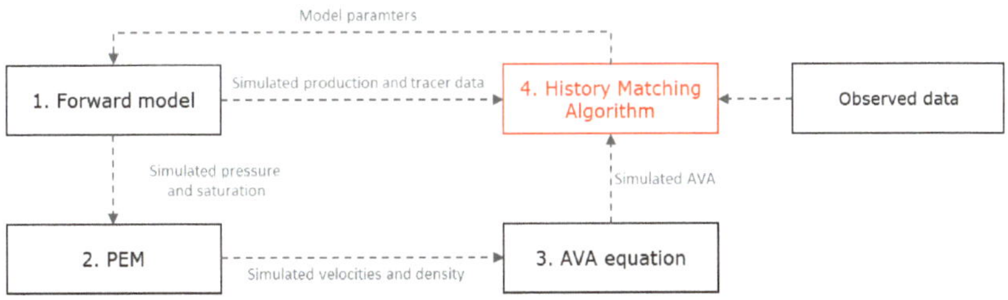

Figure 2. Integrated history matching workflow with multiple types of field datasets.

To mitigate this problem, one can either introduce a certain regularization term in the form of a least-square problem, or increase the number of field data in history matching [3]. Here we consider both aforementioned ideas, by including both a regularization term into a relevant cost function (cf. Equation (3)), and multiple types of field data into history matching.

Practical challenges may arise with multiple types of field data in history matching. For instance, different from production and tracer data, 4D seismic data are less frequent in time, but with a much larger size. Thus, when assimilating large datasets, computational issues regarding storage memory and CPU time may emerge. Meanwhile, large datasets could have a dominant influence on model updates in history matching. Moreover, with large field datasets, ensemble-based history matching methods are often prone to ensemble collapse (meaning that an ensemble of reservoir models also collapses into a single one, with few varieties among reservoir models), and thus tend to under-estimate model uncertainties [15]. Our approaches to handling the aforementioned issues include conducting a sparse representation of seismic data for dimension reduction, adjusting the relative weights among different types of field data to balance their influence on history matching, and conducting correlation-based localization to mitigate the issue of under-estimated uncertainties, which will be elaborated in Section 3 later.

2.1. Iterative Ensemble Smoother Based on a Regularized Levenburg–Marquardt Algorithm

As history matching problems are typically non-linear, a certain iterative method is needed to deal with the non-linearity. In the current work, we adopt one such method, called iterative ensemble smoother based on the regularized Levenburg–Marquardt algorithm (IES-RLM) [26].

In IES-RLM, one aims to find an ensemble $\mathcal{M}^{i+1} \equiv \{\mathbf{m}_j^{i+1}\}_{j=1}^{N_e}$ of reservoir models that approximately minimizes a nonlinear-least-squares (NLS) cost function, as follows:

$$\begin{aligned} &\underset{\{\mathbf{m}_j^{i+1}\}_{j=1}^{N_e}}{\arg\min} \frac{1}{N_\mathrm{e}} \sum_{j=1}^{N_\mathrm{e}} L_j^{i+1}\left(\mathbf{m}_j^{i+1}\right), \\ &L_j^{i+1}\left(\mathbf{m}_j^{i+1}\right) \equiv \left(\mathbf{d}_j^o - \mathbf{g}\left(\mathbf{m}_j^{i+1}\right)\right)^\mathrm{T} \mathbf{C}_\mathrm{d}^{-1} \left(\mathbf{d}_j^o - \mathbf{g}\left(\mathbf{m}_j^{i+1}\right)\right) + \gamma^i \left(\mathbf{m}_j^{i+1} - \mathbf{m}_j^i\right)^\mathrm{T} \left(\mathbf{C}_\mathrm{m}^i\right)^{-1} \left(\mathbf{m}_j^{i+1} - \mathbf{m}_j^i\right), \end{aligned} \tag{3}$$

where i stands for the index of iteration step, j for the index of ensemble member, N_e for the ensemble size, and $\mathbf{C}_\mathrm{m}^i$ for the sample model error-covariance matrix with respect to the prior ensemble $\mathcal{M}^i \equiv \{\mathbf{m}_j^i\}_{j=1}^{N_e}$, in the form of $\mathbf{C}_\mathrm{m}^i = \mathbf{S}_\mathbf{m}^i (\mathbf{S}_\mathbf{m}^i)^\mathrm{T}$, with the squared-root matrix $\mathbf{S}_\mathbf{m}^i$ defined in Equation (5).

The NLS-type cost function in Equation (3) contains two terms: the first one (called data mismatch term) calculates the difference between observed and simulated data, and the second one (called regularization term) imposes a constraint that aims to prevent a large deviation of the updated model $\mathbf{m}_j^{i+1}$ from its predecessor $\mathbf{m}_j^i$. Here, the regularization form is known as Tikhonov regularization [30], but one can also use other types of regularization [27]. Meanwhile, $\gamma^i > 0$ in Equation (3) is a coefficient influencing the relative weight between the data mismatch and regularization terms, whose value changes over the iteration steps [26]. In this optimization problem, the goal is to minimize the average of an ensemble of cost functions at each iteration step, until a certain stopping criterion is reached. The stopping criteria used in this work consist of the following three ones: (1) the maximum number of iteration is reached; (2) the relative change of data mismatch values in-between two consecutive iteration steps is less than 0.01%; (3) the data mismatch value at a given iteration step is lower than the number of observed data points (N_d) times a factor (1 in this study).

By minimizing the cost function in Equation 3, one can obtain the following approximate model update formula:

$$\mathbf{m}_j^{i+1} = \mathbf{m}_j^i + \mathbf{S}_\mathbf{m}^i \left(\widetilde{\mathbf{S}}_\mathbf{d}^i\right)^\mathrm{T} \left(\widetilde{\mathbf{S}}_\mathbf{d}^i \left(\widetilde{\mathbf{S}}_\mathbf{d}^i\right)^\mathrm{T} + \gamma^i \mathbf{I}\right)^{-1} \Delta\widetilde{\mathbf{d}}_j^i, \tag{4}$$

where the square root matrix $\mathbf{S}_\mathbf{m}^i$ in the model space is defined as

$$\mathbf{S}_\mathbf{m}^i = \frac{1}{\sqrt{N_e - 1}} \left[\mathbf{m}_1^i - \overline{\mathbf{m}}^i, \cdots, \mathbf{m}_{N_e}^i - \overline{\mathbf{m}}^i\right]; \tag{5}$$

$$\overline{\mathbf{m}}^i = \frac{1}{N_e} \sum_{j=1}^{N_e} \mathbf{m}_j^i. \tag{6}$$

Similar to Equation (5), one can define a square root matrix $\mathbf{S}_\mathbf{d}^i$ in the data space as

$$\mathbf{S}_\mathbf{d}^i = \frac{1}{\sqrt{N_e - 1}} \left[\mathbf{g}(\mathbf{m}_1^i) - \mathbf{g}(\overline{\mathbf{m}}^i), \cdots, \mathbf{g}(\mathbf{m}_{N_e}^i) - \mathbf{g}(\overline{\mathbf{m}}^i)\right]. \tag{7}$$

In many practical history matching problems, the observations may contain different types of field data in different orders of magnitudes. To mitigate potential issues related to these imbalanced magnitudes, one can introduce a normalization procedure to quantities in the observation space, so that $\mathbf{S}_\mathbf{d}^i$ and $\left(\mathbf{d}_j^o - \mathbf{g}(\mathbf{m}_j^i)\right)$ are normalized by a square root $\mathbf{C}_\mathbf{d}^{-1/2}$ of $\mathbf{C}_\mathrm{d}$ ($\mathbf{C}_\mathrm{d} = \mathbf{C}_\mathbf{d}^{1/2}(\mathbf{C}_\mathbf{d}^{1/2})^\mathrm{T}$) to arrive at the following formulations:

$$\widetilde{\mathbf{S}}_\mathbf{d}^i = \mathbf{C}_\mathbf{d}^{-1/2} \mathbf{S}_\mathbf{d}^i, \tag{8}$$

$$\Delta\widetilde{\mathbf{d}}_j^i = \mathbf{C}_\mathbf{d}^{-1/2} \left(\mathbf{d}_j^o - \mathbf{g}(\mathbf{m}_j^i)\right). \tag{9}$$

In a practical implementation of the model update formula Equation (4), numerical stability could be an issue when inverting the matrix $(\widetilde{\mathbf{S}}_\mathbf{d}^i (\widetilde{\mathbf{S}}_\mathbf{d}^i)^\mathrm{T} + \gamma^i \mathbf{I})$ (especially when using 4D seismic data), since the data size N_d is typically much larger than the ensemble size N_e. Therefore, a common procedure introduced in the literature to obtain a numerically more stable algorithm is to carry out an inversion for a certain matrix with a lower dimension (less than the ensemble size N_e) by applying a Truncated Singular Value Decomposition (TVSD) to the matrix $\widetilde{\mathbf{S}}_\mathbf{d}^i$ [24,31]. Suppose that, through the TSVD, one obtains

$$\widetilde{\mathbf{S}}_\mathbf{d}^i \approx \hat{\mathbf{U}}^i \hat{\mathbf{\Sigma}}^i (\hat{\mathbf{V}}^i)^\mathrm{T}, \tag{10}$$

where $\hat{\mathbf{U}}^i \equiv [\mathbf{u}_1^i, \cdots, \mathbf{u}_{\mathrm{sv}}^i]$ and $\hat{\mathbf{V}}^i \equiv [\mathbf{v}_1^i, \cdots, \mathbf{v}_{\mathrm{sv}}^i]$ are unitary matrices consisting of the kept left and right eigenvectors of $\widetilde{\mathbf{S}}_\mathrm{d}$, respectively, and $\hat{\mathbf{\Sigma}} \in \mathbb{R}^{\mathrm{sv} \times \mathrm{sv}}$ is a rectangular diagonal matrix containing in its diagonal a set of retained leading singular values. The integer sv is chosen as suggested by [31] to keep the leading singular values that add up to at least 99% of the total sum of squared singular values.

Inserting Equation (10) into Equation (4), one obtains a modified update formula

$$\mathbf{m}_j^{i+1} = \mathbf{m}_j^i + \mathbf{S}_{\mathbf{m}}^i \hat{\mathbf{V}}^i (\hat{\boldsymbol{\Sigma}}^i)^{\mathrm{T}} \left(\hat{\boldsymbol{\Sigma}}^i \left(\hat{\boldsymbol{\Sigma}}^i \right)^{\mathrm{T}} + \gamma^i \mathbf{I} \right)^{-1} (\hat{\mathbf{U}}^i)^{\mathrm{T}} \Delta \widetilde{\mathbf{d}}_j^i. \tag{11}$$

In the current work, the regularization parameter γ^i in Equation (11) is chosen as

$$\gamma^i = \eta^i \times Tr\left((\hat{\boldsymbol{\Sigma}}^i)^{\mathrm{T}} \hat{\boldsymbol{\Sigma}}^i \right) / N_{\mathrm{sv}}, \tag{12}$$

where η^i is a positive number that starts with a value of 1 and varies as a function of the iteration step, following the rule in [26]; and the operator Tr calculates the trace of a matrix.

2.2. Correlation-Based Automatic and Adaptive Localization

In practical applications of ensemble-based methods, a limited number of reservoir models is typically adopted to reduce computational cost. Under this setting, certain numerical issues may arise when the number of reservoir models N_{e} is significantly smaller than the sizes of reservoir model N_{m} and field data N_{d}. For instance, in the context of ensemble-based history matching, the limited number of reservoir models can produce spurious correlations between uncorrelated reservoir model parameters and observation data, leading to unsatisfactory history matching performance due to the excessive reduction of ensemble variability (e.g., ensemble collapse). To mitigate the negative effects of the limited number of reservoir models, it is common to adopt an auxiliary technique, called (Kalman gain) localization [32–35] for ensemble history matching algorithms.

For an easier explanation of the main idea behind localization, one can rearrange Equation (11) as

$$\mathbf{m}_j^{i+1} = \mathbf{m}_j^i + \mathbf{K}^i \Delta \widetilde{\mathbf{d}}_j^i, \tag{13}$$

with the Kalman gain matrix $\mathbf{K}^i \in \mathbb{R}^{N_{\mathrm{m}} \times N_{\mathrm{d}}}$ defined as

$$\mathbf{K}^i = \mathbf{S}_{\mathbf{m}}^i \hat{\mathbf{V}}^i (\hat{\boldsymbol{\Sigma}}^i)^{\mathrm{T}} \left(\hat{\boldsymbol{\Sigma}}^i \left(\hat{\boldsymbol{\Sigma}}^i \right)^{\mathrm{T}} + \gamma^i \mathbf{I} \right)^{-1} (\hat{\mathbf{U}}^i)^{\mathrm{T}}. \tag{14}$$

To conduct localization in Equation (13), one can replace the Kalman gain $\mathbf{K}^i$ by a Schur product (element-wise product) between $\mathbf{K}$ and a localization matrix $\mathbf{C}$, as in

$$\mathbf{m}_j^{i+1} = \mathbf{m}_j^i + \left(\mathbf{C} \circ \mathbf{K}^i \right) \Delta \widetilde{\mathbf{d}}_j^i. \tag{15}$$

Here, the localization matrix $\mathbf{C}$ is implemented as a tapering matrix, introduced to assign different weights (tapering coefficients) for different combinations of reservoir model parameters and field data points. Note that, in its conventional form, the Kalman gain matrix needs to be computed at each iteration step, which could be a very high-dimensional matrix, especially with big observation data (e.g., 4D seismic data) and big reservoir models. To deal with the high dimensionality of $\mathbf{K}^i$, one can choose to sparsely represent big observation data in another domain, so as to obtain a much smaller number of data points used in history matching, as will be explained later.

Equivalently, Equation (15) can be expressed as:

$$\begin{aligned} \mathrm{m}_{j,k}^{i+1} &= \mathrm{m}_{j,k}^i + \sum_{s=1}^{N_{\mathrm{d}}} (\mathrm{c}_{\mathrm{ks}} \mathrm{k}_{\mathrm{ks}}^i) \Delta \widetilde{\mathrm{d}}_{j,s}^i \\ &= \mathrm{m}_{j,k}^i + \sum_{s=1}^{N_{\mathrm{d}}} \mathrm{k}_{\mathrm{ks}}^i (\mathrm{c}_{\mathrm{ks}} \Delta \widetilde{\mathrm{d}}_{j,s}^i). \end{aligned} \tag{16}$$

where $c_{ks} \in [0,1]$ and k^i_{ks} stand for the elements of $\mathbf{C}$ and $\mathbf{K}^i$ on the k-th row and the s-th column, respectively, $\Delta\tilde{d}^i_{j,s}$ is the s-th element of $\Delta\tilde{\mathbf{d}}^i_j$, and $m^{i+1}_{j,k}$, and $m^i_{j,k}$ the k-th element of the vectors $\mathbf{m}^{i+1}_j$, and $\mathbf{m}^i_j$, respectively.

There are different ways to compute the tapering coefficients, c_{ks}, in the literature. Among them, a common approach appears to be distance-based localization [32,33,36], in which the observations and reservoir model variables are assumed to have physical locations, so that one can compute the distance between the physical locations of a model variable and a field data point. This approach can work well in many situations. However, there are some notable issues. For instance, the observation data need to have associated physical locations, the tapering function is not adaptive to reservoir heterogeneities, and the difficulty to "localize" non-local observations when there is a long range of correlations between model variables and field data.

To deal with the aforementioned issues, ref. [35] proposed a correlation-based adaptive localization scheme. In this approach, the authors computed the tapering values c_{ks} dependent on the sample correlation ρ_{ks} between the k-th element of the initial ensemble $m^0_{j,k}$ and the corresponding initial ensemble of the s-th innovation element $\Delta\tilde{d}^0_{j,s}$ $(j = 1, 2, \cdots, N_e)$. Then, they defined a threshold value θ_s dependent on the noise level (standard deviation, STD) of the sampling errors. To compute the noise level, one can assume that the members of the initial ensemble $\mathbf{m}^0_j$ $(j = 1, 2, \cdots, N_e)$ are independent and identically distributed (i.i.d). Due to the independence assumption, one can obtain a new ensemble $\hat{\mathbf{m}}^0_j$ by randomly shuffling the indices j of $\mathbf{m}^0_j$. After that, one can compute the sampling errors $\hat{\boldsymbol{\rho}}_s$, which represents the correlation field between the new ensemble $\hat{\mathbf{m}}^0_j$ $(j = 1, 2, \cdots, N_e)$ and the ensemble of the s-th innovation-data point $\Delta\tilde{d}^i_{j,s}$ $(j = 1, 2, \cdots, N_e)$. Under the i.i.d assumption, $\hat{\mathbf{m}}^0_j$ and $\mathbf{m}^0_j$ are independent. As a result, one can take the correlation field $\hat{\boldsymbol{\rho}}_s$ as a realization of sampling errors of the sample correlations between all model parameters and the s-th innovation-data point. Then, for each type of petro-physical parameter and each field data point, one can estimate the noise level $\hat{\sigma}_s$ with respect to the sampling errors $\hat{\boldsymbol{\rho}}_s$ as follows [35]:

$$\hat{\sigma}_s = \frac{\text{median}(\text{abs}(\hat{\boldsymbol{\rho}}_s))}{0.6745}. \tag{17}$$

After that, one can further compute the threshold value θ_s by

$$\theta_s = \hat{\sigma}_s \sqrt{2\ln(\#\hat{\boldsymbol{\rho}}_s)}, \tag{18}$$

where $\#\hat{\boldsymbol{\rho}}_s$ is the number of elements in $\hat{\boldsymbol{\rho}}_s$. Note that one should perform this procedure for each type of petro-physical parameter and each field data point.

Finally, one can generate a smooth tapering function as in [35]

$$z = \frac{1 - \text{abs}(\rho_{ks})}{1 - \theta_s}, \tag{19}$$

where the k-th element ρ_{ks} of $\boldsymbol{\rho}_s$ is the sample correlation between a k-th model variable and the s-th field data point. Then, the variable z is used in the Gaspari–Cohn function (Gaspari and Cohn, 1999)

$$c_{ks} = f_{GC}(z) = \begin{cases} -\frac{1}{4}z^5 + \frac{1}{2}z^3 + \frac{5}{8}z^3 - \frac{5}{3}z^2 + 1, & \text{if } 0 \le z \le 1 \\ -\frac{1}{12}z^5 - \frac{1}{2}z^4 + \frac{5}{8}z^3 + \frac{5}{3}z^2 - 5z + 4 - \frac{2}{3}z^{-1}, & \text{if } 1 < z \le 2 \\ 0, & \text{if } z > 2, \end{cases} \tag{20}$$

to finally obtain the tapering coefficient c_{ks}.

3. Application to the Brugge Field Case Study

The Brugge field model is a benchmark case based on a North sea field and developed by [29], and it has the characteristics and complexities of reservoir models used in real field case studies. The Brugge field consists of two zones, one with higher permeability and porosity, and the other with lower permeability and porosity. The zone with high permeability and porosity is located on layers 1–2 and layers 6–8, while the other with low permeability and porosity is located on layers 3–5 and 9.

The dimensions of the reservoir model are 139 × 48 × 9, summing up to 60,048 grid-block, in which 44,550 are active. Figure 3 shows the location of the field wells, including 30 wells (20 inner producers and 10 injectors). The producers are located more centrally in the red area and labelled as BR-P-1-BR-P-20, while the injectors are located on the border and labelled as BR-I-1-BR-I-10, with water being the only injected fluid.

The initial ensemble of this benchmark contains 104 realizations of reservoir models, from which we take the first realization (#1) as our reference (true) case to generate the observed data. As for reservoir model variables, we consider porosity (PORO) and permeability along x-, y-, and z- directions (denoted by PERMX, PERMY, and PERMZ), summing up to 178,200 (4 × 44,550) uncertain petro-physical parameters to estimate in history matching.

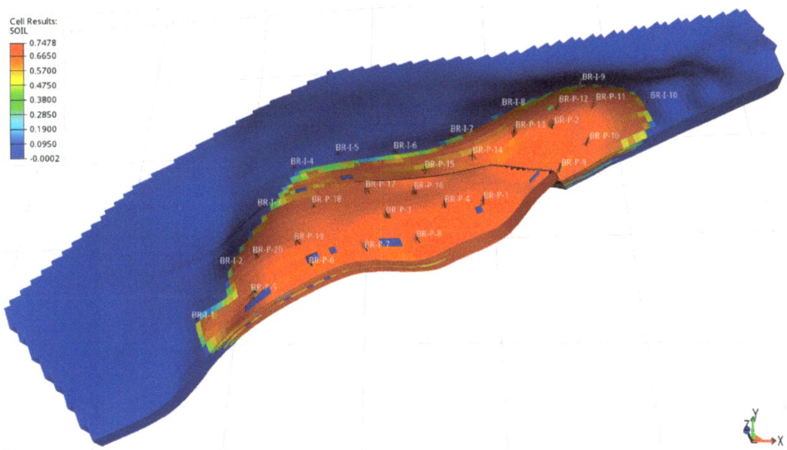

Figure 3. A numerical model of the Brugge field with initial oil saturation and distribution of the wells.

For illustration, Figures 4 and 5 show the distributions of the porosity and the log permeability (along the x-direction), respectively, on each layer from realization number two (#2) of the initial ensemble. As reported in [29], the permeability and porosity are linearly correlated, and the spatial distribution of the permeability is anisotropic, meaning that permeability distributions along different directions may not be the same.

In history matching, we condition reservoir models on three types of field data, namely, production, tracer, and 4D seismic data. The historical production data cover a period of 3647.5 days, with production forecasts from 3647.5 to 9869.5 days, where reservoir simulations are conducted using a black oil simulator (ECLIPSE©). The production data include well bottom hole pressure (WBHP), well oil and water rates (WOPR, and WWPR), summing up to 1400 data points. In addition, an injection of a water passive tracer pulse of 1.0 lb per STB is performed from day 749 to day 1812 in the injector well BR-I-6 (WTPCW06), summing up to 400 data points. To mimic the presence of measurement noise in real production data, here we also introduce certain zero-mean Gaussian white noise to the reference production and tracer data, for which the STD of the noise is set as 10% of the magnitude of each production or tracer data point, except for WBHP (for which the noise STD is set as 50 psi for each data point instead).

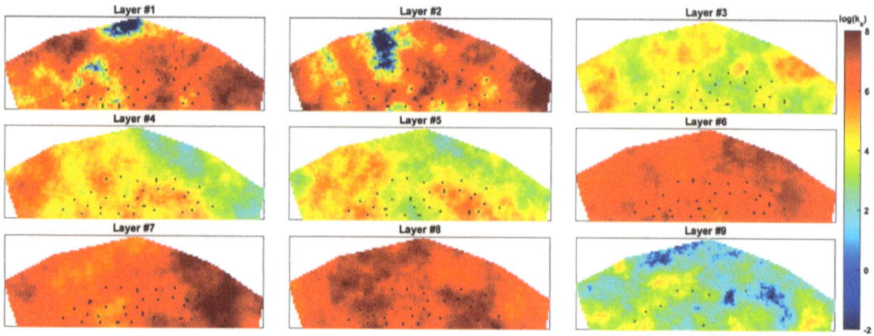

Figure 4. Log permeability along the x-direction, with respect to realization #2 of the initial ensemble from the Brugge field dataset.

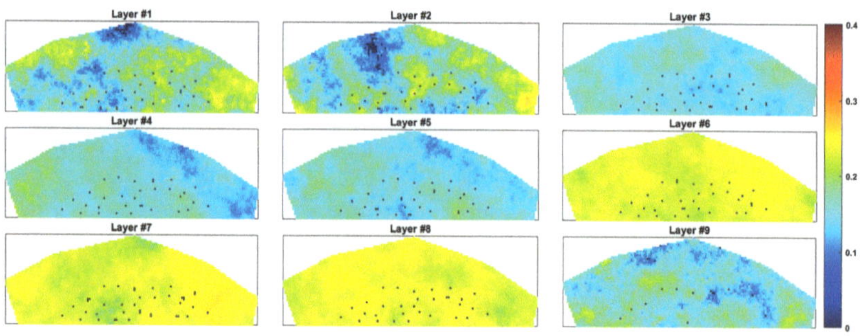

Figure 5. Porosity with respect to realization #2 of the initial ensemble from the Brugge field dataset.

The 4D seismic data are obtained from three surveys: base (day 1), monitor #1 (day 991), and monitor #2 (day 2999). From each seismic survey, the seismic attributes (AVA data) are obtained using two different offset angles, near (10°) and mid (20°). The AVA data are in a seismic domain and do not possess physical locations in the reservoir coordinate system, meaning that the AVA data are in a different dimension compared to the dimension of the reservoir model. In this work, the dimension of the AVA data is $139 \times 48 \times 176$ for each seismic survey, summing up to 7,045,632 data points in total [15]. We also introduce certain zero-mean Gaussian white noise to the AVA data, with the STD of the noise set to 30% of the magnitude as in [15].

In order to deal with the issue of big seismic data in history matching, and thereby the required memory, we choose to sparsely represent the AVA data by first applying 3D discrete wavelet transform (DWT) to the seismic data, and then (through a thresholding operation) selecting a small set of the resulted leading wavelet coefficients as the observations in history matching. For brevity, we skip the details of the DWT-based sparse data representation procedure. Readers are referred to [15,37] for more information.

From the algorithmic perspective, in the presence of the DWT-based sparse data representation procedure, the effective observation system with respect to 4D seismic data becomes

$$\mathcal{T}(\mathcal{W}(\mathbf{d}^o)) = \mathcal{T}(\mathcal{W}(\mathbf{g}^s(\mathbf{m}_{\text{true}})) + \mathcal{W}(\delta)), \tag{21}$$

where $\mathcal{W}$ and $\mathcal{T}$ denote wavelet transformation and thresholding operators, respectively, and $\mathbf{g}^s$ corresponds to the seismic forward simulator, as discussed in Appendix A. By using the aforementioned formulation, we are able to represent the original 4D seismic (7,045,632 data points) by a much smaller set of 1896 wavelet coefficients, as elaborated in [15].

Experiment Settings

In our previous work [3], we have shown that adopting both production and tracer data in the Brugge benchmark helps improve the performance of history matching, in comparison to the choice of using production data only. The current work can be considered as a follow-up study, in which we aim to further examine the impacts of 4D seismic data on history matching, and show the complexity of the joint history matching problem in the presence of multiple types of field data. To this end, in the current work we conduct two experiments complementary to those in [3]:

- **Case 1:** using production and 4D seismic data;
- **Case 2:** using production, tracer, and 4D seismic data.

In both cases, we apply the DWT-based sparse data representation procedure to the seismic attributes obtained at three surveys for improved computational efficiency.

We use IES-RLM as the history matching algorithm, and equip it with the correlation-based adaptive localization scheme described in Section 2.2. We start the IES-RLM algorithm with $\eta^0 = 1$, with the reduction and increment factor being, 0.9 and 2, respectively. More specifically, if the average data mismatch is reduced at the current iteration step, then the η value at the next iteration step is set to 0.9 times that at the current step; otherwise, the next η value is set to 2 times the value of the current iteration step, and an inner loop is then activated to search for lower data mismatch.

To simultaneously history match multiple types of field data, we introduce a normalization procedure to adjust the relative weights assigned to different types of field data, as in [2]. Following the formulation in [3], we substitute the measurement error-covariance matrix $\mathbf{C}_\mathrm{d}$ in Equation (3) by the Schur product $\mathrm{diag}(\mathbf{w}) \circ \mathbf{C}_\mathrm{d}$, where $\mathbf{w}$ is a vector to be specified later, and the operator $\mathrm{diag}(\mathbf{w})$ stands for a diagonal matrix whose diagonal elements are taken from the vector $\mathbf{w}$.

For production and tracer data, the elements of $\mathbf{w}$ are determined using the following rule:

$$w_k = (max(\mathrm{type}(\mathbf{d}_k^o)) \times p(\mathrm{type}(\mathbf{d}_k^o))/\sigma_k)^2, \text{ for } k = 1, 2, \cdots, N_d, \tag{22}$$

where the operator $max(\mathrm{type}(\mathbf{d}_k^o))$ takes the maximum value of a given type of field data, and $\mathrm{type}(\mathbf{d}_k^o)$ stands for the type (WBHP, WOPR, WWPR or WTPCW06) of the k-th field-data point. Note that the value $max(\mathrm{type}(\mathbf{d}_k^o))$ is normalized by the measurement-noise STD σ_k associated with the data point $\mathbf{d}_k^o$, aiming to mitigate potential problems caused by the different orders of magnitudes of the field data. The notation p represents a percentage value, which is set to 5% if $\mathbf{d}_k^o$ is a type of production data (WBHP, WOPR, or WWPR), and to 20% instead if $\mathbf{d}_k^o$ is a tracer data point (WTPCW06).

For seismic data, the elements of $\mathbf{w}$ are determined using the following rule:

$$w_k = \left(\left(\frac{median(abs(\hat{\mathcal{W}}(\mathbf{d}_k^o)))}{0.6745} \right) / \sigma_k \right)^2, \tag{23}$$

where $\hat{\mathcal{W}}(\mathbf{d}_k^o))$ is the wavelet coefficient belonging to the finest sub-band of $\mathcal{W}(\mathbf{d}_k^o)$.

As aforementioned, 4D seismic data usually contain a large number of data points, which can dominate the updates during history matching. To overcome this noticed issue, a certain scaling factor s is introduced to adjust the relative weights assigned to production, tracer and 4D seismic data during history matching. The scaling factor is determined using data mismatch of the initial ensemble, which is computed using the following formula:

$$\mathrm{HM}_j^i = (\Delta\widetilde{\mathbf{d}}_j^i)^T \Delta\widetilde{\mathbf{d}}_j^i. \tag{24}$$

By using Equation (24), one can compute the scaling factor as follows:

$$s = \overline{HM_1^0} / \overline{HM_2^0}, \tag{25}$$

where the overline denotes the mean value over the ensemble members, $\overline{HM_1^0}$ corresponds to the mean data mismatch value of 4D seismic data, and $\overline{HM_2^0}$ to that of production data (Case 1) or that of both production and tracer data (Case 2). Note that in Case 2, we group production and tracer data together to compute the mean of data mismatch. To construct the scaled observational data, one needs to multiply the observation data from group 2 by the scalar s, so that $\Delta\widetilde{\mathbf{d}}_{j,2}^{i}$ is replaced by $[\Delta\widetilde{\mathbf{d}}_{j,2}^{i} \times s]$.

To assess uncertainty reduction for each type of petro-physical parameter (PERMX, PERMY, PERMZ, and PORO) in the experiments, we use the Sum of Normalized Variance (SNV) [38], as in

$$\text{SNV} = \sum_{k=1}^{N_\text{m}} \frac{\text{var}(m_k^f)}{\text{var}(m_k^0)}, \tag{26}$$

where $\text{var}(m_k^0)$ and $\text{var}(m_k^f)$ denote the variances of a particular type of petro-physical parameter distributed on the k-th active reservoir gridblock of the initial and final ensembles, respectively.

Finally, to assess and compare history matching performance, in terms of the discrepancy between an estimated reservoir model $\mathbf{m}$ and the true one $\mathbf{m}_\text{true}$, we use the Root Mean Squared Error (RMSE), as in

$$\text{RMSE} = \frac{\| \mathbf{m} - \mathbf{m}_\text{true} \|_2}{\sqrt{N_\text{m}}}. \tag{27}$$

4. Results and Discussions

This section focuses on illustrating the performance of the history matching of the two aforementioned experiments, in terms of data mismatch values during history matching (which is calculated through Equation (24)) and forecast periods, and also uncertainty quantification that is represented by mean, standard deviation, and RMSE of the final ensemble.

Table 1 summarizes the values of data mismatch ($mean \pm STD$) and RMSE ($mean \pm STD$) with respect to the initial ensemble and the final ensembles of the two experiments' settings (Cases 1 and 2). In comparison with the initial ensemble, both experiments exhibit better history matching performance, in terms of both lower data mismatch and RMSE values. Comparing the results between the two different experiments shows that we achieve slightly lower data mismatch for production data (−0.2389%), but at the cost of a slightly higher data mismatch for 4D seismic data (+0.9862%), when jointly history matching production, tracer, and 4D seismic data (Case 2). In terms of RMSE, Case 2 contains lower average values for PERMY and PERMZ (−1.0565% and −3.3100%, respectively), but slightly higher ones for PERMX and PORO (+1.6433% and +0.7905%, respectively). In addition, the average RMSE values for all parameters together in Case 2 is slightly lower in comparison with Case 1, which indicates that the overall history matching performance in terms of RMSE tends to improve by adding more types of field data, but the complexity of the history matching process tends to increase. Note that in Case 2, only inter-well tracer from injector BR-I-6 was included in the history matching workflow, which provides additional information regarding well-to-well connectivity for the producers (BR-P-3, BR-P-4, BR-P-14, BR-P-15, BR-P-16, and BR-P-17), where tracer is detected. However, one cannot expect a substantial reduction in the overall data mismatch and RMSE values due to the use of the tracer data, as the inter-well tracer is usually detected in a limited number of production wells in the reservoir and tracer from a single injector well may not be sufficient to provide information to improve the connectivity of the entire reservoir. We discuss more about reservoir connectivity in the forecast analysis.

Table 1. Values of data mismatch and RMSE with respect to the initial ensemble, and the final ensembles of Cases 1 and 2.

	Initial Ensemble	Case 1	Case 2
Production data mismatch	$(7.7861 \pm 4.5592) \times 10^3$	$1.1586 \times 10^3 \pm 1.4297 \times 10^2$ (-85.1196%)	$1.1400 \times 10^3 \pm 1.1465 \times 10^2$ (-85.3585%)
Tracer data mismatch	$(3.9620 \pm 2.6721) \times 10^1$		$6.0112 \pm 9.3504\ (-84.8279\%)$
4D Seismic data mismatch	$(9.2270 \pm 3.4727) \times 10^3$	$(6.3050 \pm 1.6124) \times 10^3$ (-31.6679%)	$(6.3960 \pm 1.6680) \times 10^3$ (-30.6817%)
RMSE (PERMX)	1.5578 ± 0.5356	$1.0284 \pm 0.2160\ (-33.9838\%)$	$1.0540 \pm 0.2332\ (-32.3405\%)$
RMSE (PERMY)	1.5524 ± 0.5315	$1.0686 \pm 0.2280\ (\ 31.1646\%)$	$1.0522 \pm 0.2286\ (-32.2211\%)$
RMSE (PERMZ)	1.7523 ± 0.5454	$1.2992 \pm 0.2620\ (-25.8574\%)$	$1.2412 \pm 0.2276\ (-29.1674\%)$
RMSE (PORO)	0.0253 ± 0.0039	$0.0240 \pm 0.0034\ (-5.1383\%)$	$0.0242 \pm 0.0035\ (-4.3478\%)$
RMSE (all parameters)	1.4067 ± 0.4634	$0.9866 \pm 0.2011\ (-29.8642\%)$	$0.9700 \pm 0.1961\ (-31.0443\%)$

For illustration, Figure 6 reports the ensemble of data mismatch values obtained at each iteration step, in the form of box plots in Cases 1 and 2. In both cases, data mismatch values tend to decrease over the iteration steps for all types of field data. In addition, one can observe that main reductions of data mismatch values take place within the first few iteration steps for production data, and afterwards the changes of data mismatch values are less substantial. These main reductions in data mismatch are related to the changes of well bottom hole pressure (WBHP), which is strongly reduced in the first few iterations steps. Furthermore, the IES-RLM algorithm is close to convergence after 10 iterations steps, which is the maximum number of iterations used in the case studies.

Similarly, Figure 7 shows box plots of RMSE values in the two case studies, for PERMX, PERMY, PERMZ and PORO, respectively. As one can see, the average RMSE values also tend to decrease over the iterations steps. Consistent with Table 1, the results show that jointly history matching production, tracer and 4D seismic data (Case 2) tends to result in lower mean RMSE values for PERMY and PERMZ, but slightly higher ones for PERMX and PORO, in comparison with the choice of history matching both production and 4D seismic data (Case 1).

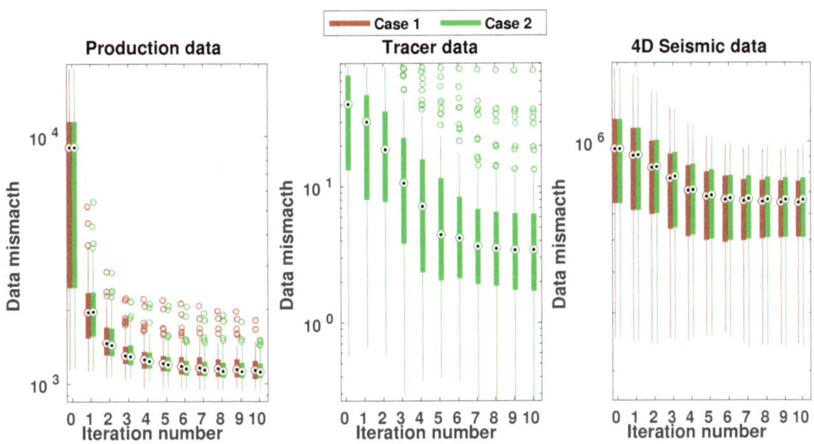

Figure 6. Box plots of data mismatch with respect to production (**left**), tracer (**middle**), and 4D seismic (**right**) data at different iteration steps, where the vertical axes are in the logarithmic scale. The experiment results are represented by a colour scheme, in which the results of Case 1 correspond to red color; whereas those of Case 2 to green. Note that Case 1 does not use tracer data. Therefore, there is no corresponding result (in red) in the middle sub-figure.

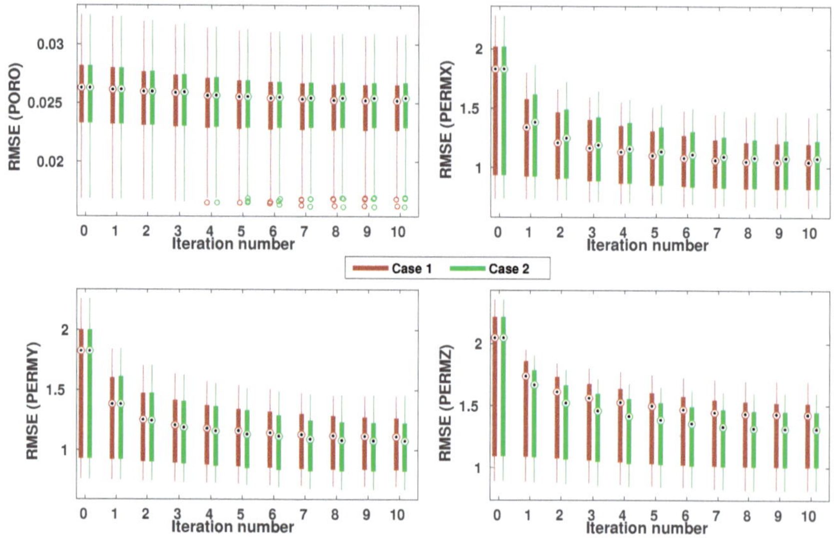

Figure 7. Box plots of RMSE values at different iteration steps, with respect to porosity (**upper left**), and permeability in x-, y-, and z-directions (**upper right**, **bottom left**, and **bottom right**, respectively).

Figure 8 presents the profiles of WOPR and WWPR from well BR-P-15 and BR-P-16, with respect to the initial and final ensembles in the two case studies. In comparison to the initial ensemble, both cases indicate significant improvements in terms of reduced data mismatch and uncertainty (in terms of ensemble variability). For instance, compared to the choice of history-matching production and 4D seismic data (Case 1), one can see that the inclusion of tracer in Case 2 tends to slightly reduce the variability of the final ensemble of WWPR in BR-P-16, which means that the final ensemble follows the observed data better.

Similar to Figure 8, Figure 9 shows the profiles of tracer data injected in well BR-I-6 and recorded at wells BR-P-15 and BR-P-16, with respect to the initial and final ensembles in the two case studies. Here, the history matching performance is reasonably good in terms of data mismatch, e.g., for both production wells, history matching helps reduce data mismatch of the final ensembles, which is consistent with Table 1.

In addition, Table 2 provides an assessment of parameter uncertainty reduction (in terms of SNV) in the two aforementioned experiments. As one can see, in both case studies, the final ensembles maintain substantial variability, meaning that the localization scheme is useful for preserving ensemble variability. Consistent with Table 1, Case 2 contains slightly lower SNV values for PERMY and PERMZ, while slightly higher ones for PORO and PERMX, in comparison with Case 1. As tracer data are more correlated to permeability, one may expect more uncertainty reduction for permeability than porosity. This is partially reflected by the results in Table 2, where there is a relatively stronger uncertainty reduction for PERMY and PERMZ (−0.51% and −3.09%, respectively).

Table 2. Parameter uncertainty reduction assessment in the different experiments.

Experiments	SNV (PORO)	SNV (PERMX)	SNV (PERMY)	SNV (PERMZ)
Case 1	92.43%	69.13%	69.79%	75.02%
Case 2	92.94%	69.27%	69.28%	71.93%

Another important aspect is to inspect the spatial distributions and standard deviations of the petro-physical parameters after history matching. Figures 10 and 11 report mean and standard deviation maps of the petro-physical parameters on layer 2, with respect to the initial and final ensembles in Cases 1 and 2. After history matching, one can see that there are minor changes among different experiment results. Relative to the mean maps, our history matching workflow is able to capture the main aspects of the initial ensemble in both experiments. Meanwhile, the standard deviation maps of the two experiments indicate substantial variability after history matching, which means that the ensemble collapse is avoided.

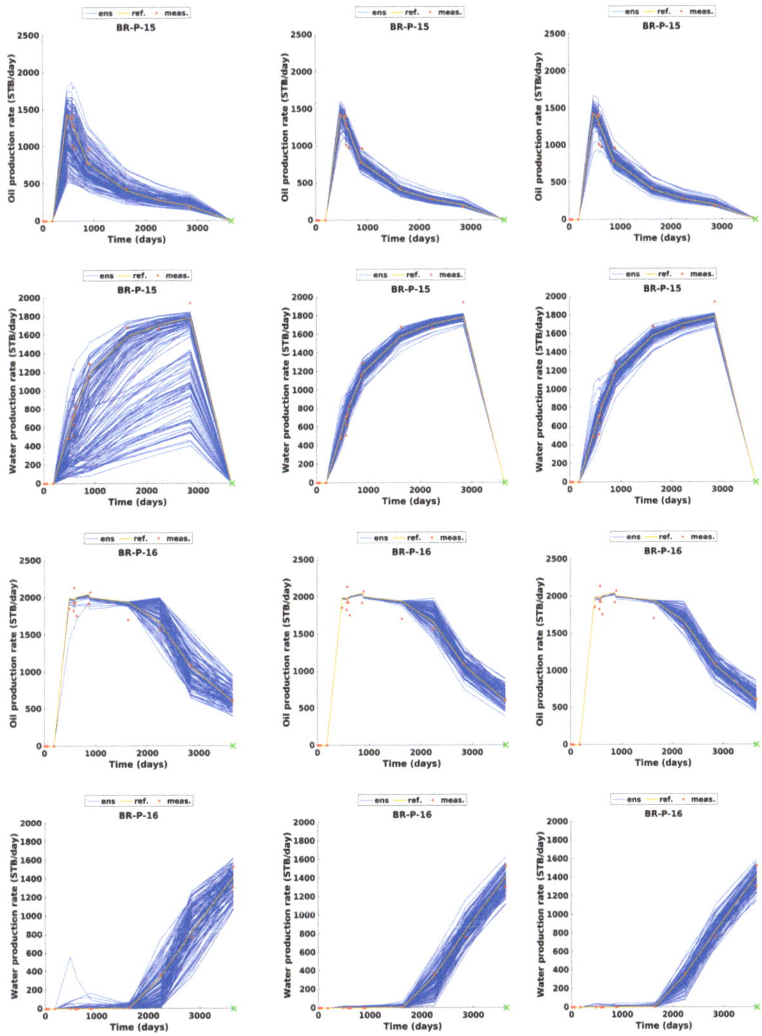

Figure 8. Profiles of WOPR (**first and third rows**) and WWPR (**second and fourth rows**) from well BR-P-15 and BR-P-16, with respect to the initial ensemble (**first column**), and the final ensembles obtained in Case 1 (**second column**) and Case 2 (**third column**), respectively. In all sub-figures, the orange curves correspond to the production data (without measurement noise) generated by the reference model, red dots to the noisy measurements used in history matching, and blue curves to the forecast production data of respective ensembles of reservoir models.

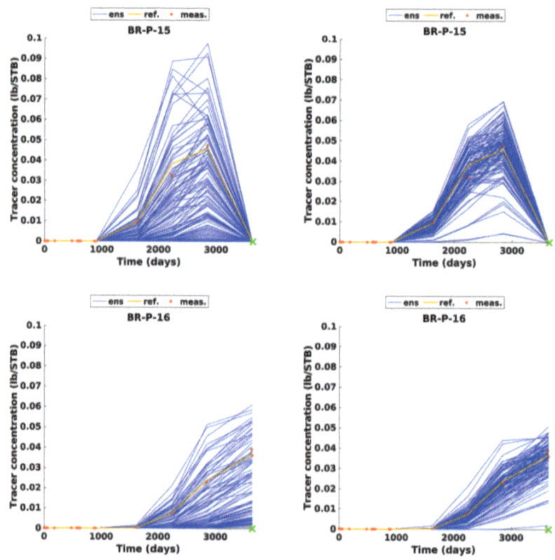

Figure 9. Similar to Figure 8, but for the profiles of tracer data (WTPCW06) in wells BR-P-15 and BR-P-16, with respect to the initial ensemble (**left**) and the final ensembles obtained in Case 2 (**right**), respectively.

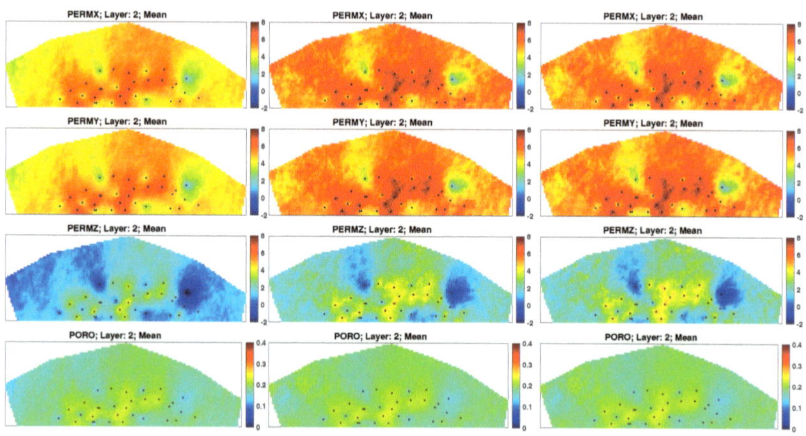

Figure 10. Mean maps for permeability in x-, y-, and z-directions (**first**, **second**, and **third rows**, respectively) and porosity (**last row**) with respect to the initial ensemble, and the final ensembles obtained in Case 1 and Case 2, (**first**, **second**, and **third columns**, respectively), on Layer 2 of the reservoir model.

Finally, we forecast the production data beyond the history matching period using the final ensembles obtained in Cases 1 and 2. For comparison, we calculate the forecast data mismatch for WOPR and WWPR, which is the difference between simulated data of the reference model and an estimated ensemble of reservoir models at the forecast period, normalized by the STD of the WOPR and WWPR.

As aforementioned, we observed better history matching performance in terms of production data mismatch and RMSE values for PERMY and PERMZ parameters, when using production, tracer, and 4D seismic data (Case 2) during the history matching period. In terms of forecast mismatch, the comparison becomes a bit more complicated. As one can see in Figure 12, in some wells (e.g., BR-P-3, and BR-P-16), we observe lower forecast mismatch for the experiment in Case 2. However, in some other wells (e.g., BR-P-6), the situation is the opposite, and higher forecast mismatch is observed instead. The complexity of performance comparison may be partially due to the geological structure of the reservoir model. In this benchmark case, the presence of a fault could have a substantial influence on intra-well fluid flows, which means that it could be relatively easy for the tracer to flow to wells close to the fault and relatively difficult to more distant wells. By checking the distances between the injector–producer pairs, one can obtain a clue for a possible explanation for these noticed differences in the forecast performance. For instance, BR-P-6 is located close to the south-east boundary of the reservoir, so its communication to the other parts of the reservoir may appear weaker, since tracer data from BR-I-6 is not detected. Consequently, it could lead to lower correlations between production in BR-P-6 and the petro-physical parameters in a large part of the reservoir, which makes it more difficult to improve the qualities of estimated parameters close to BR-P-6, hence leading to higher forecast mismatch.

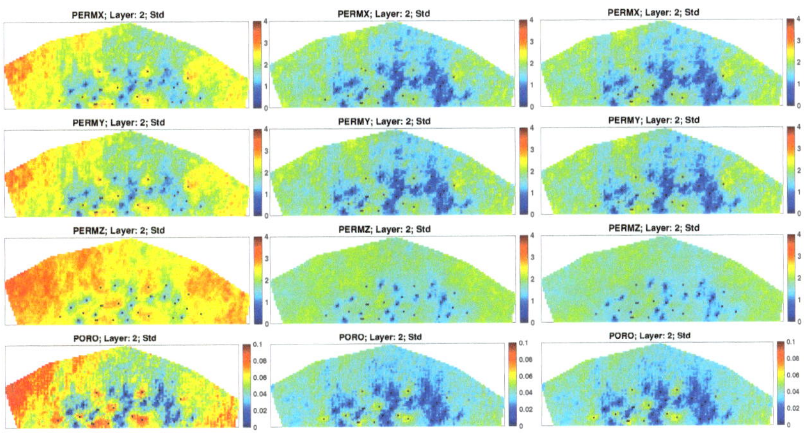

Figure 11. As in Figure 10, but for STD maps instead.

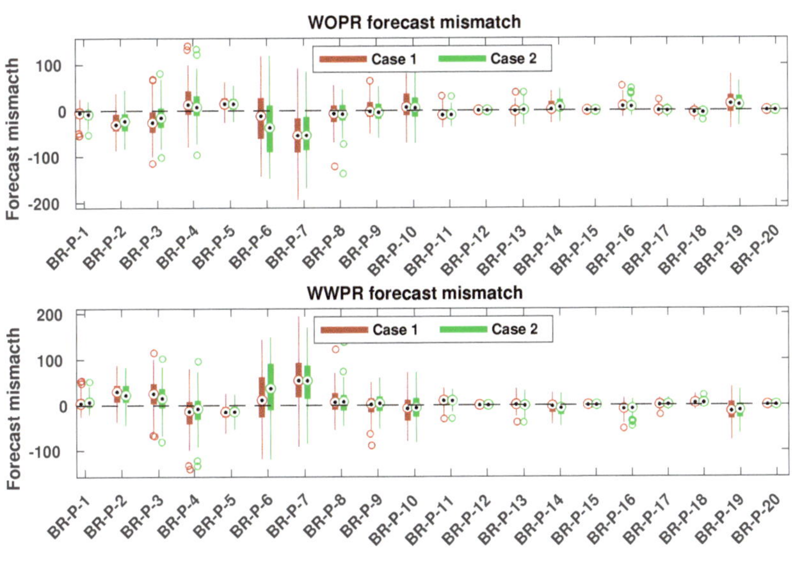

Figure 12. Forecast data mismatch of WOPR and WWPR in production wells (P1–P20) with respect to the two different experiments.

Other than the history matching performance, we also report in Appendix B the overall wall-clock time consumed in Cases 1 and 2. As one can see there, including seismic and/or tracer data into the history matching workflow substantially increases the computational time in our research environment. Nevertheless, the overall computational time is still at a reasonable and acceptable level, meaning that our proposed workflow has the potential for real field case studies.

5. Summary and Conclusions

In this work, we investigated a joint history matching problem with multiple types of field data and proposed a coherent way to integrate production, tracer, and 4D seismic data into a history matching workflow. The workflow is demonstrated in the Brugge benchmark case using the IES-RLM algorithm, with two different experiment settings (Cases 1 and 2).

The main idea behind adding multiple types of field data is to improve the reliability of reservoir models and consequently, the forecast performance. Through two different experiments, it is shown that jointly history matching production, tracer, and 4D seismic data results in lower data mismatch for production data and lower RMSE for PERMY and PERMZ, but at the cost of slightly higher data mismatch for 4D seismic data and RMSE for PORO and PERMX. In terms of the averaged RMSE over all estimated petro-physical parameters, the reliability of the reservoir models is improved in Case 2 in comparison with Case 1.

In both cases, adopting correlation-based adaptive localization helps to maintain substantial ensemble variability even in the presence of multiple types of field data, and ensemble collapse of reservoir models is avoided. Also shown in [3], the localization scheme appears beneficial to achieve a better performance during the forecast period. By a well-to-well analysis of the data mismatch during the forecast period, it is observed that both experiments appear to have good data mismatch. Nevertheless, inter-well tracer data seem to be helpful for further reducing data mismatch, which indicates a better inter-well communication in the reservoir.

Although a better history matching performance (in terms of RMSE) is achieved with the inclusion of tracer data in Case 2, the complexity of the history matching process is increased by adding more types of field data. In the proposed workflow, we adopted a scaling procedure to adjust the relative weights among production, tracer, and 4D seismic data, which appears to work reasonably well. However, to further improve the history matching performance, it could be beneficial to develop more sophisticated methods to better balance the influence of multiple types of field data in ensemble-based history matching. Other possible lines of future research would be to consider the use of multiple tracer injectors at different reservoir locations, and to extend our proposed workflow to hydraulically or naturally fractured reservoirs [39,40]. Such investigations will be considered in our future work.

Author Contributions: Methodology, W.C.C., X.L. and K.R.P.; validation, W.C.C. and X.L.; formal analysis, W.C.C. and X.L.; investigation, W.C.C. and X.L.; data curation, W.C.C. and X.L.; writing—original draft preparation, W.C.C.; writing—review and editing, W.C.C., X.L. and K.R.P.; visualization, W.C.C.; supervision, X.L. and K.R.P. All authors have read and agreed to the published version of the manuscript.

Funding: This research was partially funded by the Research Council of Norway (project numbers: 230303 and 331644) and industry partners.

Acknowledgments: The authors would like to thank reviewers for their constructive and insightful comments and suggestions, which help to significantly improve the quality of this work. William C. Cruz acknowledges financial support from the National IOR Center of Norway (project number: 230303), which is funded by the RCN and industry partners ConocoPhillips, Aker BP, Vår Energi, Equinor, Neptune Energy, Lundin, Halliburton, Schlumberger, and Wintershall Dea. Xiaodong Luo acknowledges partial financial support from the National Centre for Sustainable Subsurface Utilization of the Norwegian Continental Shelf (NCS2030), Norway, which is funded by the Research

Council of Norway (project number: 331644) and industry partners. The authors would also like to thank Schlumberger for providing academic licenses to ECLIPSE©.

Conflicts of Interest: The authors declare no conflict of interest.

Nomenclature

$\mathbf{g}$	Forward model
N_{d}	Number of observed data points
N_{e}	Number of reservoir models
$\mathbf{d}^{sim}$	Vector of simulated data
$\mathbf{m}$	Vector of reservoir model parameters
$\mathbf{d}^{o}$	Vector of observed data
$\mathbf{m}_{\mathrm{true}}$	Vector of ground-truth model parameter
$\overline{\mathbf{m}}$	Vector of mean reservoir model parameters
δ	Vector of contamination noise
$\mathbf{C}_{\mathrm{d}}$	Measurement error-covariance matrix
N_{m}	Number of reservoir parameter datapoints
$\mathcal{M}$	Ensemble of reservoir parameters
$\mathbf{C}_{\mathrm{m}}$	Model error-covariance matrix
$\mathbf{S}_{\mathbf{m}}$	Squared-root matrix of reservoir parameters
$\mathbf{S}_{\mathbf{d}}$	Squared-root matrix of simulated data
$\widetilde{\mathbf{S}}_{\mathbf{d}}$	Normalized squared-root matrix of simulated data
$\mathbf{I}$	Identity matrix
$\Delta\widetilde{\mathbf{d}}$	Normalized innovation
$\hat{\mathbf{U}}$	Matrix of the left singular vectors of $\mathbf{S}_{\mathbf{d}}$
$\hat{\mathbf{V}}$	Matrix of the right singular vectors of $\mathbf{S}_{\mathbf{d}}$
$\hat{\mathbf{\Sigma}}$	Matrix of the kept leading singular values
sv	Kept leading singular values
N_{sv}	Number of kept singular values
$\mathbf{K}$	Kalman gain matrix
$\mathbf{C}$	Localization matrix
γ	Regularization parameter
c_{ks}	Elements of $\mathbf{C}$ at the k-th row and the s-th column
k_{ks}	Elements of $\mathbf{K}$ at the k-th row and the s-th column
ρ_{ks}	Sample correlation of k-th element of the initial ensemble and s-th element of $\Delta\widetilde{\mathbf{d}}$
θ_{s}	Threshold value
$\hat{\rho}_{s}$	Sampling error of correlations
i	Index of iteration step
j	Index of ensemble member
K_{HM}	Effective bulk of the reservoir rock
μ_{HM}	Effective shear of the reservoir rock
ϕ_{c}	Critical porosity
μ_{s}	Gran shear modulus
v_{s}	Poisson's ration
P_{eff}	Effective stress
C_{p}	Average number of contacts per sphere
n	degree of root
K_{eff}	Effective dry bulk modulus
μ_{eff}	Effective shear modulus
ϕ	Porosity values
K_{s}	Grain bulk modulus
K_{f}	Effective fluid bulk modulus
K_{w}	Bulk modulus of water
K_{o}	Bulk modulus of oil
S_{w}	Saturation of water

S_o	Saturation of oil
ρ_{sat}	Saturated rock density
ρ_m	Mineral density
ρ_w	Water density
ρ_o	Oil density
V_p	P-wave velocity
V_s	S-wave velocity

Appendix A. Forward Simulation of 4D Seismic Data

The forward simulation of 4D seismic data goes through a few steps: reservoir simulation, petro-elastic model (PEM), and computation of seismic traces for a given time window. Starting from the initial ensemble containing the uncertain petro-physical parameters, i.e., porosity and permeability, one performs reservoir simulation to compute pore pressure and fluid saturations. Once one generates the dynamic reservoir properties, the next step is to compute the seismic elastic attributes, e.g., P-wave velocity, S-wave velocity, and formation density, by using a PEM. Finally, one has to transform the acquired seismic elastic data into another domain, e.g., amplitude-versus-angle (AVA), by using the AVA equation.

The PEM provides an important link to the reservoir model which governs the dynamics of fluid flow, and seismic elastic attributes which govern wave propagation. This connection is needed to convert pore pressure and fluid saturation into seismic elastic attributes for seismic interpretation or inversion. By far the most widely used method to establish this connection is the soft-sand model [41], in which the first step is to estimate the effective bulk (K_{HM}) and shear modulus (μ_{HM}) of the reservoir rock through Hertz–Minlin theory [42] as in

$$K_{\mathrm{HM}} = \left[\frac{C_p^2(1-\phi_c)^2\mu_s^2}{18\pi^2(1-\mathrm{v}_s)^2}P_{\mathrm{eff}}\right]^{1/n}, \tag{A1}$$

and

$$\mu_{\mathrm{HM}} = \frac{5-4\mathrm{v}_s}{5(2-\mathrm{v}_s)}\left[\frac{3C_p^2(1-\phi_c)^2\mu_s^2}{2\pi^2(1-\mathrm{v}_s)^2}P_{\mathrm{eff}}\right]^{1/n}, \tag{A2}$$

where ϕ_c, μ_s, v_s, and P_{eff} represent critical porosity, gran shear modulus, Poisson's ratio, and effective stress (e.g., lithostatic pressure minus pore pressure), respectively. In this work, the coordinate number C_p, which denotes the average number of contacts per sphere, is set to 9, the degree of root n is set to 3, and ϕ_c is set to 36%.

The modified Hashin–Shrikman model [43] is used to calculate the effective dry bulk modulus (K_{eff}) and the effective shear modulus (μ_{eff}) for porosity values (ϕ), as in

$$K_{\mathrm{eff}} = \left[\frac{\phi/\phi_c}{K_{\mathrm{HM}}+\frac{4}{3}\mu_{\mathrm{HM}}} + \frac{1-\phi/\phi_c}{K_s+\frac{4}{3}\mu_{\mathrm{HM}}}\right]^{-1} - \frac{4}{3}\mu_{\mathrm{HM}}, \tag{A3}$$

and

$$\mu_{\mathrm{eff}} = \left[\frac{\phi/\phi_c}{\mu_{\mathrm{HM}}+\frac{\mu_{\mathrm{HM}}}{6}\left(\frac{9K_{\mathrm{HM}}+8\mu_{\mathrm{HM}}}{K_{\mathrm{HM}}+2\mu_{\mathrm{HM}}}\right)} + \frac{1-\phi/\phi_c}{\mu_s+\frac{\mu_{\mathrm{HM}}}{6}\left(\frac{9K_{\mathrm{HM}}+8\mu_{\mathrm{HM}}}{K_{\mathrm{HM}}+2\mu_{\mathrm{HM}}}\right)}\right]^{-1} - \frac{\mu_{\mathrm{HM}}}{6} + \frac{9K_{\mathrm{HM}}+8\mu_{\mathrm{HM}}}{K_{\mathrm{HM}}+2\mu_{\mathrm{HM}}}, \tag{A4}$$

where K_s is the grain bulk modulus.

Once one calculates the effective dry bulk modulus and the effective shear modulus for each reservoir gridblock, the next step is to compute the saturated bulk modulus and shear modulus (K_{sat} and μ_{sat}, respectively) by including the saturation effect with the Gausmmann model [44], as in

$$K_{sat} = K_{eff} + \frac{\left(1 - \frac{K_{eff}}{K_s}\right)^2}{\frac{\phi}{K_f} + \frac{1-\phi}{K_s} - \frac{K_{eff}}{K_s^2}}, \tag{A5}$$

and

$$\mu_{sat} = \mu_{eff}, \tag{A6}$$

where K_f is the effective fluid bulk modulus, in which the two-phase fluid mixture is given by

$$K_f = \left(\frac{S_w}{K_w} + \frac{S_o}{K_o}\right)^{-1}, \tag{A7}$$

where K_w, K_o, S_w, S_o, are bulk modulus of water, bulk modulus of oil, saturation of water, and saturation of oil, respectively. Further, one calculates the saturated rock density (ρ_{sat}), as in [41]

$$\rho_{sat} = (1-\phi)\rho_m + \phi S_w \rho_w + \phi S_o \rho_o, \tag{A8}$$

where ρ_m, ρ_w, and ρ_o are the mineral density, water density, and oil density, respectively. Finally, one can obtain P-wave and S-wave velocities (V_p and V_s), which can be expressed as in [41]

$$V_p = \sqrt{\frac{K_{sat} + \frac{4}{3}\mu_{sat}}{\rho_{sat}}}, \tag{A9}$$

and

$$V_s = \sqrt{\frac{\mu_{sat}}{\rho_{sat}}}. \tag{A10}$$

After the seismic elastic attributes are calculated based on the outputs of reservoir simulation, we can generate synthetic seismogram by using the Zoeppritz equation [41], in which the reflection terms between two adjacent layers are defined as a function of travel time. Then, one can obtain the desired seismic AVA data by applying a convolution between the reflectivity series and a Ricker wavelet (with a dominant frequency of 45 Hz). Here, the AVA attributes are obtained without involving any inversion process, which avoids the introduction of biases and extra uncertainties into seismic history matching later [45]. For more information regarding the procedure to generate AVA data, readers are referred to, e.g., [37].

Appendix B. Comparison of CPU Time in Different Experiments

Table A1 reports the computational cost in terms of the total wall-clock time to run the ensemble-based workflow in different experiments, with Intel(R) Core(TM) i9-10900K CPU @ 3.70GHz and 64 GB memory. For the purpose of comparison, we include the CPU time in a base case with only production data (referred to as Base case hereafter), in addition to the two cases (Cases 1 and 2) performed in this work. The results shows that including 4D seismic data (Case 1) or tracer and 4D seismic data (Case 2) considerably increase the computational time in the case studies. As one can see, in comparison to the base case, the CPU time in Cases 1 and 2 are more than doubled (+111.08 % and +120.38%, respectively). Meanwhile, in comparison to Case 1, one can also notice that the inclusion of tracer data in Case 2 slightly increases the computational cost. Overall, all case studies are finished within 30 h (108,000 s) in wall-clock time. Given that the Brugge benchmark is a field-scale reservoir model, we conclude that the computational costs in all case studies are reasonable, and that the application of the proposed workflow to other case studies at a similar scale should be affordable in general. It is important to stress that in the current work we carry out forward simulations and model updates only using a single processor. The proposed history matching workflow is, however, highly parallelizable. This means that with the aid of a high-performance-computing (HPC) facility and an optimization of our research code, one can substantially improve the computational efficiency of our proposed workflow.

Table A1. CPU time test through different experiments settings.

Experiments	Base	Case 1	Case 2
CPU time (in seconds)	43,308	91,417 (+111.08%)	95,444 (+120.38%)

References

1. Jansen, J.D.; Brouwer, R.; Douma, S.G. Closed loop reservoir management. In Proceedings of the SPE Reservoir Simulation Symposium, Woodlands, TX, USA, 2–4 February 2009; OnePetro: Richardson, TX, USA, 2009.
2. Lorentzen, R.J.; Bhakta, T.; Grana, D.; Luo, X.; Valestrand, R.; Nævdal, G. Simultaneous assimilation of production and seismic data: Application to the Norne field. *Comput. Geosci.* **2020**, *24*, 907–920. [CrossRef]
3. Cruz, W.C.; Luo, X.; Petvipusit, K.R. Joint history matching of production and tracer data through an iterative ensemble smoother: A 3D field-scale case study. In Proceedings of the International Petroleum Technology Conference, Riyadh, Saudi Arabia, 21–23 February 2022; OnePetro: Richardson, TX, USA, 2022.
4. Tayyib, D.; Al-Qasim, A.; Kokal, S.; Huseby, O. Overview of tracer applications in oil and gas industry. In Proceedings of the SPE Kuwait Oil & Gas Show and Conference, Mishref, Kuwait, 14 October 2019; Society of Petroleum Engineers: Mishref, Kuwait, 2019.
5. Huseby, O.K.; Andersen, M.; Svorstol, I.; Dugstad, O. Improved Understanding of Reservoir Fluid Dynamics in the North Sea Snorre Field by Combining Tracers, 4D Seismic and Production Data. In Proceedings of the SPE Middle East Oil and Gas Show and Conference, Manama, Bahrain, 11 March 2007; OnePetro: Richardson, TX, USA, 2007.
6. Allison, S.; Pope, G.A.; Sepehrnoori, K. Analysis of field tracers for reservoir description. *J. Pet. Sci. Eng.* **1991**, *5*, 173–186. [CrossRef]
7. Datta-Gupta, A.; Lake, L.W.; Pope, G.A. Characterizing heterogeneous permeable media with spatial statistics and tracer data using sequential simulated annealing. *Math. Geol.* **1995**, *27*, 763–787. [CrossRef]
8. Iliassov, P.A.; Datta-Gupta, A.; Vasco, D. Field-scale characterization of permeability and saturation distribution using partitioning tracer tests: The Ranger field, Texas. In Proceedings of the SPE Annual Technical Conference and Exhibition, New Orleans, Louisiana, 30 September 2001; OnePetro: Richardson, TX, USA, 2001.
9. Valestrand, R.; Sagen, J.; Nævdal, G.; Huseby, O. The effect of including tracer data in the ensemble-Kalman-filter approach. *SPE J.* **2010**, *15*, 454–470. [CrossRef]
10. Chen, H.; Poitzsch, M.E. Improved reservoir history matching and production optimization with tracer data. In Proceedings of the SPE Annual Technical Conference and Exhibition, Dallas, TX, USA, 24 September 2018; OnePetro: Richardson, TX, USA, 2018.
11. Skjervheim, J.A.; Evensen, G.; Aanonsen, S.I.; Ruud, B.O.; Johansen, T.A. Incorporating 4D seismic data in reservoir simulation models using ensemble Kalman filter. In Proceedings of the SPE Annual Technical Conference and Exhibition, Dallas, TX, USA, 9–12 October 2005; OnePetro: Richardson, TX, USA, 2005.
12. Emerick, A.A. Analysis of the performance of ensemble-based assimilation of production and seismic data. *J. Pet. Sci. Eng.* **2016**, *139*, 219–239. [CrossRef]
13. Obidegwu, D.; Chassagne, R.; MacBeth, C. Seismic assisted history matching using binary maps. *J. Nat. Gas Sci. Eng.* **2017**, *42*, 69–84. [CrossRef]
14. Davolio, A.; Schiozer, D.J. Probabilistic seismic history matching using binary images. *J. Geophys. Eng.* **2018**, *15*, 261–274. [CrossRef]
15. Luo, X.; Bhakta, T.; Jakobsen, M.; Nævdal, G. Efficient big data assimilation through sparse representation: A 3D benchmark case study in petroleum engineering. *PloS ONE* **2018**, *13*, e0198586. [CrossRef]
16. Al-Mudhafar, W.J. How is multiple-point geostatistics of lithofacies modeling assisting for fast history matching? A case study from a sand-rich fluvial depositional environment of Zubair formation in South Rumaila oil field. In Proceedings of the Offshore Technology Conference, Houston, TX, USA, 30 April–3 May 2018; OnePetro: Richardson, TX, USA, 2018.
17. Thanh, H.V.; Sugai, Y. Integrated modelling framework for enhancement history matching in fluvial channel sandstone reservoirs. *Upstream Oil Gas Technol.* **2021**, *6*, 100027. [CrossRef]
18. Soares, R.V.; Luo, X.; Evensen, G.; Bhakta, T. Handling Big Models and Big Data Sets in History-Matching Problems through an Adaptive Local Analysis Scheme. *SPE J.* **2021**, *26*, 973–992. [CrossRef]
19. Evensen, G. Sequential data assimilation with a nonlinear quasi-geostrophic model using Monte Carlo methods to forecast error statistics. *J. Geophys. Res. Ocean.* **1994**, *99*, 10143–10162. [CrossRef]
20. Nævdal, G.; Johnsen, L.M.; Aanonsen, S.I.; Vefring, E.H. Reservoir monitoring and continuous model updating using ensemble Kalman filter. In Proceedings of the SPE Annual Technical Conference and Exhibition, Denver, CO, USA, 5–8 October 2003; Society of Petroleum Engineers: Denver, CO, USA, 2003. SPE-84372-MS.
21. Aanonsen, S.; Nævdal, G.; Oliver, D.; Reynolds, A.; Vallès, B. The Ensemble Kalman Filter in Reservoir Engineering: A Review. *SPE J.* **2009**, *14*, 393–412. [CrossRef]
22. Skjervheim, J.A.; Evensen, G. An ensemble smoother for assisted history matching. In Proceedings of the SPE Reservoir Simulation Symposium, The Woodlands, TX, USA, 21–23 February 2011. SPE-141929-MS.

23. Van Leeuwen, P.J.; Evensen, G. Data assimilation and inverse methods in terms of a probabilistic formulation. *Mon. Wea. Rev.* **1996**, *124*, 2898–2913. [CrossRef]
24. Chen, Y.; Oliver, D. Levenberg-Marquardt forms of the iterative ensemble smoother for efficient history matching and uncertainty quantification. *Comput. Geosci.* **2013**, *17*, 689–703. [CrossRef]
25. Emerick, A.A.; Reynolds, A.C. Ensemble smoother with multiple data assimilation. *Comput. Geosci.* **2013**, *55*, 3–15. [CrossRef]
26. Luo, X.; Stordal, A.; Lorentzen, R.; Nævdal, G. Iterative ensemble smoother as an approximate solution to a regularized minimum-average-cost problem: Theory and applications. *SPE J.* **2015**, *20*, 962–982. [CrossRef]
27. Luo, X. Novel Iterative Ensemble Smoothers Derived from A Class of Generalized Cost Functions. *Comput. Geosci.* **2021**, *25*, 1159–1189. [CrossRef]
28. Ma, X.; Hetz, G.; Wang, X.; Bi, L.; Stern, D.; Hoda, N. A robust iterative ensemble smoother method for efficient history matching and uncertainty quantification. In Proceedings of the SPE Reservoir Simulation Conference, Montgomery, TX, USA, 20–22 February 2017; OnePetro: Richardson, TX, USA, 2017.
29. Peters, L.; Arts, R.; Brouwer, G.; Geel, C.; Cullick, S.; Lorentzen, R.J.; Chen, Y.; Dunlop, N.; Vossepoel, F.C.; Xu, R.; et al. Results of the Brugge benchmark study for flooding optimization and history matching. *SPE Reserv. Eval. Eng.* **2010**, *13*, 391–405. [CrossRef]
30. Engl, H.W.; Hanke, M.; Neubauer, A. *Regularization of Inverse Problems*; Springer: Berlin/Heidelberg, Germany, 2000.
31. Evensen, G. *Data Assimilation: The Ensemble Kalman Filter*; Springer: Berlin/Heidelberg, Germany, 2009; Volume 2.
32. Emerick, A.A.; Reynolds, A.C. History matching a field case using the ensemble Kalman filter with covariance localization. *SPE Reserv. Eval. Eng.* **2011**, *14*, 443–452. [CrossRef]
33. Chen, Y.; Oliver, D.S. Cross-covariances and localization for EnKF in multiphase flow data assimilation. *Comput. Geosci.* **2010**, *14*, 579–601. [CrossRef]
34. Luo, X.; Bhakta, T.; Nævdal, G. Correlation-based adaptive localization with applications to ensemble-based 4D seismic history matching. *SPE J.* **2018**, *23*, 396–427. [CrossRef]
35. Luo, X.; Bhakta, T. Automatic and adaptive localization for ensemble-based history matching. *J. Pet. Sci. Eng.* **2020**, *184*, 106559. [CrossRef]
36. Raniolo, S.; Dovera, L.; Cominelli, A.; Callegaro, C.; Masserano, F. History match and polymer injection optimization in a mature field using the ensemble Kalman filter. In Proceedings of the IOR 2013-17th European Symposium on Improved Oil Recovery, Saint Petersburg, Russia, 16–18 April 2013; European Association of Geoscientists & Engineers: Saint Petersburg, Russia 2013; p. cp-342.
37. Luo, X.; Bhakta, T.; Jakobsen, M.; Nævdal, G. An ensemble 4D-seismic history-matching framework with sparse representation based on wavelet multiresolution analysis. *SPE J.* **2017**, *22*, 985–1010. [CrossRef]
38. Oliver, D.S.; Reynolds, A.C.; Liu, N. *Inverse Theory for Petroleum Reservoir Characterization and History Matching* ; Cambridge University Press: Cambridge, UK, 2008.
39. Li, J.; Pei, Y.; Jiang, H.; Zhao, L.; Li, L.; Zhou, H.; Zhao, Y.; Zhang, Z. Tracer flowback based fracture network characterization in hydraulic fracturing. In Proceedings of the Abu Dhabi International Petroleum Exhibition & Conference, Abu Dhabi, United Arab Emirates, 7–10 November 2016; OnePetro: Richardson, TX, USA, 2016.
40. Salishchev, M.V.; Shirnen, A.A.; Khamadaliev, D.M.; Chaplygin, D.A. Logging, Well Testing and Microseismic Fracture Geometry Investigations: Mistakes, Lessons Learnt and Challenges. In Proceedings of the SPE Russian Petroleum Technology Conference, Virtual, 26 October 2020; OnePetro: Richardson, TX, USA, 2020.
41. Mavko, G.; Mukerji, T.; Dvorkin, J. *The Rock Physics Handbook: Tools for Seismic Analysis of Porous Media*, 2nd ed.; Cambridge University Press: Cambridge, CA, USA, 2009. [CrossRef]
42. Mindlin, R.D. Compliance of elastic bodies in contact. *J. Appl. Mech.* **1949**, *16*, 259–268. [CrossRef]
43. Hashin, Z.; Shtrikman, S. A variational approach to the theory of the elastic behaviour of multiphase materials. *J. Mech. Phys. Solids* **1963**, *11*, 127–140. [CrossRef]
44. Gassmann, F. *Über die Elastizität Poröser Medien*; Zürich, Inst. für Geophysik an der ETH: Zurich, Germany 1951.
45. Bhakta, T.; Luo, X.; Nævdal, G. Ensemble based 4D seismic history matching using a sparse representation of AVA data. In Proceedings of the 2016 SEG International Exposition and Annual Meeting, Dallas, TX, USA, 16–21 October 2016; OnePetro: Richardson, TX, USA, 2016.

energies

Article

A Fully Coupled Hydro-Mechanical Approach for Multi-Fracture Propagation Simulations

Yinghao Deng [1,2,3], Di Wang [1,2], Yan Jin [3,*] and Yang Xia [3,*]

1 State Key Laboratory of Shale Oil and Gas Enrichment Mechanisms and Effective Development, Beijing 102206, China
2 State Energy Center for Shale Oil Research and Development, Beijing 100728, China
3 State Key Laboratory of Petroleum Resources and Prospecting, Beijing 102249, China
* Correspondence: jiny@cup.edu.cn (Y.J.); xiayang@vip.126.com (Y.X.)

Abstract: Hydraulic fracturing is a complex nonlinear hydro-mechanical coupled process. Accurate numerical simulation is of great significance for reducing fracturing costs and improving reservoir development benefits. The aim of this paper is to propose an efficient numerical simulation method for the fracturing-to-production problem under a unified framework that has good convergence and accuracy. A hydro-mechanical coupled fracturing model (HMFM) is established for poroelastic media saturated with a compressible fluid, and the local characteristics of the physical field are fully considered. Each fracture is explicitly characterized using the discrete fracture model (DFM), which can better reflect the physical characteristics near fractures. Based on the extended finite element method (XFEM) and the Newton–Raphson method, a fully coupled approach named Unified Extended Finite Element (UXFEM) is developed, which can solve the nonlinear system of equations that describe the solution under a unified framework. UXFEM can accurately capture the local physical characteristics of different physical fields on the orthogonal structured grids. It realizes the grid-fracture decoupling, and fractures can propagate in any direction, which shows greater flexibility in simulating fracture propagation. The fully coupled approach can better reflect the essential relationship between pressure, stress, and fracture, which is beneficial to studying hydro-mechanical coupled problems. To validate the UXFEM, UXFEM is compared with the classical KGD model, analytic solution, and COMSOL solution. Finally, based on UXFEM, the interference phenomenon and fracturing-to-production study are carried out to prove the broad practical application prospect of this new fully coupled approach.

Keywords: hydro-mechanical; fracture propagation; production; fully coupled; stress interference

Citation: Deng, Y.; Wang, D.; Jin, Y.; Xia, Y. A Fully Coupled Hydro-Mechanical Approach for Multi-Fracture Propagation Simulations. *Energies* **2023**, *16*, 1601. https://doi.org/10.3390/en16041601

Academic Editor: Reza Rezaee

Received: 4 January 2023
Revised: 27 January 2023
Accepted: 1 February 2023
Published: 5 February 2023

1. Introduction

Fracturing, as an effective technical means, can significantly improve the development efficiency of low-permeability and ultra-low-permeability reservoirs, such as shale and tight sandstone. The reservoir often develops various discontinuous structures such as natural fractures, faults, and caves, resulting in significant multi-scale characteristics of the reservoir [1–3]. Moreover, all physical processes in the subsurface are highly nonlinear and complex multiphysics problems [4,5], which poses considerable challenges to studying fracture propagation and other related research.

As the most traditional research method, experiments have played a considerable role in promoting the study of fracture propagation [2,3,6–10]. However, the limitations of physical simulation are also prominent, especially for complex multiphysics coupling problems. The experimental cost is relatively high, and the results are often obtained at the laboratory scale, which cannot accurately reflect the underground situation. Furthermore, some complex problems are often unable to be experimentally studied. In this context, numerical simulation technology is applied to the design before fracturing, monitoring during fracturing, and post-fracturing dynamic production studies. Many scholars have conducted extensive and meaningful research on the fracture propagation problem in petroleum engineering and proposed some numerical simulation methods. Those methods mainly include the finite element method (FEM) and its derivative methods, the boundary element method (BEM), and the discrete element method (DEM).

FEM is a flexible, effective, and widely used numerical method [11–17]. Fractures are highly coupled with grids, and fractures must be set along the boundary of mesh grids. Therefore, fractures must propagate along the grid boundary or continuously reconstruct mesh grids, which requires tremendous calculation. At the same time, the shape function of FEM is continuous. It is often difficult to accurately describe various discontinuities when describing the fluid pressure and the solid displacement fields. Thus, it is often necessary to use local mesh refinement to achieve high-accuracy calculations, which also increases the calculation burden. The characteristics of fracture geometry and physical field are challenging to characterize efficiently and accurately in traditional FEM, and the limitations of FEM are significant.

Some FEM-based extension methods have been proposed successively, such as node-splitting FEM [18,19], generalized FEM [20,21], and XFEM [22–27]. Node-splitting FEM allows FEM to describe the fracture width simply. Still, the fracture propagation path must follow the grid boundary, which cannot accurately describe the fracture propagation and interaction. XFEM is a method based on the partition of unity method (PUM), which uses enrichment functions to capture the physical characteristics of fracture walls and fracture tips. XFEM is an efficient method for solving discontinuous problems, which can well solve strong and weak discontinuity problems. The core idea is to capture various discontinuities with the help of enrichment functions constructed based on analytical solutions or asymptotic analytical solutions, and fractures and grids are independent of each other. XFEM, based on the discrete fracture model, can explicitly characterize each fracture. Many scholars have constructed enrichment functions to describe fracture surfaces, fracture tips, and intersecting fractures based on physical field characteristics and analytical solution characteristics, which can efficiently deal with complex fracture mechanical behavior. However, most of the current XFEM models cannot accurately consider the fluid flow and exchange between fractures and matrix, usually ignoring the fluid leak-off term during fracturing stimulation or replacing it with an empirical leak-off equation, which cannot meet the simulation requirements. Moreover, XFEM also has considerable troubles in the numerical simulation of complex fracture flow, and it is difficult to achieve a fully coupled approach under a unified numerical framework.

BEM [28–34] is a dimensionality reduction research method based on Green's formula. It can calculate fracture aperture and stress more accurately and conveniently than FEM and is suitable for simulating fracture propagation with complex topologies. However, this method faces two main problems. One is that it cannot accurately describe the influence of fractures on fluid/solid physical fields. The other one is that it is difficult to consider the leak-off during fracture propagation [35,36].

DEM [37] is a numerical simulation method specially used to solve the problem of discontinuous media. It can deal with fracture intersecting, branching, merging, and kinking problems. However, calibrating/updating particle properties remains a complex technical challenge. The particle number needs to be large enough to achieve sufficient accuracy, which prevents this technique from being widely used in large-scale models [36].

Different methods have different advantages and different defects. In order to synthesize their advantages and compensate for each other's weaknesses, some hybrid numerical simulation methods are also widely used. Recently, Li et al. [36,38–40] developed a meaningful thermal/hydro-mechanical (THM) model. They solved it using the hybrid numerical simulation method, realizing the two-dimensional and three-dimensional simulation of the construction of a complex fracture network. Settgast et al. [41] developed a fully coupled finite element/finite volume approach for simulating field-scale hydraulically driven fractures in three dimensions. Guo et al. [42] adopted the mixed finite element/displacement discontinuity method to solve for the spatial–temporal evolutions of pore pressure and in situ stress because of parent-well production and injection and models the fracture propagation during infill-well completion based on updated heterogeneous in situ stresses. Recently, a coupled simulation strategy combining the embedded discrete fracture method (EDFM) and the XFEM has been developed to simulate the fluid-driven fracture propagation process in porous media. EDFM and XFEM are used to simulate fracture-related fluid mechanics and solid mechanics, respectively, with information exchanged under the iterative numerical coupling scheme, and it realizes two-dimensional and three-dimensional hydraulic fracturing fracture propagation simulation [43–45]. Liu et al. [46] developed a hydro-mechanical model for non-planar hydraulic fracture propagation in ductile formations, which is solved by the hybrid extended finite element/finite volume method. Zhang et al. [47] combined the extended finite element/phase-field method to solve the discontinuous and continuous hydraulic fracturing formulations.

However, there are still many insurmountable problems in developing realistic simulation tools for the hydraulic fracturing process. The problems of numerical simulation of fracture propagation are mainly reflected in the contradiction between calculation efficiency and calculation accuracy, reflected explicitly in several aspects: (1) There are strict requirements for grid division; (2) The fracture propagation path is not arbitrary, and the fracture morphology of simulation results is distorted; (3) Matrix flow, fracture flow, and solid deformation coupling are challenging, and the leak-off term during the fracturing process is often ignored or simplified; (4) The nonlinearity is strong, and the fully coupled approach is scarce; (5) There are still risks and challenges in the convergence and stability of numerical calculation. The reasons for this contradiction are complex and diverse, such as rock heterogeneity, unclear mechanism of fracture propagation, and aggravation of heterogeneity caused by the intersection of fractures and artificial stimulation measures. In recent years, some new methods and models have been proposed and achieved good results, such as Peridynamics [48–51].

This paper uses the discrete fracture model (DFM) to characterize each fracture [52–56]. A fully coupled numerical approach for solving nonlinear hydro-mechanical coupled models is established based on XFEM. In this method, the solid rock deformation obeys quasi-static linear elasticity, characterized by the stress balance equation. The matrix flow follows Darcy's law, and the fracture fluid flow obeys the cubic law. In the derivation, this paper uses the normal flow velocity discontinuity term in the weak form of matrix and fracture flow equations to automatically characterize the fluid leak-off term. This paper uses XFEM to compile a fully coupled UXFEM solver for the abovementioned HMFM. This approach solves the fluid pressure field and the solid displacement field under the same framework, realizing the unification and full coupling of the model and solving.

According to the different physical characteristics of solid deformation and fluid flow, UXFEM adopts different enrichment functions and successfully realizes the numerical simulation of hydro-mechanical fracture propagation. The advantages of XFEM are entirely inherited in UXFEM. Fractures are decoupled from mesh grids, and fractures are allowed to propagate in any direction and can be deflected at any angle without mesh reconstruction or other special processing. In this paper, the maximum circumferential tensile stress criterion is used as the fracture propagation criterion, the displacement-based method is used to calculate the stress intensity factors (SIFs), and fractures always propagate along the direction of the maximum circumferential stress. When UXFEM solves nonlinear mathematical models, it adopts Newton–Raphson iterative linearization to deal with nonlinear systems. With the advantages of XFEM, UXFEM can achieve the balance between calculation efficiency and accuracy to a certain extent, which has practical significance for the actual engineering-scale fracturing simulation research.

2. Unified Extended Finite Element Method

2.1. Fracture Description

This paper uses pair of orthogonal level set functions $f(x)$, $g(x)$ to describe each fracture explicitly, and some enrichment functions are constructed based on level set functions. The fracture surface is defined as $f(x) = 0$, and the fracture tip is defined as $f(x) = 0$, $g(x) = 0$.

The core of XFEM is to capture the discontinuity of the physical field by using enrichment functions with discontinuous properties and enriched degree of freedoms (DOF) on the basis of FEM, which makes the description of the physical field characteristics independent of grids, bringing a lot of conveniences. The fracture is a strong discontinuity for the displacement field, while the fracture is a weak discontinuity for the fluid pressure field. That means the pressure on the two sides of the fracture is continuous, but the pressure derivative is discontinuous. In this paper, the set of all nodes is denoted as N_{all}, and the enriched nodes, including surface nodes and tip nodes, are denoted as N_s and N_t, respectively (Figure 1).

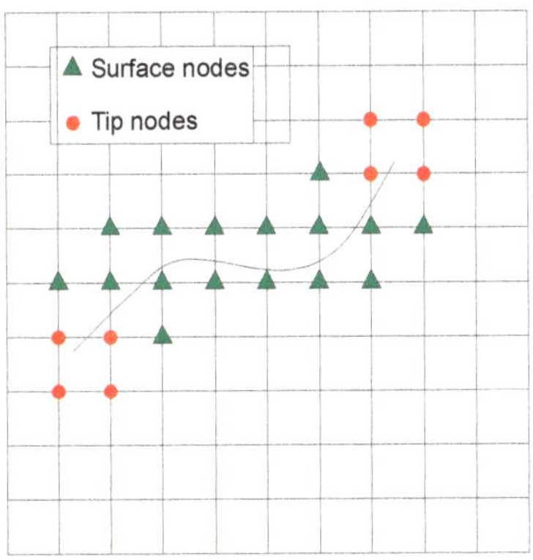

Figure 1. Enriched node-set.

2.2. Solid Displacement Field Approximation and Enrichment Functions

In the XFEM framework, the displacement approximation is written as:

$$u(\boldsymbol{x}) = \sum_{i \in N_{all}} N_i(\boldsymbol{x})u_i + \sum_{j \in N_s} N_j(\boldsymbol{x})(H(f(\boldsymbol{x})) - H((f(\boldsymbol{x}_j)))a_j + \sum_{q \in N_t} N_q(\boldsymbol{x})\sum_{l=1}^{4}(F_l(\boldsymbol{x}) - F_l(\boldsymbol{x}_q))b_q^l \tag{1}$$

where $N_i(\boldsymbol{x})$, $N_j(\boldsymbol{x})$, $N_q(\boldsymbol{x})$ are standard FEM shape function; u_i is standard displacement degrees of freedom (DDOF) for nodes N_{all}; a_j^n, b_{ql}^m are added enriched DDOFs for nodes N_s and N_t respectively; $H(\cdot)$ is the Heaviside step function; $F_l(\cdot)$ is fracture tip displacement enrichment function.

For expression brevity, we combine the last two terms in Equation (1) and rewrite Equation (1) as:

$$u(\boldsymbol{x}) = \sum_{i \in N_{all}} N_i(\boldsymbol{x})u_i + \sum_{j \in N_{enr}} \Psi_j(\boldsymbol{x})\widetilde{u}_j = \boldsymbol{N} \cdot \boldsymbol{U} \tag{2}$$

where $N_{enr} = \{N_s, N_t \times 4\}$; Ψ_j and $\widetilde{u}_j$ denote enrichment functions and its added enriched DDOFs; $\boldsymbol{N}$ is the combination of standard FEM shape function N_i and enrichment function Ψ_j; and $\boldsymbol{U}$ is the combination of standard DDOFs and added enriched DDOFs.

For the element completely penetrated by the fracture, the displacement field is discontinuous on both sides, which is a strong discontinuity. The Heaviside step function is used to enrich the surface nodes.

$$H(f(\boldsymbol{x})) = \begin{cases} 1, f(\boldsymbol{x}) \geq 0 \\ 0, f(\boldsymbol{x}) < 0 \end{cases} \tag{3}$$

For fracture tips, the stress has the singularity of $O(r^{-\frac{1}{2}})$. For isotropic linear elastic materials, tip nodes can be enriched by the fracture tip displacement enrichment function [22,57,58]:

$$[F_l(r,\theta),\ l=1,2,3,4] = [\sqrt{r}\sin\frac{\theta}{2}\sqrt{r}\cos\frac{\theta}{2}\sqrt{r}\sin\frac{\theta}{2}\sin\theta\sqrt{r}\cos\frac{\theta}{2}\sin\theta] \quad (4)$$

where r, θ are polar coordinates with origin at the fracture tip and x_1 axis oriented into the body and parallel to the fracture surfaces (Figure 2).

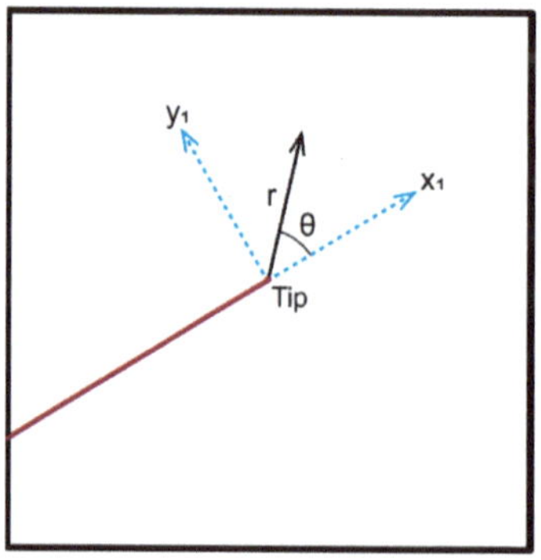

Figure 2. Polar coordinate of a fracture tip.

2.3. Fluid Pressure Field Approximation and Enrichment Functions

In the XFEM framework, the pressure approximation is written as:

$$p(\boldsymbol{x}) = \sum_{i\in N_{all}} N_i(\boldsymbol{x})p_i + \sum_{a\in N_s} N_a(\boldsymbol{x})(\phi_s(x)-\phi_s(x_a))\widetilde{p}_a + \sum_{b\in N_t} N_b(\boldsymbol{x})(\phi_t(x)-\phi_t(x_b))\widetilde{p}_b^m \quad (5)$$

where $N_i(\boldsymbol{x})$, $N_a(\boldsymbol{x})$, $N_b(\boldsymbol{x})$ are standard FEM shape function; p_i is standard pressure degrees of freedom (PDOFs) for node N_{all}; a_j^n, b_{ql}^m are added enriched PDOFs for nodes N_s and N_t respectively; $\phi_s(\cdot)$ is the modified level set absolute value function; $\phi_t(\cdot)$ is fracture tip pressure enrichment function.

For expression brevity, we combine the last two terms in Equation (5) and rewrite Equation (5) as:

$$p(\boldsymbol{x}) = \sum_{i\in N_{all}} N_i(\boldsymbol{x})p_i + \sum_{j\in N_{enr}} \Phi_j(\boldsymbol{x})\widetilde{p}_j = \boldsymbol{H}\cdot\boldsymbol{P} \quad (6)$$

where $N_{enr} = \{N_s, N_t\}$; Ψ_j and $\widetilde{u}_j$ denote enrichment functions and their added enriched PDOFs; $\boldsymbol{H}$ is the combination of standard FEM shape function N_i and enrichment function Φ_j; and $\boldsymbol{P}$ is the combination of standard PDOFs and added enriched PDOFs.

For the fracture surface, the pressure is continuous, but the pressure gradient is discontinuous. Moës [59] uses the modified level set absolute value function to capture the fluid pressure characteristics, avoiding the appearance of blending elements.

$$\phi_s(\boldsymbol{x}) = \sum_j |f(\boldsymbol{x}_j)|N_j(\boldsymbol{x}) - \left|\sum_j f(\boldsymbol{x}_j)N_j(\boldsymbol{x})\right| \quad (7)$$

According to the study by Chen et al. [60], the pressure gradient has the singularity of $O(r^{-\frac{1}{2}})$. A new enrichment function was constructed based on the asymptotic analytic solution [61–63].

$$\phi_t(\boldsymbol{x}) = \sqrt{r}\cos\frac{\theta}{2} \quad (8)$$

2.4. XFEM Discretization

This paper only considers the quasi-static process with infinitesimal strain in porous media, which is homogeneous, isotropic, and linearly elastic. The porous medium domain Ω, with boundary $\Gamma, \Gamma_t, \Gamma_u$ (Figure 3). The fracture Γ_f consisting of two surfaces is embedded in the domain Ω. In this paper, fluid flow and matrix deformation are coupled based on poroelasticity [64], matrix and fracture flow are coupled based on discontinuous flow on the fracture surface, and fracture deformation and fracture flow are coupled using the cubic law. The coupled governing equations are Equation (A4) and (A21) in Appendix A. For the detailed formula derivation, please refer to the Appendix A.

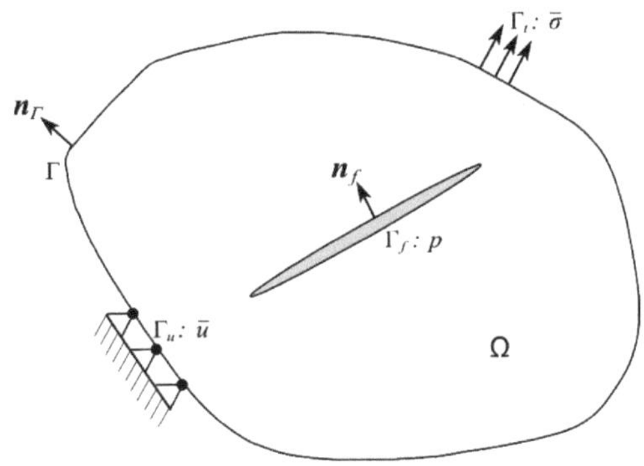

Figure 3. Illustration of a domain.

According to the basic theory of XFEM, the fracture opening can be expressed as:

$$w_f = [\![u]\!] \cdot \boldsymbol{n}_f = \left(\sum_{i \in N_s} N_i(x) a_i + 2 \sum_{q \in N_t} \sqrt{r} N_q(x) b_q \right) \cdot \boldsymbol{n}_f = [\![N]\!] \cdot \boldsymbol{U} \cdot \boldsymbol{n}_f \tag{9}$$

where $[\![\boldsymbol{u}]\!]$ denotes the displacement jump across the fracture; a_i and b_q are discontinuous enriched DDOFs; $[\![N]\!]$ denotes the discontinuous shape function matrix.

By substituting the displacement and pressure approximation expressions into the weak form of the governing equations, the discrete calculation format can be obtained as follows:

$$\begin{bmatrix} 0 & 0 & 0 & 0 \\ 0 & 0 & 0 & 0 \\ M_{uu} & M_{u\tilde{u}} & M_{pp} & M_{p\tilde{p}} \\ M_{\tilde{u}u} & M_{\tilde{u}\tilde{u}} & M_{\tilde{p}p} & M_{\tilde{p}\tilde{p}} \end{bmatrix} \begin{bmatrix} \dot{U} \\ \dot{\tilde{U}} \\ \dot{P} \\ \dot{\tilde{P}} \end{bmatrix} + \begin{bmatrix} K_{uu} & K_{u\tilde{u}} & K_{pp} & K_{p\tilde{p}} \\ K_{\tilde{u}u} & K_{\tilde{u}\tilde{u}} & K_{\tilde{p}p} & K_{\tilde{p}\tilde{p}} \\ 0 & 0 & KK_{pp} & KK_{pp} \\ 0 & 0 & KK_{pp} & KK_{pp} \end{bmatrix} \begin{bmatrix} U \\ \tilde{U} \\ P \\ \tilde{P} \end{bmatrix} = \begin{bmatrix} F_1 \\ \tilde{F}_1 \\ F_2 \\ \tilde{F}_2 \end{bmatrix} \tag{10}$$

Equation (10) can be simplified to the following form:

$$\begin{bmatrix} 0 & 0 \\ \boldsymbol{M}_u & \boldsymbol{M}_p \end{bmatrix} \begin{bmatrix} \dot{\boldsymbol{U}} \\ \dot{\boldsymbol{P}} \end{bmatrix} + \begin{bmatrix} \boldsymbol{K}_u & \boldsymbol{K}_p \\ 0 & \boldsymbol{KK}_p \end{bmatrix} \begin{bmatrix} \boldsymbol{U} \\ \boldsymbol{P} \end{bmatrix} = \begin{bmatrix} \boldsymbol{F}_1 \\ \boldsymbol{F}_2 \end{bmatrix} \tag{11}$$

where

$$\boldsymbol{K}_u = \int_\Omega \boldsymbol{B}^T : \boldsymbol{C} : \boldsymbol{B} d\Omega$$

$$\boldsymbol{K}_p = -\int_\Omega \boldsymbol{B}^T \alpha \varsigma^T \boldsymbol{H}^T d\Omega - \int_{\Gamma_f} [\![N]\!] \boldsymbol{n}_f \boldsymbol{H}^T d\Gamma$$

$$\boldsymbol{F}_1 = \int_{\Gamma_t} \boldsymbol{N} \cdot \overline{\sigma} d\Gamma - \int_\Omega \boldsymbol{N} \cdot (\sigma_0 + \alpha p_0 \boldsymbol{I}) d\Omega$$

$$M_u = \int_{\Omega} \alpha \rho \boldsymbol{H} \varsigma \boldsymbol{B} d\Omega + \int_{\Gamma_f} \rho \boldsymbol{H} \boldsymbol{n}_f [\![N]\!]^T d\Gamma$$

$$M_p = \int_{\Omega} \rho Q^{-1} \boldsymbol{H}\boldsymbol{H}^T d\Omega + \int_{\Gamma_f} \frac{\rho w}{K_l} \boldsymbol{H}\boldsymbol{H}^T d\Gamma$$

$$KK_p = \int_{\Omega} \frac{\rho k_m}{\mu} \boldsymbol{D}^T \boldsymbol{D} d\Omega + \int_{\Gamma_f} \frac{\rho w^3}{12\mu} \boldsymbol{D}_s \boldsymbol{D}_s{}^T d\Gamma$$

$$F_2 = \int_{\Gamma_f} HQ_m d\Gamma - \int_{\Gamma_q} \boldsymbol{H}\bar{q} d\Gamma$$

$$\varsigma = [1,1,0]$$

$$\boldsymbol{N} = [\boldsymbol{N}^u \boldsymbol{N}^{\tilde{u}}]^T, \ \boldsymbol{B} = [\nabla \boldsymbol{N}^u \ \nabla \boldsymbol{N}^{\tilde{u}}]$$

$$\boldsymbol{H} = [\boldsymbol{N}^p \ \boldsymbol{N}^{\tilde{p}}]^T, \ \boldsymbol{D} = [\nabla \boldsymbol{N}^p \ \nabla \boldsymbol{N}^{\tilde{p}}]$$

where $\boldsymbol{H}_s$ is the total shape function at the point source, $\boldsymbol{D}_s$ is the directional derivative of $\boldsymbol{H}_s$ along the tangential direction of the fracture, i.e., $D_s = \partial H_s/\partial s$. The superscripts "$u$" and "$\tilde{u}$" correspond to standard DDOFs and enriched DDOFs, respectively, and the superscripts "p" and "$\tilde{p}$" correspond to standard PDOFs and enriched PDOFs, respectively. Equation (11) is a set of coupled nonlinear equations, which are solved by the Newton–Rapson method.

3. Fracture Propagation and Solution Strategy

3.1. Fracture Propagation Criterion

In practice, the hydraulic fracture usually propagates in a mixed mode. The maximum circumferential stress criterion is adopted as the fracture propagation criterion. It is assumed that the fracture propagates when the effective stress intensity factor K_e along that direction reaches the fracture toughness K_{IC}, and it will deflect by an angle of [65]:

$$\theta = 2\arctan\left(\frac{-2K_{II}/K_I}{1+\sqrt{1+8(K_{II}/K_I)^2}}\right) \tag{12}$$

The effective stress intensity factor K_e is expressed as:

$$K_e = \cos\frac{\theta}{2}\left(K_I \cos^2\frac{\theta}{2} - 1.5K_{II}\sin\theta\right) \geq K_{IC} \tag{13}$$

The displacement-based approach is used to calculate the SIFs. K_I and K_{II} are determined by the discontinuous displacement of the fracture tip element. The relational equation is [66].

$$\begin{cases} K_I = \frac{0.806\sqrt{\pi}ED_n}{4(1-v^2)\sqrt{r}} \\ K_{II} = \frac{0.806\sqrt{\pi}ED_s}{4(1-v^2)\sqrt{r}} \end{cases} \tag{14}$$

where E is Young's modulus, v is Poisson's ratio, D_n and D_s are normal and tangential discontinuous displacements of fractures. It should be noted that the calculation models in this paper are all under the plane strain condition.

3.2. HMFM Fracturing Simulation Process

Fracture propagation is a highly nonlinear hydro-mechanical coupled process. In this paper, the fully coupled approach is implemented. The fully coupled approach forms a single large system of equations that solve for all of the displacement/pressure unknowns at once. Figure 4 summarizes the fully coupled implementation of the solution of the HMFM.

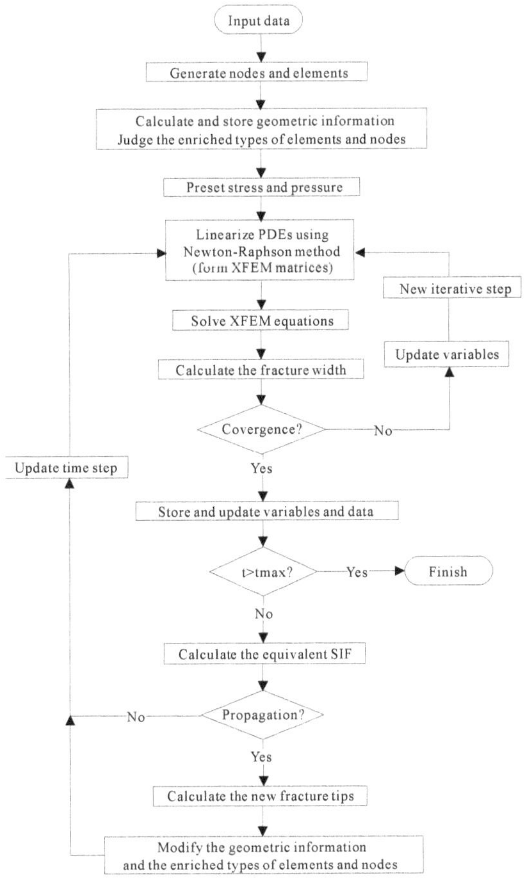

Figure 4. UXFEM flow chart.

4. Results and Discussions

4.1. Validation 1: Comparison between UXFEM and KGD Model

The KGD model is one of the most commonly used fracturing models. The viscosity-dominated hydraulic pressure is calculated using the approximate solution of the KGD model derived by Detournay [67]. Table 1 lists the parameters used by the model. The fracture length and width at the wellbore position are compared with those calculated by UXFEM and plane strain KGD model, as shown in Figure 5. It can be seen that the evolution of fracture length and width calculated by UXFEM in this paper can well fit the KGD model. The calculation error increases with the fracture length because the infinite reservoir assumption cannot be satisfied with the fracture propagation.

Table 1. Parameters for validation 1.

Parameters	Value	Unit
Young modulus	60	GPa
Poisson ratio	0.25	/
Fracture toughness	1.2	MPa·m$^{0.5}$
Fluid viscosity	50	mPa·s
Injection rate	0.01	m^2/s
Convergence tolerance	0.01	/

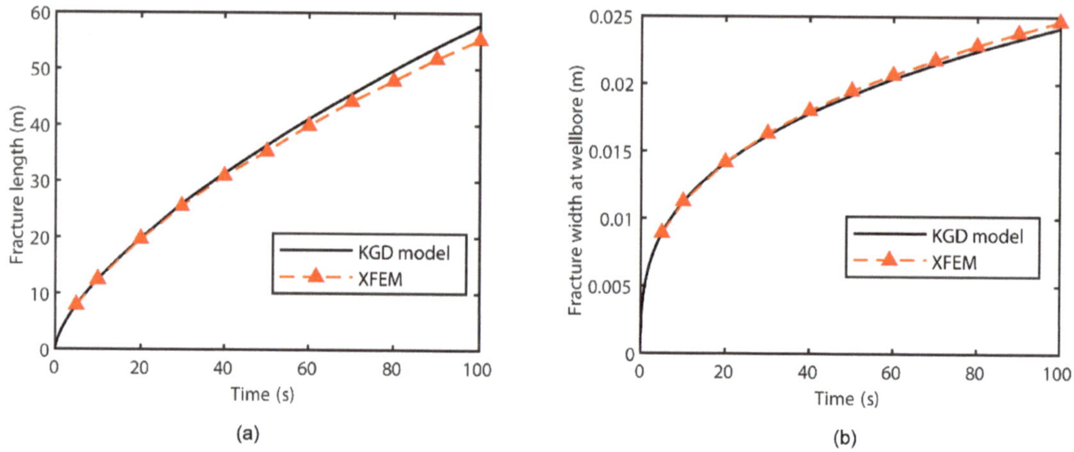

Figure 5. Comparison of KGD and UXFEM, (**a**) Comparison of fracture length, and (**b**) Comparison of fracture width at the wellbore.

4.2. Validation 2: Comparison of KI and KII in Numerical Simulation and Analytical Solution

Rice [68] deduces the analytical expressions of the SIFs at the fracture tip under hydrostatic pressure and horizontal stress. For the single fracture propagation in an infinite domain, the SIFs are expressed as:

$$\begin{cases} K_I = \sqrt{\pi L_f}\left[p - \left(\sigma_H \sin^2\beta + \sigma_h \cos^2\beta\right)\right] \\ K_{II} = \frac{\sqrt{\pi L_f}}{2}[\sigma_H - \sigma_h]\sin 2\beta \end{cases} \tag{15}$$

where L_f is the half length of the fracture; p is hydrostatic pressure in the fracture; S is the inclination angle of fracture and maximum horizontal principal stress. σ_H and σ_h are the maximum horizontal stress and the minimum horizontal stress, respectively.

In order to approximate the infinite formation and eliminate the influence of the boundary, the model size is set to 200 × 200 m, the initial half-length of the fracture is 5 m, and the domain is discretized into structured grids (159 × 159). Table 2 lists the parameters used by the model. Calculate the SIFs (K_I and K_{II}) of different inclination angles, and the results are compared, as shown in Figure 6. By comparison, the SIFs calculated by UXFEM in this paper can well fit the analytical solution. The maximum error of KI is not more than 2%, which proves the method's accuracy.

Table 2. Parameters for validation 2.

Parameters	Value	Unit
Young's modulus	60	GPa
Poisson's ratio	0.25	/
Maximum horizontal in-situ stress	60	MPa
Minimum horizontal in-situ stress	55	MPa
Hydrostatic pressure applied to the fracture surface	65	MPa
Half-length of the fracture	5	m

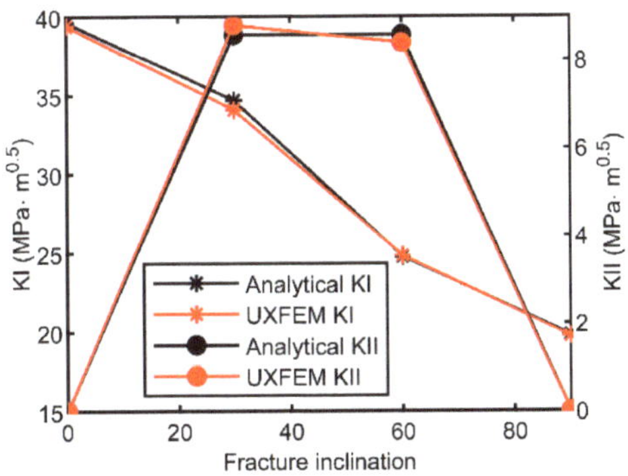

Figure 6. Calculation results comparison of stress intensity factor between UXFEM and analytical solution.

4.3. Validation 3: Comparison between UXFEM and COMSOL Multiphysics® 5.6

The purpose of this section is to show the accuracy and feasibility of our fully coupled approach. We consider a quasi-static mud loss case of a shale oil reservoir. The simulation parameters are shown in Table 3. The geometry (Figure 7) is built to carry out the simulation process. The domain is discretized into 25,860 unstructured grids in COMSOL and 3481 structured grids in UXFEM. Figure 8 shows the displacement field, effective stress component σ_y', and pressure field calculated by COMSOL and UXFEM. In Figure 8, the results calculated by UXFEM and COMSOL are in good agreement. From Figure 8a,d, the displacement at the injection point is the largest, implying that the fracture width is the largest at the injection point. Figure 8b,e shows that there is an obvious stress concentration at the fracture tip. Figure 8c,f shows that the pressure is higher along the fracture, and a small low-pressure area appears around the tips. Figure 9 shows that fracture widths calculated by UXFEM and COMSOL can be matched. Calculating the L_2 error norm:

$$\left\|F - F^h\right\|_{L_2} = \left[\int_\Omega (F - F^h) d\Omega\right]^{0.5} \tag{16}$$

where F^h is the COMSOL solution and F is the UXFEM solution.

The displacement L_2 error, effective stress component σ_y' L_2 error, and pressure L_2 error are 0.0083%, 0.041%, and 0.078%, respectively. A very good agreement between the results proves the effectiveness and accuracy of this fully coupled approach.

Table 3. Parameters for Validation 3.

Parameters	Value	Unit
Initial reservoir pressure	10	MPa
Pressure at injection point	25	MPa
Maximum horizontal in-situ stress	20	MPa
Minimum horizontal in-situ stress	15	MPa
Young's modulus	40	GPa
Poisson's ratio	0.2	/
Matrix permeability	1×10^{-18}	m^2
Initial matrix porosity	0.15	/
Bulk modulus of the solid	50	GPa
Bulk modulus of the fluid	2.5	GPa
Fluid density	1000	kg/m^3
Fluid viscosity	10	mPa·s
Biot coefficient	0.85	/

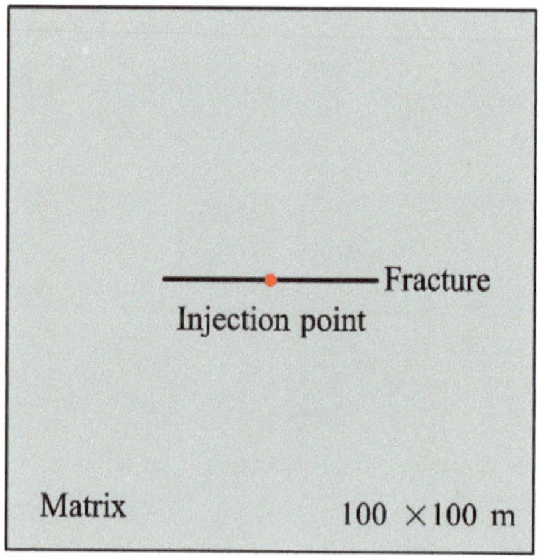

Figure 7. Geometric model for the comparison between UXFEM and COMSOL.

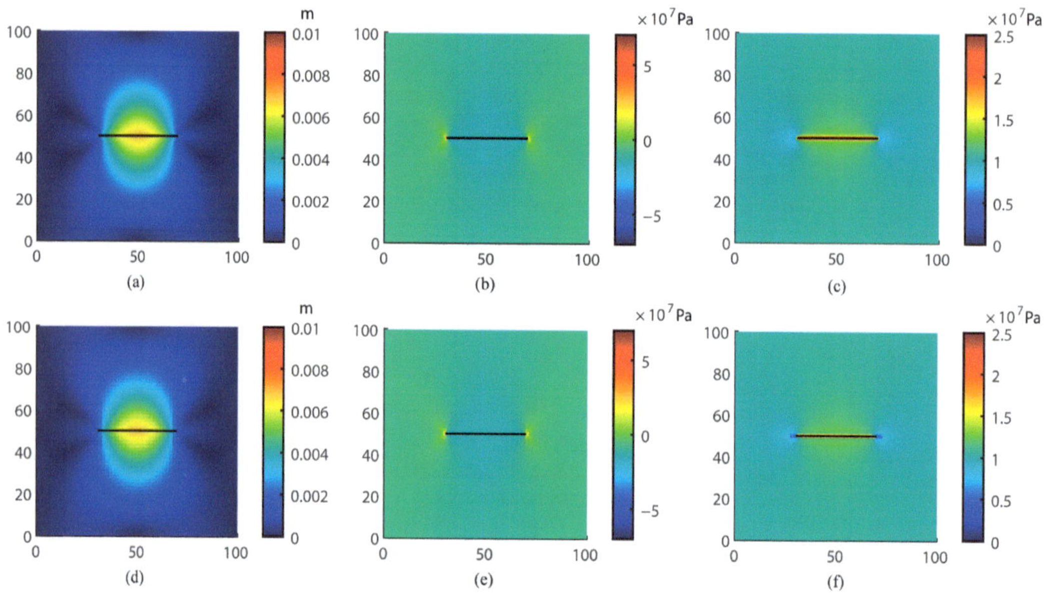

Figure 8. Distributions of the displacement, Effective stress component σ'_y and pressure at 1000 s, (**a**) Displacement distribution calculated by COMSOL, (**b**) Effective stress component σ'_y distribution calculated by COMSOL, (**c**) Pressure distribution calculated by COMSOL, (**d**) Displacement distribution calculated by UXFEM, (**e**) Effective stress component σ'_y distribution calculated by UXFEM, (**f**) Pressure distribution calculated by UXFEM.

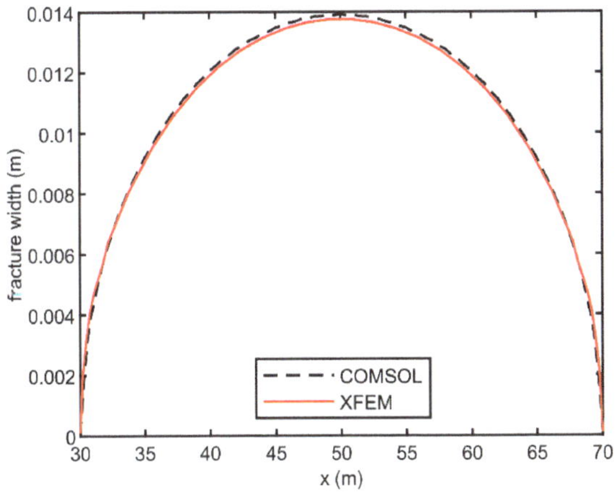

Figure 9. Comparison of the fracture width between UXFEM and COMSOL.

4.4. Case Study1: Propagation of Three Parallel Fractures

A 200 m × 200 m two-dimensional model is established to study the fracture deflection propagation caused by the stress shadow effect. The initial half-length of the fracture was 5 m, the distance between the three fractures was 14 m or 21 m, and the midpoints of the three fractures are the injection points. The injection rate at injection points is 0.01 m^2/s. Other parameters are listed in Table 4.

Table 4. Parameters for case study.

Parameters	Value	Unit
Initial reservoir pressure	50	MPa
Maximum horizontal in situ stress	60	MPa
Minimum horizontal in situ stress	55	MPa
Matrix permeability	1×10^{-17}	m^2
Initial matrix porosity	0.1	/
Initial fracture length	10	m
Injection rate at injection points	0.01	m^2/s
Bulk modulus of the fluid	2.5	GPa
Fluid density	1000	kg/m^3
Fluid viscosity	50	mPa·s
Fracture toughness K_{IC}	5	$MPa{\cdot}m^{0.5}$
Biot coefficient	0.85	/

The final morphology of the simulated fracture, displacement field, and effective stress component σ'_y are shown in Figure 10. For the model of three fractures, the middle fracture is squeezed and suppressed by the same on both sides. The middle fracture almost propagates along a straight line, and its length is significantly shorter than the other two fractures. Due to the stress interference between fractures, the side fractures deflect in the opposite direction, away from the middle fracture [69]. Fractures will actively choose to propagate along the path with the least resistance. As the fracture spacing is smaller, the fracture deflection angle is larger, indicating that the fracture spacing is an important factor affecting the stress interference intensity.

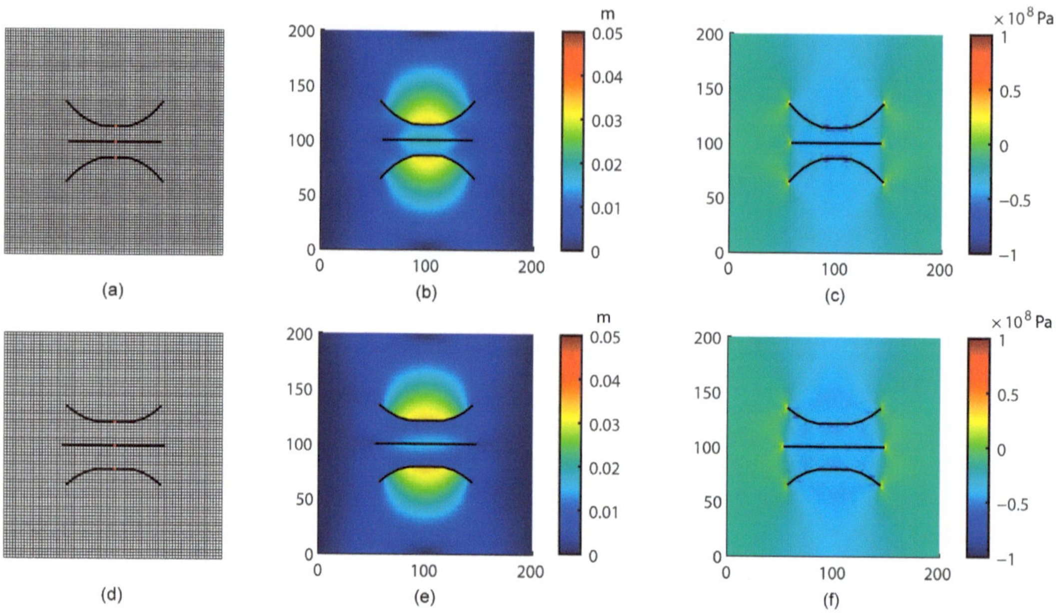

Figure 10. Simulation results with fracture spacing = 14 or 21 m after 280 s, (**a**) Fracture final morphology of spacing = 14 m, (**b**) Displacement of spacing = 14 m, and (**c**) Effective stress component σ'_y of spacing = 14 m, (**d**) Fracture final morphology of spacing = 21 m, (**e**) Displacement of spacing = 21 m, and (**f**) Effective stress component σ'_y of spacing = 21 m.

4.5. Case Study2: Propagation in Multiple En Échelon Fractures

En échelon fractures are common in geology and are caused by the mechanical interaction between their near-tip stress fields [36,70]. To study the evolution of En échelon fractures, a 200 × 200 m geometry model is built. The model contains three prefabricated fractures, with an initial fracture length of 10 m and a vertical spacing of 16 m. The initial parameters of the three fractures are completely the same. The injection point is located at the midpoint of the fracture (The red points in Figure 11), and the injection rate is set to 0.01 m^2/s. Other parameters of the model are the same as those in 5.4 (Table 4). See Figure 11 for the calculation results. It can be seen that after the fractures overlap, the propagation of the middle fracture is restrained, and the tendency of mutual attraction between adjacent fractures is becoming more and more obvious. Enhancement or impediment of the growth ahead of an individual fracture front leads to asymmetric fracture geometry [36].

The stress interference of parallel fractures has been studied previously, and the fractures tend to mutually repel each other. However, for such parallel fractures with incomplete left-right overlap, the stress shadow effect will cause the stress direction to change, and the overlapping tips will appear as a phenomenon of mutual attraction. In the process of hydraulic fracturing, we hope that hydraulic fractures can form complex shapes. However, this phenomenon may lead to fracture collision and reduce the fracturing effect, which needs to be avoided. Therefore, the optimization of the overlapping length and spacing of fractures is an important job.

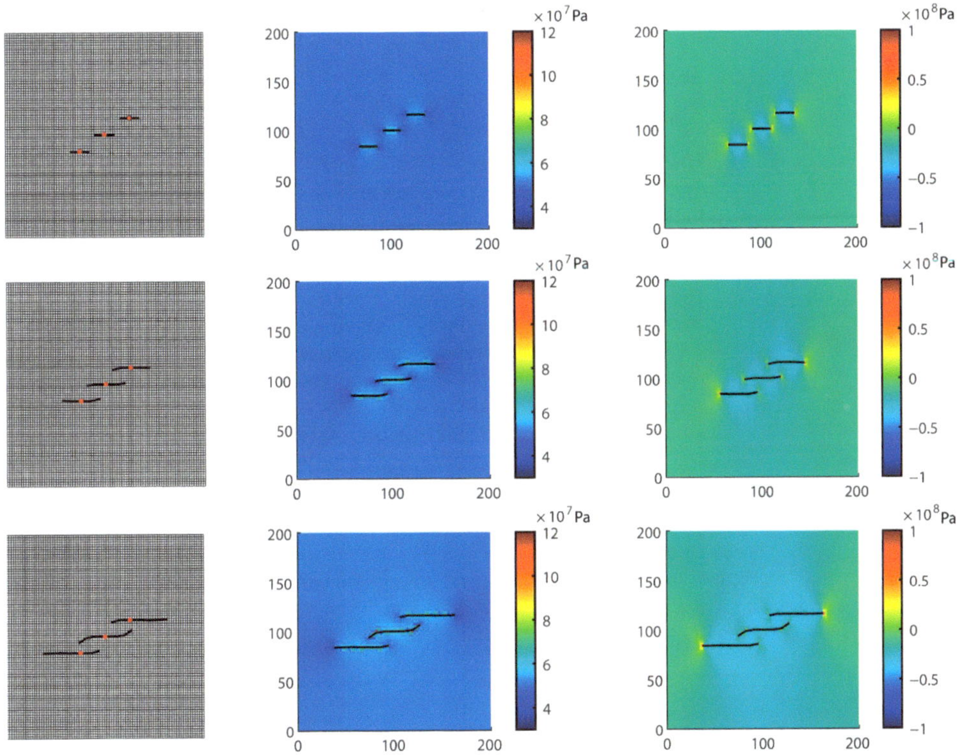

Figure 11. Sequential growth of En échelon fractures: trajectory, pressure, and effective stress component σ'_y after 20 s, 70 s, and 170 s.

4.6. Case Study3: Multi-Well and Multi-Cluster Simultaneous Fracturing and Production

The problem considered here is the fracturing-production problem of 7 hydraulic fractures generated by two horizontal wells (#1 and #3) and one vertical well (#2), numbered 1 to 7, respectively. The schematic diagram of the initial fractures and the model is shown in Figure 12. The size of the model is 200 m × 100 m. Two horizontal wells are along the straight-line x = 60 m and x = 140 m, respectively. The vertical well is located in the middle of the model. Table 4 lists the detailed parameters.

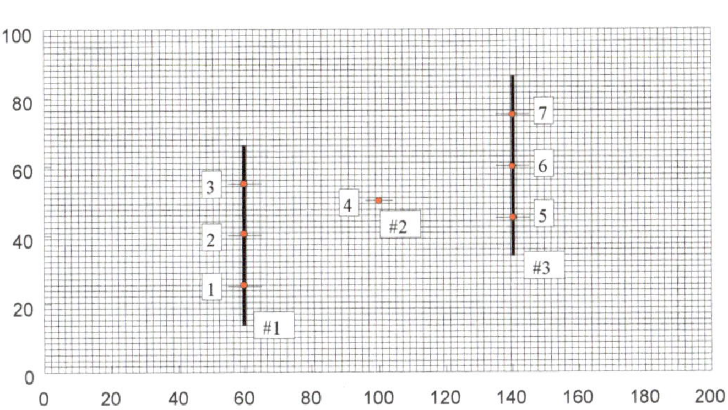

Figure 12. Multi-well and multi-cluster simultaneous fracturing model.

This study aims to analyze the interaction between evolved multi-fractures and local stress-field-pressure fields and help explore the interference phenomenon's influence on fracture morphology and production. Taking well #1 as an example, it can be seen that the fractures are subject to obvious stress interference from Figures 13 and 14, and the stress interference is manifested in two points:

(1) The fractures will undergo obvious diversion. For well #1, Frac. 1–3 have different degrees of deflection. The deflection of the two Frac. 1 and 3 are more prominent. In the early stage, the middle fracture Frac. 2 does not deflect obviously due to the interference of superimposed stress. In the later stage of fracturing, Frac. 2 and 4 attract each other, showing En échelon fracture morphology [69].

(2) There are noticeable differences in fracture length and width. The length of Frac. 2 and 6 of the two horizontal wells is shorter, followed by Frac. 3 and 5, and the longest is Frac. 1 and 7. Frac. 4 cannot propagate far enough and even the front edge of the fracture tends to close. Although Frac. 1 and 7 have propagated long enough, one side of the fracture tends to close under the effect of stress shadow. These phenomena are very likely to lead to the reduction of fracturing efficiency and cost waste.

The complex morphology of fractures is the result of stress interference. Studies have found that 25–30% of the fractures do not bring production [71,72], indicating that 25–30% of the fractures are not opened or closed after being opened, which is also the effect of stress interference. Therefore, stress interference has both positive and negative effects. Numerical simulation can be used to find the best design for well placement, perforation, and fracturing so that the fractures can be fully opened and stabilized to maximize the benefits of fracturing.

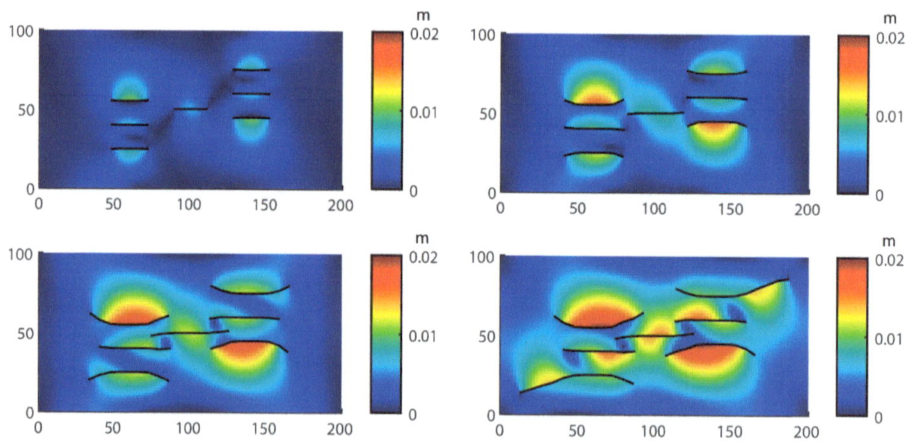

Figure 13. Displacement field after 20, 50, 70, and 100 s.

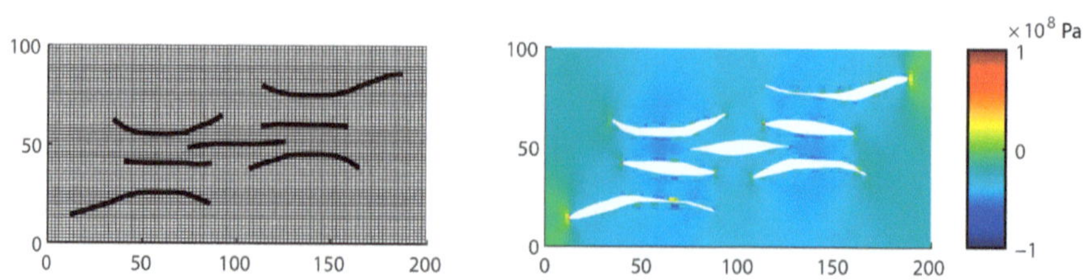

Figure 14. Fracture trajectory, effective stress component σ'_y, and fracture width distribution after fracturing.

Change the bottom hole condition to a constant pressure of 25 MP, and proppant instead of the hydraulic pressure supports the fracture. Assuming that the fracture has high conductivity and obeys Darcy's law, it is only necessary to change the fracture flow coefficient $\frac{w^2}{12\mu}$ in Equation (A4) into the hydraulic fracture permeability. The stress-sensitive parameters are brought into the governing equation for calculation during production. For the estimation method of stress-sensitive parameters, see the study of Lu et al. [73]. The pressure change in the production process is shown in Figure 15, which shows that there is almost no production interference between fractures and wells within 30 days, and obvious interference can be observed after 300 days. The calculated cumulative production is shown in Figure 16, which clearly shows the rising pattern of the cumulative production. The early production increases rapidly because the reservoir near the well is fully stimulated. However, the reservoir's extremely low permeability makes it difficult to maintain rapid production.

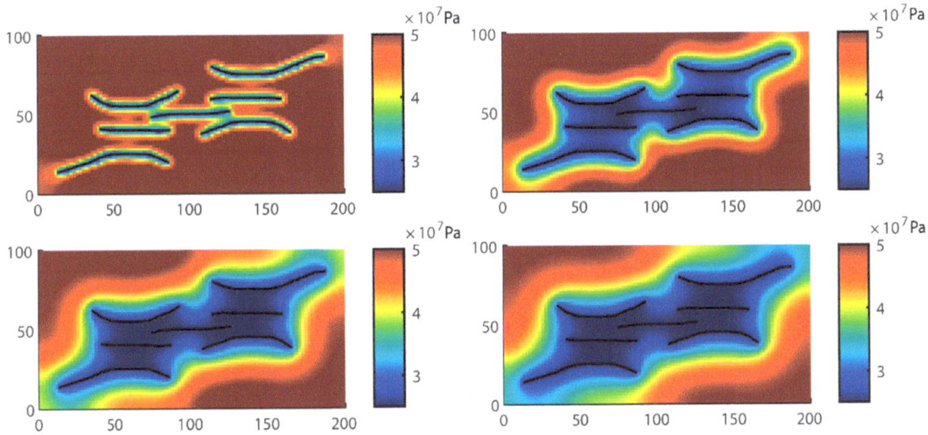

Figure 15. Pressure field after 30, 300, 600, and 900 days.

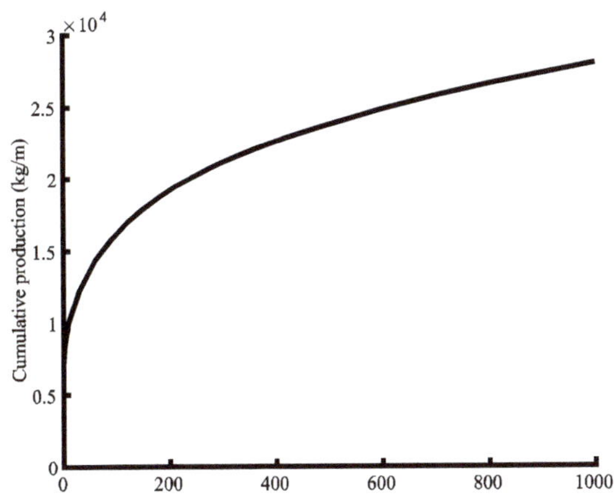

Figure 16. Cumulative production curve during the hydro-mechanical coupled production process.

Four calculation examples with different well-spacing (40 m, 60 m, 80 m, and 100 m) are simulated for comparison to verify the effect of well-spacing on fracture morphology and production. After 60 s injection, the fracture morphology and displacement field are as Figure 17. The final morphology of fractures under the stress interference is different with different well-spacing. Then, realize conversion from hydraulic fracturing to production, as shown in Figure 18, and the cumulative production curve after 2000 days is plotted in Figure 19. Figure 18 shows that small well-spacing will lead to slow pressure drop near boundaries, resulting in a low production rate. Thus, the cumulative production of small well-spacing is smaller than that of large well-spacing, which can be explained by the small well-spacing leading to a poor fracturing effect. Therefore, reasonable well-spacing is a crucial factor affecting the formation of a complex fracture network and efficient production.

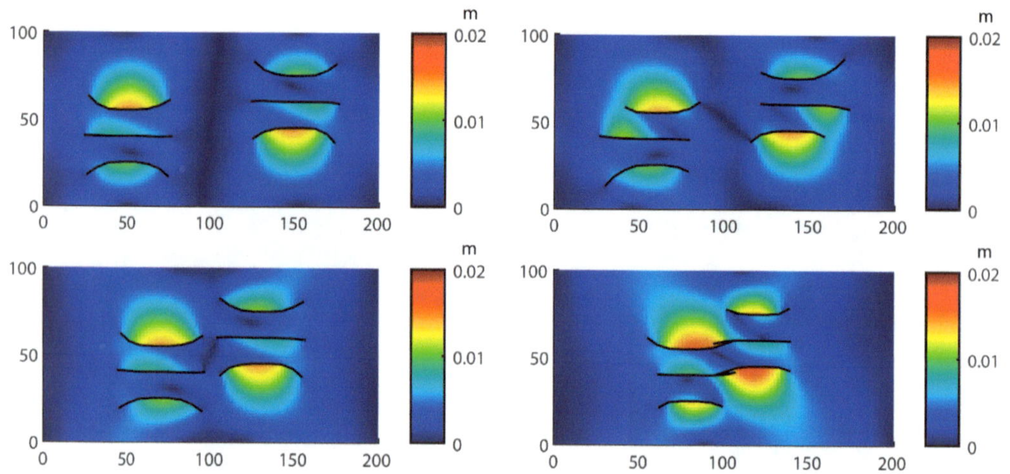

Figure 17. Displacement field after 60 s (well-spacing = 100, 80, 60, and 40 m).

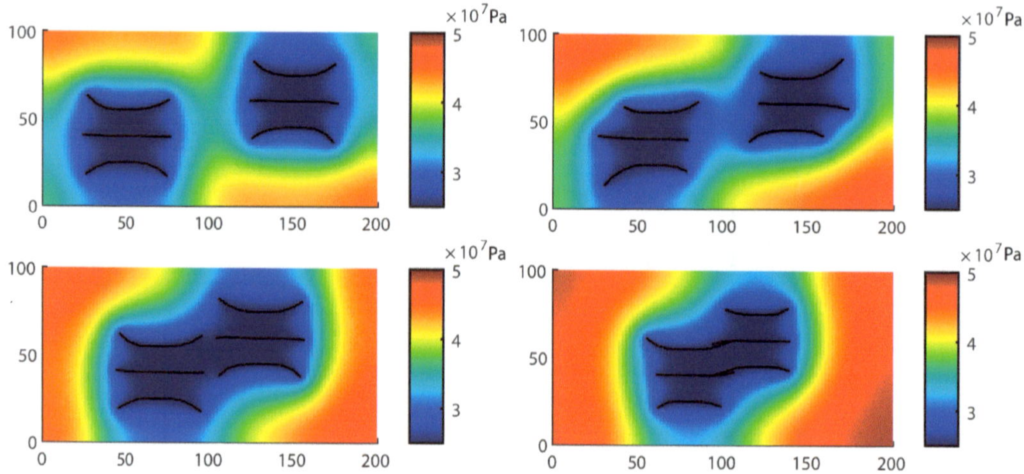

Figure 18. Pressure field after 2000 days (well-spacing = 100, 80, 60, and 40 m).

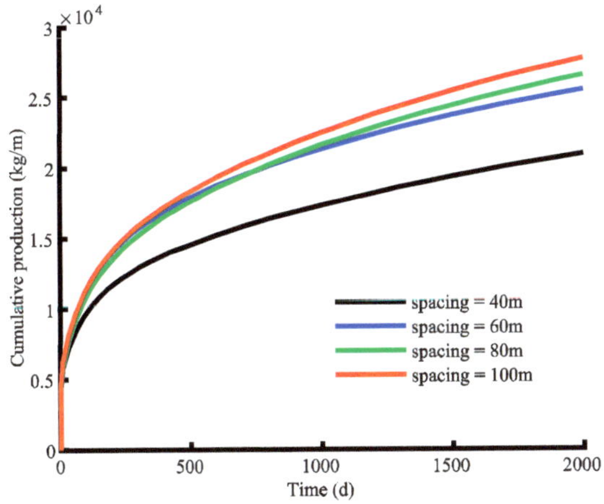

Figure 19. Comparison of cumulative production curves with different well-spacing.

5. Conclusions

Aiming at solving the fracturing-to-production problem under the unified framework, a fully coupled approach UXFEM for solving the HMFM under the same framework is established based on the extended finite element method and Newton–Rapson method. This technology inherits all the excellent characteristics of XFEM and can fully reflect the mechanical mechanism of fracture propagation. Meanwhile, it also avoids some potential problems caused by the iterative coupling method, such as convergence reduction. The effectiveness and accuracy of UXFEM in HMFM research are validated by comparing it with the results of classical KGD models, analytical solutions, and COMSOL Multiphysics® 5.6. To study the interference problem during the fracturing process, three examples are used to illustrate the effect of stress shadow on fracturing. At the same time, after simultaneous fracturing, this paper also successfully realized the fracturing-to-production analysis. The research shows that hydraulic fractures are interfered with by the stress between wells and fractures, which plays an essential role in forming the final fracture network morphology and cumulative production. Proper well location design, perforation plan, and production plan are the keys to determining whether the stress interference can be used reasonably to increase the fracturing effects and reservoir development efficiency. This research has certain practical significance for fracturing design and production prediction.

Author Contributions: Data curation, D.W.; Methodology, Y.D.; Supervision, Y.J. and Y.X.; Writing—Original draft, Y.D. All authors have read and agreed to the published version of the manuscript.

Funding: This work was financially supported by the National Natural Science Foundation of China (Grant NO. U19B6003-05 and 51874321).

Data Availability Statement: Not applicable.

Acknowledgments: The financial support provided by the National Natural Science Foundation of China is highly appreciated.

Conflicts of Interest: The authors declare that they have no known competing financial interests or personal relationships that could have appeared to influence the work reported in this paper.

Appendix A

For a quasi-static process, ignoring the effect of gravity, the stress balance equation of the rock matrix is [74]:

$$\nabla \cdot \sigma = 0 \tag{A1}$$

where σ is the Cauchy stress tensor. Considering the effective stress, the Cauchy stress tensor σ can be written as:

$$\sigma = \sigma' + \sigma_0 - \alpha(p - p_0)\boldsymbol{I} \tag{A2}$$

where σ_0 is the initial total stress; p_0 is the initial pressure; α is the Biot coefficient; $\boldsymbol{I}$ is the identity matrix; σ' is the effective stress; $\sigma' = \boldsymbol{C} : \varepsilon$. $\boldsymbol{C}$ is the elasticity tensor;ε is the strain, which can be calculated from the geometric equation.

The boundary condition is:

$$\begin{cases} \boldsymbol{u} = \overline{\boldsymbol{u}} & \text{on } \Gamma_u \\ \sigma \cdot \boldsymbol{n} = \overline{\sigma} & \text{on } \Gamma_t \\ \sigma \cdot \boldsymbol{n}_f = -p\boldsymbol{n}_f & \text{on } \Gamma_f \end{cases} \tag{A3}$$

where Γ_u, Γ_t and Γ_f are displacement boundary, traction boundary, and fracture surfaces, respectively; $\boldsymbol{n}$ and $\boldsymbol{n}_f$ are the normal direction to the related boundary.

Taking $\delta\boldsymbol{u}$ as the trial function, the weak form of the solid matrix deformation governing equation can be obtained from Equations (A1)–(A3),

$$\int_\Omega \nabla\delta\boldsymbol{u} \cdot (\boldsymbol{C} : \varepsilon + \sigma_0 - \alpha(p - p_0)\boldsymbol{I})d\Omega - \int_{\Gamma_f} [\![\delta\boldsymbol{u}]\!] \cdot p\boldsymbol{n}_f d\Gamma - \int_{\Gamma_t} \delta\boldsymbol{u} \cdot \overline{\sigma} d\Gamma = 0 \tag{A4}$$

The governing equation for mass conservation of a compressible-fluid flow is:

$$\frac{dm}{dt} + \nabla \cdot (\rho \boldsymbol{v}_m) = 0 \tag{A5}$$

where m denotes the fluid mass, $\boldsymbol{v}_m$ is the matrix flux. ρ is the fluid density.

The constitutive relations for poroelasticity material are written as [64]

$$\delta m = \rho(\alpha\delta e + \frac{\delta p}{Q}) \tag{A6}$$

where e is the volumetric strain, $e = \nabla \cdot \boldsymbol{u}$; Q is the Biot modulus, $Q = \left[\frac{\phi}{K_l} + \frac{\alpha - \phi}{K_s}\right]^{-1}$. where K_l is the fluid modulus; ϕ is the porosity of the matrix.

Combine Equation (A5) and (A6), the matrix fluid flow governing equation can express as:

$$\alpha\rho\nabla \cdot \dot{u} + \frac{\rho}{Q}\frac{\partial p}{\partial t} + \nabla \cdot (\rho \boldsymbol{v}_m) = 0 \tag{A7}$$

The boundary condition is:

$$\begin{cases} \rho \boldsymbol{v}_m \cdot \boldsymbol{n} = \overline{q} & \text{on } \Gamma_q \\ [\![\rho \boldsymbol{v}_m \cdot \boldsymbol{n}_f]\!] = \rho v_d & \text{on } \Gamma_f \end{cases} \tag{A8}$$

where Γ_q is the flow boundary, $\overline{q}$ is the mass flux; $[\![\rho \boldsymbol{v}_m \cdot \boldsymbol{n}_f]\!]$ denotes the normal discontinuous flow on the fracture surface; v_d denotes the fluid transfer from the fracture into the matrix, which can be interpreted as the leak-off effect.

Taking δp as the trial function, the weak form of the matrix flow governing equation can be obtained from Equations (A7) and (A8),

$$\int_\Omega \alpha\rho\delta p\nabla \cdot \dot{\boldsymbol{u}} d\Omega + \int_\Omega \rho Q^{-1}\delta p\dot{p} d\Omega + \int_{\Gamma_q} \delta p\overline{q} d\Gamma - \int_{\Gamma_f} \delta p\rho v_d d\Gamma - \int_\Omega \nabla(\delta p)\rho \boldsymbol{v}_m d\Omega = 0 \tag{A9}$$

Matrix flow obeys Darcy's law, $\boldsymbol{v}_m = -\frac{k_m}{\mu}\nabla p$. Thus, Equation (A9) can be rewritten as:

$$\int_{\Gamma_f} \delta p \rho v_d d\Gamma = \int_{\Omega} \alpha \rho \delta p \nabla \cdot \dot{\boldsymbol{u}} d\Omega + \int_{\Omega} \rho Q^{-1} \delta p \dot{p} d\Omega + \int_{\Gamma_q} \delta p \bar{q} d\Gamma + \int_{\Omega} \frac{\rho k_m}{\mu} \nabla(\delta p) \nabla p d\Omega \tag{A10}$$

Compressible fluid flow in fracture satisfies mass conservation and cubic law [75]:

$$\frac{\partial(\rho w)}{\partial t} + \nabla \cdot (\rho \boldsymbol{q}_f) = Q_m \tag{A11}$$

$$\boldsymbol{q}_f = -\frac{w^3}{12\mu} \frac{\partial p}{\partial s} = \boldsymbol{v}_f w \tag{A12}$$

where $\boldsymbol{q}_f$ is the fluid flux in the fracture; w is the fracture aperture; $\boldsymbol{v}_f$ is the average fluid velocity of fracture section; and Q_m is the mass source term in the fracture.

$$\frac{\partial(\rho w)}{\partial t} = \frac{\partial \rho}{\partial t} w + \rho \frac{\partial w}{\partial t} = \frac{\rho w}{K_l} \frac{\partial p}{\partial t} + \rho \frac{\partial w}{\partial t} \tag{A13}$$

where $\frac{1}{K_l} = \frac{1}{\rho} \frac{\partial \rho}{\partial p}$. Combine Equation (A11) and (A13):

$$\frac{\rho w}{K_l} \frac{\partial p}{\partial t} + \rho \frac{\partial w}{\partial t} + \nabla \cdot \left(\rho \boldsymbol{q}_f \right) = Q_m \tag{A14}$$

The boundary condition is:

$$\boldsymbol{q}_f \cdot \boldsymbol{n}_f = w \boldsymbol{v}_d \tag{A15}$$

Taking δp as the trial function, the weak form of the fracture flow governing equation can be obtained from Equation (A14),

$$\int_{\Omega_f} \delta p \frac{\rho w}{K_l} \frac{\partial p}{\partial t} d\Omega + \int_{\Omega_f} \delta p \rho \frac{\partial w}{\partial t} d\Omega + \int_{\Omega_f} \delta p \nabla \cdot (\rho \boldsymbol{q}_f) d\Omega - \int_{\Omega_f} \delta p Q_m d\Omega = 0 \tag{A16}$$

The third term in Equation (A16) can be rewritten as:

$$\int_{\Omega_f} \delta p \nabla \cdot (\rho \boldsymbol{q}_f) d\Omega = \int_{\Omega_f} \nabla \cdot (\delta p \rho \boldsymbol{q}_f) d\Omega - \int_{\Omega_f} \nabla \delta p \cdot \rho \boldsymbol{q}_f d\Omega = \int_{\Gamma_f} \delta p \rho \boldsymbol{q}_f \cdot \boldsymbol{n}_f d\Gamma + \int_{\Omega_f} \frac{\partial(\delta p)}{\partial s} \cdot \frac{\rho w^3}{12\mu} \frac{\partial p}{\partial s} d\Omega \tag{A17}$$

For fluid flow in the fracture, the fluid exchange (leak-off) term can be written as:

$$\int_{\Gamma_f} \delta p \rho \boldsymbol{q}_f \cdot \boldsymbol{n}_f d\Gamma = \int_{\Gamma_f} \delta p \rho w v_d d\Gamma \tag{A18}$$

Thus,

$$\int_{\Gamma_f} \delta p \rho w v_d d\Gamma = -\int_{\Omega_f} \frac{\partial(\delta p)}{\partial s} \cdot \frac{\rho w^3}{12\mu} \frac{\partial p}{\partial s} d\Omega - \int_{\Omega_f} \delta p \frac{\rho w}{K_l} \frac{\partial p}{\partial t} d\Omega - \int_{\Omega_f} \delta p \rho \frac{\partial w}{\partial t} d\Omega + \int_{\Omega_f} \delta p Q_m d\Omega \tag{A19}$$

Since the fracture aperture is much smaller than its length, the variation of fluid pressure across the fracture width can be ignored. The fracture flow governing equation can be obtained from Equation (A19).

$$\int_{\Gamma_f} \delta p \rho v_d d\Gamma = -\int_{\Gamma_f} \frac{\partial(\delta p)}{\partial s} \cdot \frac{\rho w^3}{12\mu} \frac{\partial p}{\partial s} d\Gamma - \int_{\Gamma_f} \delta p \frac{\rho w}{K_l} \frac{\partial p}{\partial t} d\Gamma - \int_{\Gamma_f} \delta p \rho \frac{\partial w}{\partial t} d\Gamma + \int_{\Gamma_f} \delta p Q_m d\Gamma \tag{A20}$$

Based on Equation (A10) and (A20), the weak form of the coupled matrix-fracture flow governing equation is:

$$\begin{aligned}&\int_\Omega \alpha\rho\delta p\nabla\cdot\dot{u}d\Omega+\int_\Omega \rho Q^{-1}\delta p\dot{p}d\Omega+\int_{\Gamma_q}\delta p\bar{q}d\Gamma+\int_\Omega \frac{\rho k_m}{\mu}\nabla(\delta p)\nabla pd\Omega+\\&\int\limits_{\Gamma_f}\frac{\partial(\delta p)}{\partial s}\cdot\frac{\rho w^3}{12\mu}\frac{\partial p}{\partial s}d\Gamma+\int\limits_{\Gamma_f}\delta p\frac{\rho w}{K_l}\dot{p}d\Gamma+\int\limits_{\Gamma_f}\delta p\rho\frac{\partial w}{\partial t}d\Gamma-\int\limits_{\Gamma_f}\delta pQ_m d\Gamma=0\end{aligned} \tag{A21}$$

References

1. Liu, W.; Yao, J.; Chen, Z.; Liu, Y. Effect of quadratic pressure gradient term on a one-dimensional moving boundary problem based on modified Darcy's law. *Acta Mech. Sin.* **2016**, *32*, 38–53. [CrossRef]
2. Bunger, A.P.; Kear, J.; Jeffrey, R.G.; Prioul, R.; Chuprakov, D. Laboratory investigation of hydraulic fracture growth through weak discontinuities with active ultrasound monitoring. In Proceedings of the 13th ISRM International Congress of Rock Mechanics, Montreal, QC, Canada, 10–13 May 2015.
3. Guo, J.; Luo, B.; Lu, C.; Lai, J.; Ren, J. Numerical investigation of hydraulic fracture propagation in a layered reservoir using the cohesive zone method. *Eng. Fract. Mech.* **2017**, *186*, 195–207. [CrossRef]
4. Yang, D.; Zhou, Y.; Xia, X.; Gu, S.; Xiong, Q.; Chen, W. Extended finite element modeling nonlinear hydro-mechanical process in saturated porous media containing crossing fractures. *Comput. Geotech.* **2019**, *111*, 209–221. [CrossRef]
5. Olson, J.E.; Wu, K. Sequential versus simultaneous multi-zone fracturing in horizontal wells: Insights from a non-planar, multi-frac numerical model. In Proceedings of the SPE Hydraulic Fracturing Technology Conference, The Woodlands, TX, USA, 6–8 February 2012; OnePetro: Richardson, TX, USA, 2012.
6. Zoback, M.D.; Rummel, F.; Jung, R.; Raleigh, C.B. Laboratory hydraulic fracturing experiments in intact and pre-fractured rock. *Int. J. Rock Mech. Min. Sci. Geomech. Abstr.* **1977**, *14*, 49–58. [CrossRef]
7. Bunger, A.P.; Gordeliy, E.; Detournay, E. Comparison between laboratory experiments and coupled simulations of saucer-shaped hydraulic fractures in homogeneous brittle-elastic solids. *J. Mech. Phys. Solids* **2013**, *61*, 1636–1654. [CrossRef]
8. Guo, T.; Zhang, S.; Qu, Z.; Zhou, T.; Xiao, Y.; Gao, J. Experimental study of hydraulic fracturing for shale by stimulated reservoir volume. *Fuel* **2014**, *128*, 373–380. [CrossRef]
9. Hou, B.; Chang, Z.; Fu, W.; Muhadasi, Y.; Chen, M. Fracture initiation and propagation in a deep shale gas reservoir subject to an alternating-fluid-injection hydraulic-fracturing treatment. *SPE J.* **2019**, *24*, 1839–1855. [CrossRef]
10. Yang, H.; Guo, Y.; Wang, L.; Bi, Z.; Guo, W.; Zhao, G.; Yang, C. Study on the Stimulation Effectiveness Evaluation of Large-Scale Hydraulic Fracturing Simulation Experiment based on Optical Scanning Technology. *SPE J.* **2022**, *27*, 2941–2959. [CrossRef]
11. Advani, S.H.; Lee, J.K. Finite element model simulations associated with hydraulic fracturing. *SPE J.* **1982**, *22*, 209–218.
12. Devloo, P.R.; Fernandes, P.D.; Gomes, S.M.; Bravo, C.M.A.A.; Damas, R.G. A finite element model for three dimensional hydraulic fracturing. *Math. Comput. Simul.* **2006**, *73*, 142–155. [CrossRef]
13. Wangen, M. Finite element modeling of hydraulic fracturing in 3D. *Comput. Geosci.* **2013**, *17*, 647–659. [CrossRef]
14. Bao, J.Q.; Fathi, E.; Ameri, S. A coupled finite element method for the numerical simulation of hydraulic fracturing with a condensation technique. *Eng. Fract. Mech.* **2014**, *131*, 269–281. [CrossRef]
15. Chen, Z.; Jeffrey, R.G.; Zhang, X.; Kear, J. Finite-element simulation of a hydraulic fracture interacting with a natural fracture. *SPE J.* **2017**, *22*, 219–234. [CrossRef]
16. Pakzad, R.; Wang, S.; Sloan, S.W. 3D finite element modelling of fracturing in heterogeneous rock: From pure solid to coupled fluid/solid analysis. In Proceedings of the ISRM European Rock Mechanics Symposium-EUROCK, St. Petersburg, Russia, 22–26 May 2018; Society of Petroleum Engineers: Richardson, TX, USA, 2018.
17. Pezzulli, E.; Nejati, M.; Salimzadeh, S.; Matthai, S.K.; Driesner, T. Finite element simulations of hydraulic fracturing: A comparison of algorithms for extracting the propagation velocity of the fracture. *Eng. Fract. Mech.* **2022**, *274*, 108783. [CrossRef]
18. Ji, L.; Settari, A.; Sullivan, R.B. A novel hydraulic fracturing model fully coupled with geomechanics and reservoir simulation. *SPE J.* **2009**, *14*, 423–430. [CrossRef]
19. Kim, J.; Moridis, G.J. Numerical analysis of fracture propagation during hydraulic fracturing operations in shale gas systems. *Int. J. Rock Mech. Min.* **2015**, *76*, 127–137. [CrossRef]
20. Duarte, C.A.; Reno, L.G.; Simone, A. A high-order generalized FEM for through-the-thickness branched cracks. *Int. J. Numer. Methods Eng.* **2007**, *72*, 325–351. [CrossRef]
21. Gupta, V.; Kim, D.J.; Duarte, C.A. Analysis and improvements of global–local enrichments for the generalized finite element method. *Comput. Methods Appl. Mech. Eng.* **2012**, *245*, 47–62. [CrossRef]
22. Belytschko, T.; Black, T. Elastic crack growth in finite elements with minimal remeshing. *Int. J. Numer. Meth. Eng.* **1999**, *45*, 601–620. [CrossRef]
23. Belytschko, T.; Moës, N.; Usui, S.; Parimi, C. Arbitrary discontinuities in finite elements. *Int. J. Numer. Meth. Eng.* **2001**, *50*, 993–1013. [CrossRef]
24. Moës, N.; Dolbow, J.; Belytschko, T. A finite element method for crack growth without remeshing. *Int. J. Numer. Meth. Eng.* **1999**, *46*, 131–150. [CrossRef]

25. Mohammadnejad, T.; Andrade, J.E. Numerical modeling of hydraulic fracture propagation, closure and reopening using XFEM with application to in-situ stress estimation. *Int. J. Numer. Anal. Met.* **2016**, *40*, 2033–2060. [CrossRef]
26. Xia, Y.; Jin, Y.; Chen, M.; Chen, K.P. An enriched approach for modeling multi-scale discrete-fracture/matrix interaction for unconventional-reservoir simulations. *SPE J.* **2019**, *24*, 349–374. [CrossRef]
27. Zheng, H.; Pu, C.; Xu, E.; Sun, C. Numerical investigation on the effect of well interference on hydraulic fracture propagation in shale formation. *Eng. Fract. Mech.* **2020**, *228*, 106932. [CrossRef]
28. Olson, J.E. *Predicting Fracture Swarms—The Influence of Subcritical Crack Growth and the Crack-Tip Process Zone on Joint Spacing in Rock*; Geological Society, Special Publications: London, UK, 2004; Volume 231, pp. 73–88.
29. Weng, X.; Kresse, O.; Cohen, C.; Kresse, O.; Cohen, C.-E.; Wu, R.; Gu, H. Modeling of hydraulic-fracture-network propagation in a naturally fractured formation. *SPE Prod. Oper.* **2011**, *26*, 368–380.
30. Castonguay, S.T.; Mear, M.E.; Dean, R.H.; Schmidt, J.H. Predictions of the growth of multiple interacting hydraulic fractures in three dimensions. In Proceedings of the SPE Annual Technical Conference and Exhibition, New Orleans, LA, USA, 30 September–2 October 2013; Society of Petroleum Engineers: Richardson, TX, USA, 2013.
31. Hou, B.; Chen, M.; Cheng, W.; Diao, C. Investigation of hydraulic fracture networks in shale gas reservoirs with random fractures. *Arab. J. Sci. Eng.* **2016**, *41*, 2681–2691. [CrossRef]
32. Tang, H.; Winterfeld, P.H.; Wu, Y.S.; Huang, Z.Q.; Di, Y.; Pan, Z.; Zhang, J. Integrated simulation of multi-stage hydraulic fracturing in unconventional reservoirs. *J. Nat. Gas. Sci. Eng.* **2016**, *36*, 875–892. [CrossRef]
33. Kumar, D.; Ghassemi, A. Three-dimensional poroelastic modeling of multiple hydraulic fracture propagation from horizontal wells. *Int. J. Rock Mech. Min.* **2018**, *105*, 192–209. [CrossRef]
34. Shen, B.; Shi, J. A numerical scheme of coupling of fluid flow with three-dimensional fracture propagation. *Eng. Anal. Bound. Elem.* **2019**, *106*, 243–251. [CrossRef]
35. Salimzadeh, S.; Paluszny, A.; Zimmerman, R.W. Three-dimensional poroelastic effects during hydraulic fracturing in permeable rocks. *Int. J. Solids Struct.* **2017**, *108*, 153–163. [CrossRef]
36. Li, S.; Firoozabadi, A.; Zhang, D. Hydromechanical modeling of non-planar three-dimensional fracture propagation using an iteratively coupled approach. *J. Geophys. Res. Solid Earth.* **2020**, *125*, e2020JB020115. [CrossRef]
37. Cundall, P.A. A computer model for simulating progressive, large-scale movement in blocky rock system. In Proceedings of the International Symposium on Rock Mechanics, Nancy, France, 4–6 October 1971.
38. Li, S.; Feng, X.T.; Zhang, D.; Tang, H. Coupled thermo-hydro-mechanical analysis of stimulation and production for fractured geothermal reservoirs. *Appl. Energy* **2019**, *247*, 40–59. [CrossRef]
39. Li, S.; Zhang, D.; Li, X. A new approach to the modeling of hydraulic-fracturing treatments in naturally fractured reservoirs. *SPE J.* **2017**, *22*, 1064–1081. [CrossRef]
40. Li, S.; Li, X.; Zhang, D. A fully coupled thermo-hydro-mechanical, three-dimensional model for hydraulic stimulation treatments. *J. Nat. Gas. Sci. Eng.* **2016**, *34*, 64–84. [CrossRef]
41. Settgast, R.R.; Fu, P.; Walsh, S.D.; White, J.A.; Annavarapu, C.; Ryerson, F.J. A fully coupled method for massively parallel simulation of hydraulically driven fractures in 3-dimensions. *Int. J. Numer. Anal. Met.* **2017**, *41*, 627–653. [CrossRef]
42. Guo, X.; Wu, K.; An, C.; Tang, J.; Killough, J. Numerical Investigation of Effects of Subsequent Parent-Well Injection on Interwell Fracturing Interference Using Reservoir-Geomechanics-Fracturing Modeling. *SPE J.* **2019**, *24*, 1884–1902. [CrossRef]
43. Zeng, Q.; Liu, W.; Yao, J. Hydro-mechanical modeling of hydraulic fracture propagation based on embedded discrete fracture model and extended finite element method. *J. Petrol. Sci. Eng.* **2018**, *167*, 64–77. [CrossRef]
44. Wang, C.; Huang, Z.; Wu, Y.S. Coupled numerical approach combining X-FEM and the embedded discrete fracture method for the fluid-driven fracture propagation process in porous media. *Int. J. Rock. Mech. Min.* **2020**, *130*, 104315. [CrossRef]
45. Ren, G.; Younis, R.M. An integrated numerical model for coupled poro-hydro-mechanics and fracture propagation using embedded meshes. *Comput. Methods Appl. Mech. Eng.* **2021**, *376*, 113606. [CrossRef]
46. Liu, W.; Yao, J.; Zeng, Q. A numerical hybrid model for non-planar hydraulic fracture propagation in ductile formations. *J. Petrol. Sci. Eng.* **2021**, *196*, 107796. [CrossRef]
47. Zhang, J.; Yu, H.; Xu, W.; Lv, C.; Micheal, M.; Shi, F.; Wu, H. A hybrid numerical approach for hydraulic fracturing in a naturally fractured formation combining the XFEM and phase-field model. *Eng. Fract. Mech.* **2022**, *271*, 108621. [CrossRef]
48. Ouchi, H.; Katiyar, A.; Foster, J.T.; Sharma, M.M. A peridynamics model for the propagation of hydraulic fractures in heterogeneous, naturally fractured reservoirs. In Proceedings of the SPE Hydraulic Fracturing Technology Conference, The Woodlands, TX, USA, 3–5 February 2015; OnePetro: Richardson, TX, USA, 2015.
49. Rabczuk, T.; Ren, H. A peridynamics formulation for quasi-static fracture and contact in rock. *Eng. Geol.* **2017**, *225*, 42–48. [CrossRef]
50. Qin, M.; Yang, D.; Chen, W.; Xia, X. Hydraulic fracturing network modeling based on peridynamics. *Eng. Fract. Mech.* **2021**, *247*, 107676. [CrossRef]
51. Diehl, P.; Lipton, R.; Wick, T.; Tyagi, M. A comparative review of peridynamics and phase-field models for engineering fracture mechanics. *Comput. Mech.* **2022**, *69*, 1259–1293. [CrossRef]
52. Moinfar, A.; Sepehrnoori, K.; Johns, R.T.; Varavei, A. Coupled geomechanics and flow simulation for an embedded discrete fracture model. In Proceedings of the SPE Reservoir Simulation Symposium, The Woodlands, TX, USA, 18–20 February 2013; Society of Petroleum Engineers: Richardson, TX, USA, 2013.

53. Zidane, A.; Firoozabadi, A. An efficient numerical model for multicomponent compressible flow in fractured porous media. *Adv. Water Resour.* **2014**, *74*, 127–147. [CrossRef]
54. Garipov, T.T.; Karimi-Fard, M.; Tchelepi, H.A. Discrete fracture model for coupled flow and geomechanics. *Comput. Geosci.* **2016**, *20*, 149–160. [CrossRef]
55. McClure, M.W.; Babazadeh, M.; Shiozawa, S.; Huang, J. Fully Coupled Hydromechanical Simulation of Hydraulic Fracturing in 3D Discrete-Fracture Networks. *SPE J.* **2016**, *21*, 1302–1320. [CrossRef]
56. Wei, S.; Kao, J.; Jin, Y.; Shi, C.; Xia, Y.; Liu, S. A discontinuous discrete fracture model for coupled flow and geomechanics based on FEM. *J. Petrol. Sci. Eng.* **2021**, *204*, 108677. [CrossRef]
57. Dolbow, J.E. An Extended Finite Element Method with Discontinuous Enrichment for Applied Mechanics. Ph.D. Dissertation, Northwestern University, Evanston, IL, USA, December 1999.
58. Gordeliy, E.; Peirce, A. Enrichment strategies and convergence properties of the XFEM for hydraulic fracture problems. *Comput. Methods Appl. Mech. Eng.* **2015**, *283*, 474–502. [CrossRef]
59. Moës, N.; Cloirec, M.; Cartraud, P.; Remacle, J.F. A computational approach to handle complex microstructure geometries. *Comput. Methods Appl. Mech. Eng.* **2003**, *192*, 3163–3177. [CrossRef]
60. Chen, K.P.; Jin, Y.; Chen, M. Pressure-gradient singularity and production enhancement for hydraulically fractured wells. *Geophys. J. Int.* **2013**, *195*, 923–931. [CrossRef]
61. Xia, Y.; Jin, Y.; Huang, Z.; Lu, Y.; Wang, H. An Extended Finite Element Method for Hydro-Mechanically Coupled Analysis of Mud Loss in Naturally Fractured Formations. In Proceedings of the 53rd US Rock Mechanics/Geomechanics Symposium, New York, NY, USA, 23–26 June 2019; Society of Petroleum Engineers: Richardson, TX, USA, 2019.
62. Xia, Y.; Jin, Y.; Oswald, J.; Chen, M.; Chen, K. Extended finite element modeling of production from a reservoir embedded with an arbitrary fracture network. *Int. J. Numer. Methods Fluids* **2018**, *86*, 329–345. [CrossRef]
63. Liu, P.; Liu, F.; She, C.; Zhao, L.; Luo, Z.; Guan, W.; Li, N. Multi-phase fracturing fluid leakoff model for fractured reservoir using extended finite element method. *J. Nat. Gas. Sci. Eng.* **2016**, *28*, 548–557. [CrossRef]
64. Coussy, O. *Poromechanics*; John Wiley & Sons: Hoboken, NJ, USA, 2004.
65. Schöllmann, M.; Richard, H.A.; Kullmer, G.; Fulland, M. A new criterion for the prediction of crack development in multiaxially loaded structures. *Int. J. Fract.* **2002**, *117*, 129–141. [CrossRef]
66. Olson, J.E.; Taleghani, A.D. Modeling simultaneous growth of multiple hydraulic fractures and their interaction with natural fractures. In Proceedings of the SPE Hydraulic Fracturing Technology Conference, The Woodlands, TX, USA, 19–21 January 2009; Society of Petroleum Engineers: Richardson, TX, USA, 2009.
67. Detournay, E. Propagation regimes of fluid-driven fractures in impermeable rocks. *Int. J. Geomech.* **2004**, *4*, 35–45. [CrossRef]
68. Rice, J.R. Mathematical analysis in the mechanics of fracture. *Fract. Adv. Treatise* **1968**, *2*, 191–311.
69. Liu, X.; Rasouli, V.; Guo, T.; Qu, Z.; Sun, Y.; Damjanac, B. Numerical simulation of stress shadow in multiple cluster hydraulic fracturing in horizontal wells based on lattice modelling. *Eng. Fract. Mech.* **2020**, *238*, 107278. [CrossRef]
70. Schultz, R. Discontinuity Patterns and Their Interpretation. In *Geologic Fracture Mechanics*; Cambridge University Press: Cambridge, UK, 2019; pp. 170–208.
71. Miller, C.; Waters, G.; Rylander, E. Evaluation of production log data from horizontal wells drilled in organic shales. In Proceedings of the North American Unconventional Gas Conference and Exhibition, The Woodlands, TX, USA, 14–16 June 2011; Society of Petroleum Engineers: Richardson, TX, USA, 2011.
72. Lecampion, B.; Desroches, J. Simultaneous initiation and growth of multiple radial hydraulic fractures from a horizontal wellbore. *J. Mech. Phys. Solids* **2015**, *82*, 235–258. [CrossRef]
73. Lu, Y.; Wei, S.; Xia, Y.; Jin, Y. Modeling of geomechanics and fluid flow in fractured shale reservoirs with deformable multi-continuum matrix. *J. Petrol. Sci. Eng.* **2021**, *196*, 107576. [CrossRef]
74. Hughes, T.J.R. *The Finite Element Method: Linear Static and Dynamic Finite Element Analysis*; Prentice-Hall: Englewood Cliffs, NJ, USA, 1987.
75. Witherspoon, P.A.; Wang, J.S.; Iwai, K.; Gale, J.E. Validity of cubic law for fluid flow in a deformable rock fracture. *Water Resour. Res.* **1980**, *16*, 1016–1024. [CrossRef]

energies

MDPI

Article

Experimental Investigation of Fracture Propagation in Clayey Silt Hydrate-Bearing Sediments

Yanjiang Yu [1,2], Kaixiang Shen [1,2,*] and Haifeng Zhao [3,*]

1 Guangzhou Marine Geological Survey, China Geological Survey, Guangzhou 511458, China; yuyanjiang2004@163.com
2 National Engineering Research Center of Gas Hydrate Exploration and Development, Guangzhou 511458, China
3 College of Petroleum Engineering, China University of Petroleum, Beijing 102249, China
* Correspondence: skxv@163.com (K.S.); zhaohf@cup.edu.cn (H.Z.)

Abstract: More than 90% of the natural gas hydrate resources are reserved as marine clayey silt sediments. It is of great significance to efficiently develop a clayey silt hydrate. At present, there are problems of low single well production and small depressurization range in its production test, which is still a long way from commercial exploitation. The combination of hydraulic fracturing technology and other methods such as depressurization method is regarded as one of the potential technical means to achieve the commercial exploitation of the hydrate. However, compared with shale gas reservoirs and coalbed methane reservoirs, clayey silt hydrate reservoirs have special mechanical properties, resulting in unique hydraulic fracturing processes. Therefore, it is necessary to study the fracture initiation and propagation laws of clayey silt hydrate reservoirs. To this end, we carried out large-scale (30 × 30 × 30 cm) true triaxial hydraulic fracturing experiments using a simulated material with similar mechanics, porosity, and permeability to clayey silt hydrate-bearing sediments. The effects of completion method, fracturing method, and fracturing fluid displacement on hydraulic fracture propagation of clayey silt hydrate-bearing sediments were studied. The results showed that a perforated completion can significantly increase the fracture reconstruction area and decrease the fracture initiation pressure compared to an open hole completion. Due to the small horizontal stress difference, it is feasible to carry out temporary plugging fracturing in clayey silt hydrate reservoirs. Temporary plugging fracturing can form steering fractures and significantly improve fracture complexity and fracture area. Increasing the fracturing fluid displacement can significantly increase the fracture area as well. When conducting fracturing in clayey silt hydrate-bearing sediments, the fracturing fluid filtration area is obviously larger than the fracture propagation area. Therefore, it is recommended to use a high-viscosity fracturing fluid to reduce the filtration of the fracturing fluid and improve the fracturing fluid efficiency. This study preliminarily explores the feasibility of temporary plugging fracturing in clayey silt hydrate reservoirs and analyzes the effect of completion methods on the propagation of fracturing fractures, which can provide a reference for the research conducted on the fracturing stimulation of clayey silt hydrate reservoirs.

Keywords: clayey silt hydrate; perforated completion; temporary plugging fracturing; fracture propagation; stimulated rock area

Citation: Yu, Y.; Shen, K.; Zhao, H. Experimental Investigation of Fracture Propagation in Clayey Silt Hydrate-Bearing Sediments. *Energies* **2024**, *17*, 528. https://doi.org/10.3390/en17020528

Academic Editor: João Fernando Pereira Gomes

Received: 25 September 2023
Revised: 9 November 2023
Accepted: 13 November 2023
Published: 22 January 2024

1. Introduction

Natural gas hydrate (NGH) is a clathrate crystalline compound generated from water molecules and hydrocarbon gas molecules under suitable conditions of low temperature and high pressure [1]. It exhibits a high energy density, an extensive distribution, and abundant reserves [2–4]. NGHs are viewed as one of the most likely clean energies to replace traditional fossil energy [5].

The existing hydrate exploitation technologies mainly include the depressurization method [6], thermal stimulation method [7], chemical inhibitor injection method [8], CO_2/N_2

replacement method [9], solid fluidization method [10], and combination method [11–13]. Depressurization and thermal stimulation methods are viewed as two kinds of methods with high feasibility, which are the most used in laboratory research and field test production. As for the NGH reservoir, effective permeability is an important property that would produce a significant impact on the pressure drop diffusion and fluid migration of the depressurization method and the heating efficiency of the thermal stimulation method [14]. Increasing reservoir permeability has a significant impact on the NGH production [15]. However, the NGH resources present as marine clayey silt sediments exceed 90% [16,17], and the stored environment is characterized by high argillaceous content and low permeability [18–20]. At the same time, the hydrate dispersed in the sediment will cause a significant decline in the reservoir permeability. These characteristics lead to an extremely low effective permeability of this kind of reservoir, which limits the efficiency of depressurization and thermal stimulation methods.

Hydraulic fracturing technology can increase the drainage area, enhance the reservoir permeability, and improve reservoir seepage conditions. It has been widely applied in unconventional low-permeability reservoirs such as coal and shale reservoirs [21–23]. It is considered as one of the potential technical means to achieve the efficient development and utilization of NGHs [24–27]. Relevant studies have proved that fracturing reconstruction is an effective technique to improve the productivity of NGH reservoirs [28–32]. Thus, it is meaningful to conduct extensive research on the hydraulic fracture initiating and propagating law of NGH reservoirs. Researchers have begun to explore the fracability and the hydraulic fracture initiation and extension law of NGH reservoirs. Too et al. [33] confirmed by the injection pressure curve that the sandy NGH reservoir has a preferable fracability when the hydrate saturation exceeds 50%. Liu et al. [34] put forward a fracability estimation model applicable to hydrate-bearing sediments based on analytic hierarchy process-entropy method. Its results suggested that a high-viscosity fracturing fluid is supposed to be used when reconstructing hydrate-bearing sediments with a low fracability index. Ito et al. [35] validated the possibility of forming hydraulic fractures in unconsolidated sand–mud interbed sediments using a true triaxial loading experimental equipment. The experimental results showed that the interface fracture between the sand and mud layers is inclined to generate hydraulic fractures. Konno et al. [36] executed hydraulic fracturing experiments based on the sandy sediment with a hydrate saturation of 72%. The fractures perpendicular to the maximum horizontal principal stress were observed through X-CT scanning. Furthermore, the gas production experiment using depressurization after fracturing detected that the hydraulic fracturing can improve the permeability of the sample, and the fracture could still maintain a high permeability during the depressurization process. Taking the Alaska hydrate test area as the geological background, Zhang et al. [37] and Liu et al. [34] synthesized the NGH reservoir skeleton with similar physical and mechanical properties to the reservoir samples in the laboratory, and then synthesized CH_4 hydrate in the skeleton. The impacts of the stress conditions and the fracturing fluid performance on the hydraulic fracturing behavior of the hydrate-bearing sediments were studied. The results showed that hydrate-bearing sediments are prone to tensile failure to form a single fracture under the low-stress and high-stress difference condition. Complex multiple fractures tend to form under the high-stress and low-stress difference condition. Under a high in situ stress, increasing the fracturing fluid viscosity is a feasible method to effectively create fractures. Yao et al. [38] established a discrete element model, studied the fracturing behavior of sandy CH_4 hydrate, and explored the effects of hydrate saturation, hydrate microscopic distribution mode, and in situ stress conditions on the fracturing behavior of hydrate-bearing sediments. The results showed that, with the increase in hydrate saturation, the initiation pressure increases gradually. When the hydrate saturation is less than 30%, the fracture-creating ability is poor, whereas when the hydrate saturation is higher than 40%, a good fracture-creating ability and a preference to form multiple main fractures are observed. As the in situ stress difference increases, the fracture morphology becomes simple. Based on the extended finite element method, a multi-cluster fracture propagation model for horizontal wells in NGH reservoirs was established to analyze the correlation between fracture propagating paths and fracture spacing, hydrate saturation, and horizontal stress difference [39]. Lu et al. [40] conducted hydraulic fracturing tests under different confining pressures for clayey silt sediments in Shenhu area, and evaluated their mechanism of fracture initiating and propagating. Ma et al. [24] built a three-dimensional hydraulic fracture propagating model on the basis of cohesive elements. They studied the effects of flow rate, cluster spacing, and fracturing methods (simultaneous fracturing and sequential fracturing) on the fracture morphology of clayey silt hydrate reservoirs. Sun et al. [41] discussed the fracturing law of hydrate-bearing clayey silt and frozen clayey silt. They evaluated the impacts of hydrate/ice saturation, stress difference, fluid viscosity, and flow rate on the fracturing behavior. The consequences indicated that the initiation pressure increases with the increase in hydrate/ice saturation, flow rate, and fluid viscosity. Increasing fluid viscosity and flow rate would result in the formation of complex fractures, and the stress difference is the main controlling parameter of fracture propagating direction. Although the NGH resources present as marine clayey silt sediments exceed 90%, few studies have explored the fracture propagating law of hydrate-bearing clayey silt, and the existing research has mainly focused on numerical simulation. There is no experimental study on conventional and temporary plugging fracturing of large size samples.

Therefore, we prepared large size clayey silt hydrate-bearing sediment samples. Subsequently, true triaxial physical simulation experiments of conventional and temporary plugging fracturing were carried out. The fracture morphology of conventional and temporary plugging fracturing under different fracturing fluid displacements was explored, and the impact of completion mode on fracture geometry was investigated. The findings could provide fundamental understanding and reasons for the hydraulic fracturing of marine clayey silt hydrate reservoirs.

2. Experiments

2.1. Experimental Sample Preparation

The large-scale true triaxial fracturing simulation test is a significant means to investigate the fracture initiating and propagating law. It has been used extensively in unconventional reservoirs such as shale and coal seam [42–44]. There are two types of samples in the experiments: one is the outcrop taken from the field, and the other is the artificial simulated sample preparation in the laboratory. However, compared with shale and coal, the clayey silt hydrate reservoirs are characterized by non-diagenesis. It is very expensive and difficult to drill large-scale in situ hydrate-bearing sediments samples. Therefore, artificially simulated samples are used to conduct indoor physical simulation experiments [41]. Tetrahydrofuran (THF) hydrate sediment samples are currently being utilized to carry out relevant studies [37]. Since the structure and mechanical properties of the THF hydrate are close to those of the NGH and it can be dissolved in water in any proportion, it is convenient to control the hydrate saturation in an experiment. The hydrate formed after mixing with the porous medium sample can basically exist evenly between the sediment particles within the sample [45]. The THF solution can synthesize a hydrate at an atmospheric pressure and at 0~4 °C [46]. Therefore, THF is usually used to replace methane to investigate the mechanical and physico-chemical properties of the generated hydrate. The THF hydrate has also been studied to comprehend its fracture initiating and propagating laws of hydrate fracturing [47]. The THF hydrate can exist stably at 0.1 MPa and 4 °C [45,48]. The THF hydrate can maintain its stability during an experiment, and it can be conveniently compacted into a sample for synthesis. The conventional fracturing equipment can be upgraded to carry out THF hydrate fracturing experiments. Therefore, we choose THF to prepare hydrate sediment samples for hydraulic fracturing experiments.

The target area of this paper is the exploration area of the GMGS1 voyage. GMGS1 voyage is a hydrate drilling project organized and implemented by Guangzhou Marine Geological Survey in 2007. Eight stations were drilled based on the GMGS1 voyage, and hydrate samples were acquired at SH2, SH3, and SH7 stations [49,50]. The research data in this paper are based on the logging data of SH2 borehole. The water depth of the SH2 station is 1235 m, and hydrate samples were successfully acquired from 188 m to 228 m below the seafloor. After testing, the porosity of the hydrate sediment is 0.38, the hydrate saturation range is 1.0~47.3%, and the inherent permeability of the reservoir is approximately 10 mD [51]. In this experiment, the hydrate saturation was set to 45%. The mineral composition of the target block is shown in Table 1 [52,53]. Based on these data, sand, silt and clay are selected as hydrate host sediments. The proportions of each part are detailed in Table 1. The results indicated that, when the mass fraction of THF is 19%, the THF-H_2O solution can be completely converted into a hydrate, and considering the possible volatility of THF, we choose a THF-H_2O solution with a 21% mass proportion [54]. The mass of deionized water and THF used in the experiments can be calculated using Equations (1) and (2) [47].

$$m_w = \frac{306}{378}\rho_h S_h V_\varphi \tag{1}$$

$$m_{THF} = \frac{306 \times 21}{378 \times 79}\rho_h S_h V_\varphi \tag{2}$$

m_w and m_{THF} are the masses (g) of H_2O and THF; ρ_h is the density of THF hydrate, 0.888 g/cm^3 in our experiment; S_h is the hydrate saturation, 45% in our experiment; V is the volume of the sample, 27,000 cm^3 in our experiment; and φ is the porosity, 0.40 in our experiment.

Table 1. The mineralogical components of clayey-silty sediments.

Grain Size (μm)	Actual Mass Proportion (%)	Average Value (%)	Experimental Mass Proportion (%)
Clay: <4	15–45	--	21
Silt: 4–63	50–80	--	75
Sand: >63	<5	--	4
Clay minerals at SH2 site	11–27	19.64	21
Montmorillonite	33–59	47.04	45
Illite	22–39	29.28	30
Chlorite	9–17	13.17	15
Kaolinite	7–14	10.51	10

The procedure for the preparation of THF hydrate-bearing sediments sample is as follows: (1) The required mass of silt, sand, and clay is calculated according to the sample mineral composition ratio, and the amount of THF and water is calculated according to the required hydrate saturation. (2) First, the silt, sand, and clay are thoroughly stirred and mixed, and then, the THF solution is sprayed on the aggregate, stirred evenly with a mixer, and then loaded into the sediment to prepare a mold. (3) The mold is placed in a low temperature incubator for sample synthesis, and it is kept frozen at −9 °C for 48 h to generate the THF hydrate, and then demolded to remove the sediment sample. The wellbore is placed in the mold before loading, so that the wellbore can be consolidated with the surrounding sediments during the sample freezing process. According to the above method, 1~5 clayey silt hydrate-bearing simulation samples (30 × 30 × 30 cm) are fabricated. To carry out the THF hydrate fracturing experiment, we upgrade the experimental equipment for conventional fracturing physical simulation and place the equipment in a sealed low and constant temperature environment. The entire fracturing experiment is conducted in a low and constant temperature chamber. Before carrying out large-scale fracturing physical simulation experiments, we first prepare a small size-standard core using the method described above and test its rock mechanics, porosity, and permeability parameters. The porosity, permeability, and mechanical parameters of the simulated samples are depicted in Table 2. The test results indicated that the mechanical, porosity, and permeability parameters of the sample prepared by the above method were in a good agreement with the values of the research block [51,55]. On this basis, large-scale hydraulic fracturing physical simulation tests were carried out.

Table 2. The porosity, permeability, and mechanical parameters of the simulated samples.

Specimen Number	Elastic Modulus (MPa)	Poisson's Ratio	Compressive Strength (MPa)	Tensile Strength (MPa)	Porosity (%)	Permeability (mD)
1	391.35	0.29	3.36	0.31	36.4	2.39
2	399.15	0.30	3.63	0.33	32.3	2.76
3	387.54	0.31	2.99	0.27	36.6	2.97
4	386.62	0.32	2.8	0.26	39.6	3.05
5	396.46	0.33	3.41	0.31	34.5	2.66

2.2. Experimental Apparatus

Hydraulic fracturing simulation device was used in the experiment (Figure 1). In the experiment, the pressure plate is pushed by an oil pressure pump set and a hydraulic cylinder to apply triaxial stress to rock samples in the core chamber. The maximum loading stress of X-axis is 10 MPa, and the maximum loading stress of Y-axis and Z-axis is 20 MPa. After the stress loading is completed, the piston pump is used to push the piston in the intermediate container to squeeze the fracturing fluid into the pipeline and enter the wellbore fracturing rock sample through the fluid injection pipeline. The maximum displacement of the piston pump is 1250 mL/min.

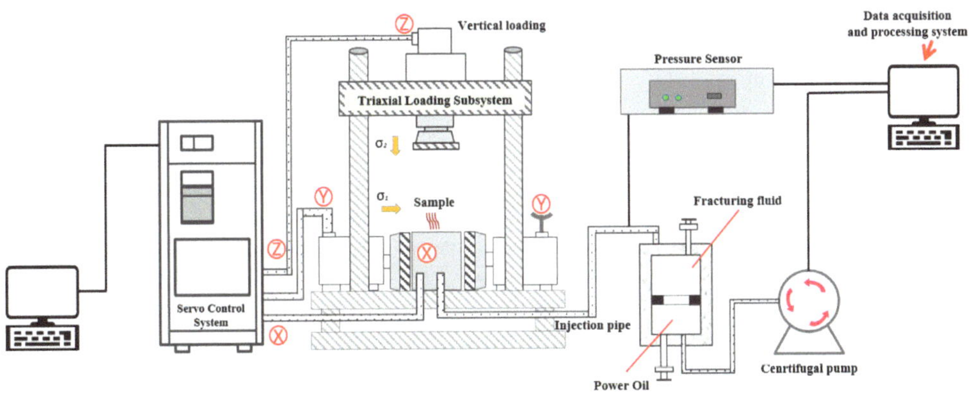

Figure 1. True triaxial hydraulic fracturing system.

2.3. Experimental Design

To analyze the effects of the completion method, fracturing method, and fracturing fluid displacement on the fracture propagating of a clayey silt hydrate reservoir, we designed the experimental scheme depicted in Table 3. Guar gum fracturing fluid is used as the fracturing fluid in conventional fracturing, and fiber fracturing fluid is used as the fracturing fluid in temporary plugging fracturing, as demonstrated in Figure 2a,b. Fiber fracturing fluid is prepared by using guar gum solution as the base fluid, adding fiber (fiber length: 6 cm) and cross-linking to form jelly. Figure 2c shows the wellbore for 4# and 5# samples with an initial perforation interval of 2.5 cm.

Table 3. Summary of experimental parameters.

Sample Number	Completion Method	Fracturing Method	($\sigma_v/\sigma_H/\sigma_h$ (MPa))	Displacement (mL/min)	Fluid Viscosity (mPa·s)
1#	open hole completion	conventional fracturing	4/2/1	100	100
2#		temporary plugging fracturing		50	100
3#		temporary plugging fracturing		100	
4#	perforated completion	conventional fracturing		50	100
5#		conventional fracturing		100	

Figure 2. Diagram of fiber fracturing fluid and wellbore: (**a**) fiber used for configuring fiber fracturing fluid; (**b**) fiber fracturing fluid with tracer added; and (**c**) wellbore in perforated mode.

2.4. Experimental Procedure

(1) Triaxial stress loading process

The sample was first placed in a triaxial stress loading chamber. To avoid the damage to the rock sample due to unbalanced loads in different directions, the stress in different directions was first applied at 1 MPa simultaneously in the experiment. Next, the vertical stress and the maximum horizontal stress continue to increase synchronously to 2 MPa, and then, the vertical stress gradually increases to 4 MPa. The specimen is stabilized under the boundary stress for 10 min to ensure the stress balance in the specimen. As the strength of the sample is much lower than that of conventional shale and sandstone samples, the stress is gradually applied at an interval of 0.5 MPa to avoid sample damage during confining pressure loading.

(2) Fracturing process with guar gum fracturing fluid injection

The guar gum fracturing fluid is pumped into the sample using a centrifugal pump, and the change in pump pressure is recorded over time. The experiment is terminated when the injection pressure drops suddenly or fracturing fluid leakage is observed.

(3) Fracturing process with fiber fracturing fluid injection

In the temporary plugging fracturing stage, a new fracturing fluid with a temporary plugging agent is pumped into the sample. The test is terminated when the injection pressure drops suddenly or the fracturing fluid leakage is observed.

(4) Observation and recording of fracture morphology

After each test, the sample was removed from the true triaxial test frame. The sample was then split along induced fractures using a hammer and chisel to observe and record fracture morphology.

1#, 4#, and 5# samples shall be subject to conventional fracturing tests, and relevant tests shall be carried out according to steps (1), (2), and (4). 2# and 3# samples were subjected to the temporary plugging fracturing test, and relevant experiments were carried out according to steps (1), (2), (3), and (4).

3. Experimental Results and Analysis

3.1. Influence of Completion Scheme

Stimulated rock area (SRA) was utilized to quantitatively evaluate the effect of hydraulic fracturing [56]. SRA refers to the total induced fracture area in a sample after fracturing in the hydraulic fracturing simulation experiment. It is believed that the larger the SRA, the larger the area of the reservoir that the hydraulic fracture can communicate with after a reservoir reconstruction, and the better the effect of the increasing and stabilizing output. In the actual hydraulic fracturing simulation experiment, the area of the entire fracture plane is recorded as 1.00 (that is, close to 30 cm $\times$ 30 cm). In the actual calculation, the distribution area of the tracer on the fracture plane is divided into grades of 0.25, 0.50, and 0.75, respectively. After the test, the fractures on the surface of the sample were used to determine the fracture propagation direction. The approximate value of the fracture volume was then determined from the distribution pattern of the tracer. Finally, the complete fracture morphology is reconstructed using a drawing software.

The purpose of 1# and 5# samples is to investigate the effect of the completion method on hydraulic fracture propagating and fracture geometry. The results demonstrated that the completion method has a significant impact on the SRA.

As shown in Figure 3, although the fracturing experiments of different completion methods finally formed approximately symmetrical double-wing vertical fractures, the fracture communication area (SRA = 0.75) produced by the perforated completion sample was significantly larger than that of the open hole completion sample (SRA = 0.5). At the same time, the breakdown pressure of the perforated completion sample was lower than that of the open hole completion sample. It was attributed to the prefabricated perforation section playing similar roles like micro-fractures and induced hydraulic fractures.

3.2. Impact of Fracturing Scheme

To evaluate the impact of fracturing mode on the hydraulic fracture propagation, the hydraulic fracturing experiment of 1# and 3# samples were conducted. The results suggested that the fracture morphology and the SRA are affected remarkably by the fracturing mode.

As depicted in Figure 4, a single vertical main fracture was formed in the conventional fracturing experiments. In the temporary plugging fracturing experiment, a vertical main fracture with an SRA of 0.75 was formed during the initial fracturing. In the subsequent temporary plugging fracturing, a diverting fracture with an angle of nearly 90° to the initial fracture was formed. Its fracture shape is obviously more complicated than that of the conventional fracturing. The SRA of the temporary plugging fracturing experiment (SRA = 1.5) is also significantly larger than that of the conventional fracturing (SRA = 0.5).

It should be pointed out that temporary plugging fracturing can increase the fracture complexity and significantly enhance the volume of reconstruction, but it requires a high treatment pressure, which is a problem. Meanwhile, it should be considered that the clayey silt hydrate reservoir is not diagenetic, its strength is low, and it can easily lose stability. Therefore, before a hydraulic fracturing construction, it is necessary to consider the reservoir conditions comprehensively, and fracturing in the layer with high hydrate saturation should be given priority.

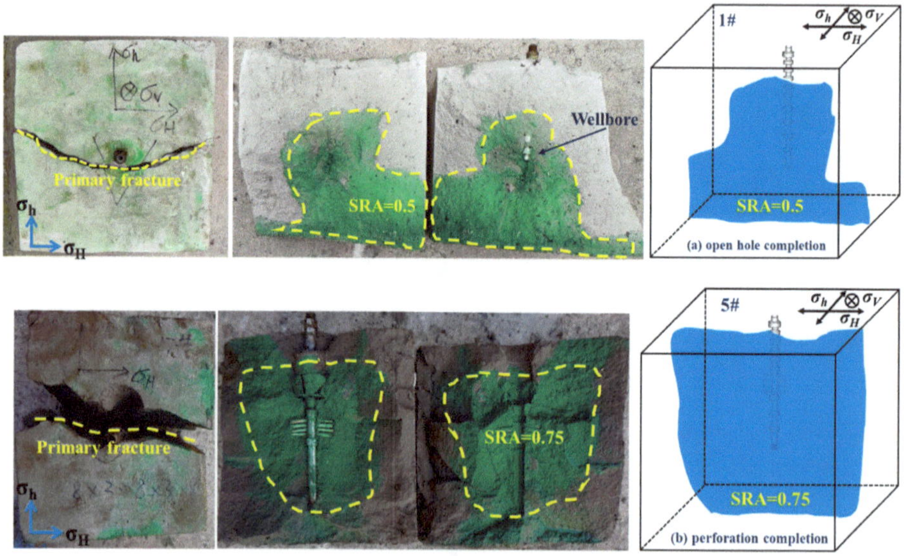

Figure 3. Fracture morphology comparison between open hole completion and perforation completion: (**a**) open hole completion (1#) and (**b**) perforation completion (5#).

(**A**)

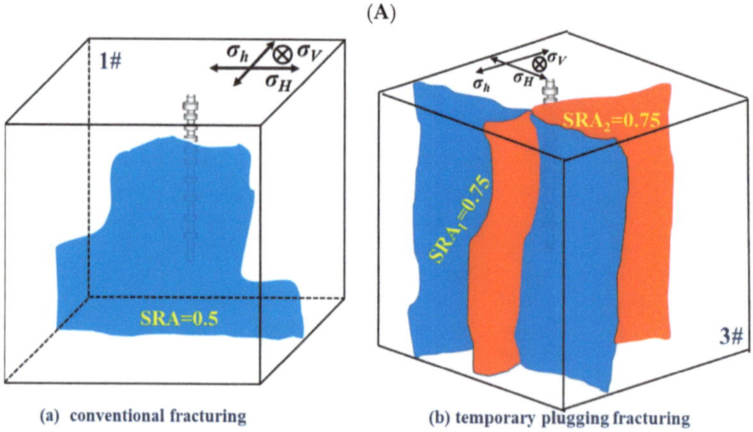

(**B**)

Figure 4. Fracture morphology comparison between conventional fracturing and temporary plugging fracturing: (**A**) fracture morphology after hydraulic fracturing (3#) and (**B**) fracture reconstruction diagram after hydraulic fracturing (1# and 3# samples).

3.3. Impact of Displacement

The objectives of 2# and 3# samples and 4# and 5# samples are to explore the impact of displacement on hydraulic fracture propagating and fracture geometry. The results indicated that the SRA was affected deeply by displacement.

As demonstrated in Figure 5B, although the complex fracture morphology of main fracture and steering fracture is formed in 2# and 3# samples after fracturing, the fracture morphology is analogous, but increasing the displacement can significantly improve the SRA. Owing to the increase in fracturing fluid displacement, the injection pressure and the pressure in the fracture would increase, and the stress at the fracture tip would increase, resulting in an increase in fracture propagation speed. Meanwhile, increasing the displacement increases the initiation pressure. This can be attributed to the high fracturing fluid displacement causing a high strain rate at the fracture tip, leading to an increase in the strength and initiation pressure of the hydrate sediments. Similar experimental phenomena are observed in the 4# and 5# samples. Although nearly symmetrical double wing vertical fractures are formed after fracturing, and the fracture morphology is similar, and increasing the displacement could significantly improve the SRA. Meanwhile, the increase in the displacement could lead to an increase in the fracture initiation pressure.

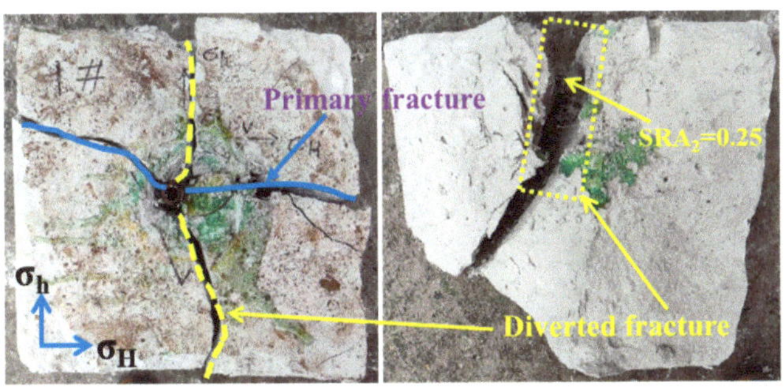

(A)

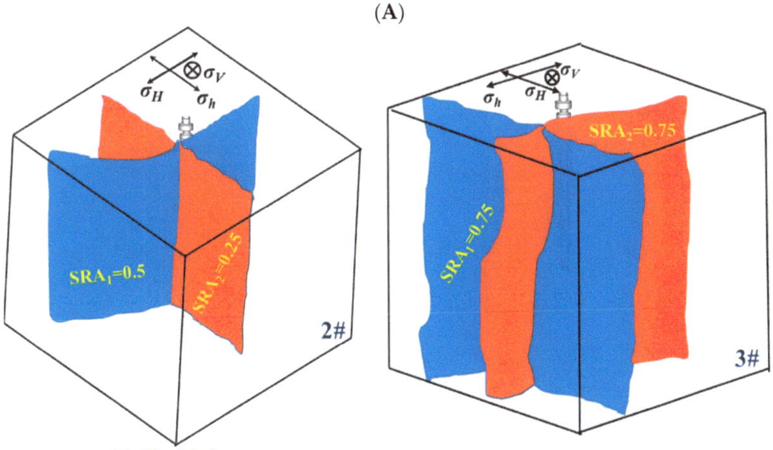

(a) 25mL/min

(b) 50mL/min

(B)

Figure 5. *Cont.*

(C)

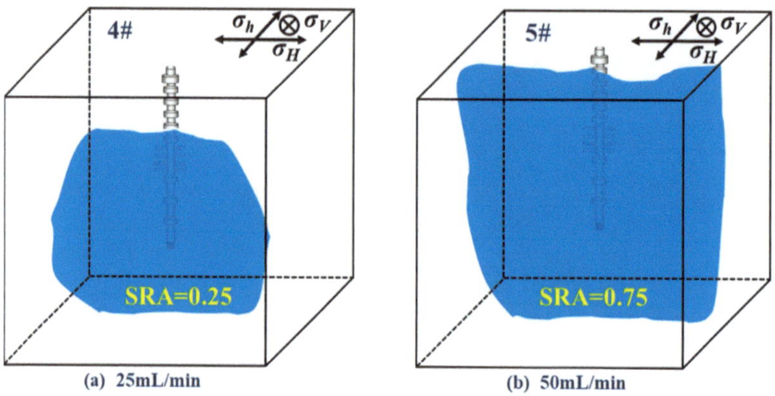

(D)

Figure 5. Fracture morphology comparison between low and high displacements: (**A**) fracture morphology after hydraulic fracturing (2#), (**B**) fracture reconstruction diagram after hydraulic fracturing (2# and 3# samples), (**C**) fracture morphology after hydraulic fracturing (4#), and (**D**) fracture reconstruction diagram after hydraulic fracturing (4# and 5# samples).

It is important to note that increasing the displacement can significantly increase the reconstruction volume, but in actual construction, a higher fracturing fluid displacement will increase the treatment pressure, thus improving the requirements for fracturing construction equipment. Meanwhile, considering that the clayey silt hydrate reservoir is not diagenetic, with a low strength and exhibiting an easy stability loss, the construction displacement must be reasonably optimized before conducting hydraulic fracturing.

3.4. Fracturing Pressure Curve

As depicted in Figure 6a, the breakdown pressure of 1#sample is obvious, about 11.78 MPa, and the propagation pressure curve is relatively stable, revealing the fracture geometry to be more regular. The result coincides with that of the single vertical fracture, which is observed by splitting the sample after the fracturing experiment (Figure 3). The breakdown pressure of 4# sample is obvious, about 7.53 MPa, and the propagation pressure is relatively stable, manifesting a more regular fracture geometry, but the breakdown pressure is significantly lower than that of 1# sample. This was due to the fact that the prefabricated perforation interval plays a similar role to that of microcracks. In addition, compared with 1# sample, 4# sample has a larger fluctuation after the breakdown point, which may be due to the difference in perforation initiation time. The pressure curve of the

5# sample resembles that of the 4# sample, but the breakdown pressure was higher than that of the 4#sample, which is 8.39 MPa.

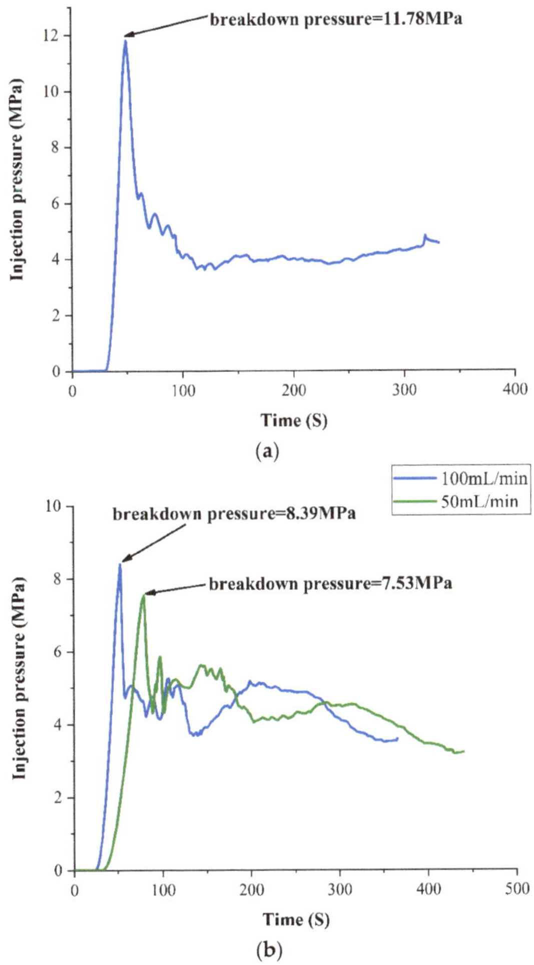

Figure 6. Pressure curves of conventional fracturing. (**a**) 1#; (**b**) 4# and 5#.

The pressure curves of temporary plugging fracturing are illustrated in Figure 7. In the fracturing process (q_1) of 2# sample pump injection without a temporary plugging agent fracturing fluid, there are obvious breakdown points, high fracture initiation pressure, and low fracture propagation pressure, and this is similar to the pressure change law in the conventional fracturing process. After pumping the fracturing fluid with a fiber temporary plugging agent (q_2), the treatment pressure gradually increases to 22.9 MPa, and the pressure fluctuation is obvious. The pressure characteristics are caused by the temporary plugging agent migrating into the fracture and compressing the new fracture. From the pressure curve, the pressure rises and then falls many times, and each such process can be regarded as a quasi-rupture process. As the fiber fracturing fluid is injected into the wellbore, the fluid is filtered along the formed fracture, and the fiber slowly accumulates at the beginning of the fracture, thereby creating a filter cake. When the filter cake reaches a certain level, the rate of fluid filtration slows down, or the liquid can no longer enter the fracture, causing a pressure rise. After the pressure has risen to a certain level, due to the formation of new fractures or the propagation of old fractures, the fluid is filtered into the sample again, resulting in a rapid drop in pressure. When the new filter loss reaches a certain extent, the fiber gathers again to form a filter cake, resulting in a new round of pressure rise, and the injection pressure decreases again after the new fracture or the old fracture propagates, and this process is repeated. The fracturing curve of 3# sample is similar. The main difference is that the fluctuation is more obvious after pumping the fracturing fluid with a fiber temporary plugging agent. The phenomenon indicates that many new fractures are generated during the fracturing process, and the fracture area of 3# sample is also significantly larger than that of 2# sample.

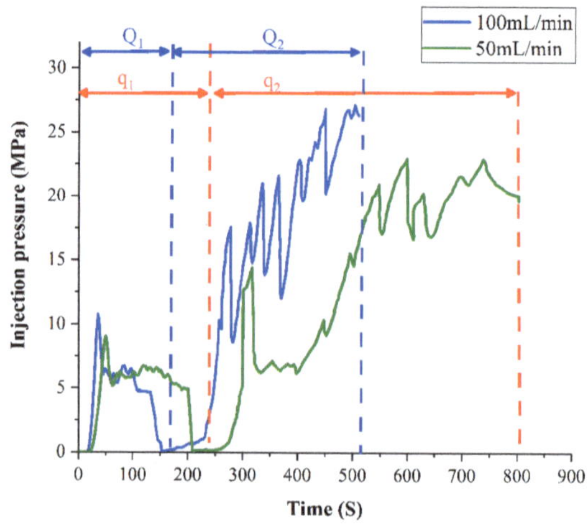

Figure 7. Pressure curves of temporary plugging fracturing.

3.5. *Analysis of Filtration Area and Fracture Propagation Area*

Fracture propagation area and fracturing fluid filtration area are be observed in the fracturing physical simulation experiment. The Figure 8 shows the distribution of fracture propagation area and fracturing fluid filtration area of 4# and 5# samples after fracturing. It can be found that the fracturing fluid filtrate area is significantly larger than the fracture propagation area under both low and high displacement conditions. Compared with tight sandstone reservoir samples [57], the difference between fracturing fluid filtration area and fracture propagation area is greater, as shown in Figure 9. The filtration area of fracturing fluid is obviously larger than the fracture propagation area. Fracturing fluid can form a large range of filtration area quickly, which may be due to the high argillaceous content of clayey silt hydrate reservoirs. However, in the actual hydraulic fracturing treatment, we should avoid the formation of a large range of filtration area, which is a waste of fracturing fluid and fracturing fluid energy. Therefore, when performing the hydraulic fracturing in clayey silt hydrate reservoirs, high-viscosity fracturing fluids should be used to reduce the fracturing fluid filtration and increase the fracturing fluid efficiency.

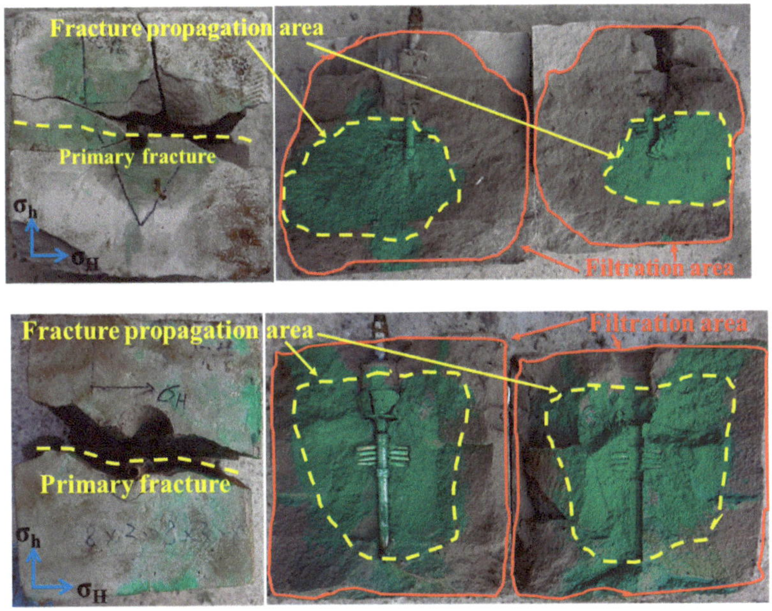

Figure 8. The distribution of filtration area and fracture propagation area after fracturing.

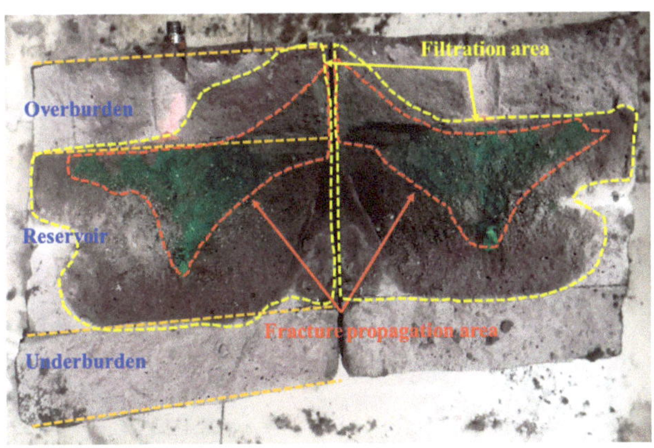

Figure 9. The distribution of filtration area and fracture propagation area after fracturing of tight sandstone sample (Orange dashed Lines representing the boundary between reservoirs and interlayer).

4. Conclusions

Based on a series of large-size (30 × 30 × 30 cm) laboratory physical simulation experiments, the effects of the completion method, fracturing method, and fracturing fluid displacement on the hydraulic fracture propagation behavior of clayey silt hydrate-bearing sediments were studied. The drawn conclusions are as follows:

(1) Compared with the open hole completion, the perforated completion can result in a significant increase in the fracture reconstruction area and a reduction in the fracture initiation pressure. As the clayey silt hydrate reservoir is not diagenetic and has a low strength, the perforated completion has obvious advantages. It is recommended to use the perforated completion in the fracturing of this type of reservoir.

(2) Due to the small horizontal stress difference, the temporary plugging fracturing is feasible in clayey silt hydrate reservoirs. Temporary plugging fracturing can form steering fractures and significantly improve fracture complexity and fracture area.

(3) The fracture area is significantly increased with the increase in fracturing fluid displacement, but a higher fracturing fluid displacement will improve the treatment pressure, thus increasing the requirements of fracturing construction equipment. Meanwhile, considering that the clayey silt hydrate reservoir is not diagenetic, its strength is low, and it can easily lose stability, the displacement needs to be reasonably optimized before fracturing.

(4) When conducting the fracturing construction in a clayey silt hydrate reservoir, the fracturing fluid filtration area obviously exceeds the fracture propagation area. Therefore, a high-viscosity fracturing fluid should be selected to reduce the filtration and improve the fracturing fluid efficiency.

(5) It should be pointed out that, in our research, we only considered the similarity of the mechanical properties and porosity and permeability properties between the simulated samples and the actual reservoir conditions. However, the actual fracturing of the NGH reservoir also involves the decomposition and phase transformation of the hydrate. Hydrate dissociation may lead to a rapid increase in pore pressure, and it also reduces the hydrate formation. The change in formation characteristics will result in the change in reservoir porosity and permeability. Those changes have an important impact on the safety and fracture propagation behavior of the hydraulic fracturing construction. Therefore, it is necessary to focus on this aspect in the future research.

Author Contributions: Methodology, Y.Y. and H.Z.; Investigation, K.S. All authors have read and agreed to the published version of the manuscript.

Funding: This work is supported by the National Natural Science Foundation of China (grant no. 51991364) and the Basic and Applied Basic Research Foundation of Guangdong Province, China (No.2020B0301030003). The authors gratefully appreciate their support.

Data Availability Statement: Data are contained within the article.

Conflicts of Interest: The authors declare no conflict of interest.

Nomenclature

NGH	natural gas hydrate
THF	Tetrahydrofuran
SRA	stimulated rock area
GMGS	Guangzhou Marine Geological Survey

References

1. Sloan, E.D. Fundamental principles and applications of natural gas hydrates. *Nature* **2003**, *426*, 353–359. [CrossRef]
2. Ruppel, C.D.; Kessler, J.D. The interaction of climate change and methane hydrates. *Rev. Geophys.* **2017**, *55*, 126–168. [CrossRef]
3. Management, M.M.J.E. *Natural Gas Hydrate: In Oceanic and Permafrost Environments*; Springer: Berlin/Heidelberg, Germany, 2003.
4. Kerr, R.A. Energy-Gas hydrate resource: Smaller but sooner. *Science* **2004**, *303*, 946–947. [CrossRef]
5. Milkov, A.V. Global estimates of hydrate-bound gas in marine sediments: How much is really out there? *Earth-Sci. Rev.* **2004**, *66*, 183–197. [CrossRef]
6. Yang, M.; Zhe, F.; Zhao, Y.; Jiang, L.; Zhao, J.; Song, Y. Effect of depressurization pressure on methane recovery from hydrate–gas–water bearing sediments. *Fuel* **2016**, *166*, 419–426. [CrossRef]
7. Tupsakhare, S.S.; Kattekola, S.; Castaldi, M.J. An Application of the Results from the Large-Scale Thermal Stimulation Method of Methane Hydrate Dissociation to the Field Tests. *Ind. Eng. Chem. Res.* **2017**, *56*, 4588–4599. [CrossRef]
8. Li, G.; Li, X.-S.; Tang, L.-G.; Zhang, Y. Experimental investigation of production behavior of methane hydrate under ethylene glycol injection in unconsolidated sediment. *Energy Fuels* **2007**, *21*, 3388–3393. [CrossRef]
9. Koh, D.-Y.; Kang, H.; Lee, J.-W.; Park, Y.; Kim, S.-J.; Lee, J.; Lee, J.Y.; Lee, H. Energy-efficient natural gas hydrate production using gas exchange. *Appl. Energy* **2016**, *162*, 114–130. [CrossRef]
10. Ma, S.; Zheng, J.-N.; Zhao, J.; Yang, M.; Song, Y. Experimental analysis on thermodynamic stability and methane leakage during solid fluidization process of methane hydrate. *Fuel* **2021**, *284*, 119020. [CrossRef]

11. Nair, V.C.; Prasad, S.K.; Kumar, R.; Sangwai, J.S. Energy recovery from simulated clayey gas hydrate reservoir using depressurization by constant rate gas release, thermal stimulation and their combinations. *Appl. Energy* **2018**, *225*, 755–768. [CrossRef]
12. Feng, Y.; Chen, L.; Suzuki, A.; Kogawa, T.; Okajima, J.; Komiya, A.; Maruyama, S. Enhancement of gas production from methane hydrate reservoirs by the combination of hydraulic fracturing and depressurization method. *Energy Convers. Manag.* **2019**, *184*, 194–204. [CrossRef]
13. Zhao, J.; Zhang, L.; Chen, X.; Zhang, Y.; Liu, Y.; Song, Y. Combined replacement and depressurization methane hydrate recovery method. *Energy Explor. Exploit.* **2016**, *34*, 129–139. [CrossRef]
14. Jin, G.; Xu, T.; Xin, X.; Wei, M.; Liu, C. Numerical evaluation of the methane production from unconfined gas hydrate-bearing sediment by thermal stimulation and depressurization in Shenhu area, South China Sea. *J. Nat. Gas Sci. Eng.* **2016**, *33*, 497–508. [CrossRef]
15. Yu, T.; Guan, G.; Abudula, A.; Yoshida, A.; Wang, D.; Song, Y. Gas recovery enhancement from methane hydrate reservoir in the Nankai Trough using vertical wells. *Energy* **2019**, *166*, 834–844. [CrossRef]
16. Boswell, R.; Collett, T.S. Current perspectives on gas hydrate resources. *Energy Environ. Sci.* **2011**, *4*, 1206–1215. [CrossRef]
17. You, K.; Flemings, P.B.; Malinverno, A.; Collett, T.S.; Darnell, K. Mechanisms of Methane Hydrate Formation in Geological Systems. *Rev. Geophys.* **2019**, *57*, 1146–1196. [CrossRef]
18. Sun, J.; Ning, F.; Zhang, L.; Liu, T.; Peng, L.; Liu, Z.; Li, C.; Jiang, G. Numerical simulation on gas production from hydrate reservoir at the 1st offshore test site in the eastern Nankai Trough. *J. Nat. Gas Sci. Eng.* **2016**, *30*, 64–76. [CrossRef]
19. Anderson, B.; Hancock, S.; Wilson, S.; Enger, C.; Collett, T.; Boswell, R.; Hunter, R. Formation pressure testing at the Mount Elbert Gas Hydrate Stratigraphic Test Well, Alaska North Slope: Operational summary, history matching, and interpretations. *Mar. Pet. Geol.* **2011**, *28*, 478–492. [CrossRef]
20. Qin, X.-W.; Lu, J.-A.; Lu, H.-L.; Qiu, H.-J.; Liang, J.-Q.; Kang, D.-J.; Zhan, L.-S.; Lu, H.-F.; Kuang, Z.-G. Coexistence of natural gas hydrate, free gas and water in the gas hydrate system in the Shenhu Area, South China Sea-ScienceDirect. *China Geol.* **2020**, *3*, 210–220. [CrossRef]
21. Liu, C.; Shen, Y.; Zhang, J.; Lu, D.; Liu, H.; Wu, H. Production analysis in shale gas reservoirs based on fracturing-enhanced permeability areas. *Sci. China-Phys. Mech. Astron.* **2019**, *62*, 104611. [CrossRef]
22. Tan, Y.; Pan, Z.; Feng, X.-T.; Zhang, D.; Connell, L.D.; Li, S. Laboratory characterisation of fracture compressibility for coal and shale gas reservoir rocks: A review. *Int. J. Coal Geol.* **2019**, *204*, 1–17. [CrossRef]
23. Zhang, R.-H.; Zhang, L.-H.; Tang, H.-Y.; Chen, S.-N.; Zhao, Y.-L.; Wu, J.-F.; Wang, K.-R. A simulator for production prediction of multistage fractured horizontal well in shale gas reservoir considering complex fracture geometry. *J. Nat. Gas Sci. Eng.* **2019**, *67*, 14–29. [CrossRef]
24. Ma, X.; Jiang, D.; Fang, X.; Wang, X. Numerical simulation of single-cluster and multi-cluster fracturing of hydrate reservoir based on cohesive element. *Eng. Fract. Mech.* **2022**, *265*, 108365. [CrossRef]
25. Yu, Y.; Liu, J.; Li, B.; Sun, Y. Analysis of the hydraulic fracturing mechanism and fracture propagation law with a new extended finite element model for the silty hydrate reservoir in the South China Sea. *J. Nat. Gas Sci. Eng.* **2022**, *101*, 104535. [CrossRef]
26. Ma, X.; Cheng, J.; Sun, Y.; Li, S. 2D Numerical Simulation of Hydraulic Fracturing in Hydrate-Bearing Sediments Based on the Cohesive Element. *Energy Fuels* **2021**, *35*, 3825–3840. [CrossRef]
27. Liu, X.; Sun, Y.; Guo, T.; Rabiei, M.; Qu, Z.; Hou, J. Numerical simulations of hydraulic fracturing in methane hydrate reservoirs based on the coupled thermo-hydrologic-mechanical-damage (THMD) model. *Energy* **2022**, *238*, 122054. [CrossRef]
28. Chen, C.; Yang, L.; Jia, R.; Sun, Y.; Guo, W.; Chen, Y.; Li, X. Simulation Study on the Effect of Fracturing Technology on the Production Efficiency of Natural Gas Hydrate. *Energies* **2017**, *10*, 1241. [CrossRef]
29. Yang, L.; Chen, C.; Jia, R.; Sun, Y.; Guo, W.; Pan, D.; Li, X.; Chen, Y. Influence of Reservoir Stimulation on Marine Gas Hydrate Conversion Efficiency in Different Accumulation Conditions. *Energies* **2018**, *11*, 339. [CrossRef]
30. Sun, J.; Ning, F.; Liu, T.; Liu, C.; Chen, Q.; Li, Y.; Cao, X.; Mao, P.; Zhang, L.; Jiang, G. Gas production from a silty hydrate reservoir in the South China Sea using hydraulic fracturing: A numerical simulation. *Energy Sci. Eng.* **2019**, *7*, 1106–1122. [CrossRef]
31. Shan, L.; Fu, C.; Liu, Y.; Qi, Y. A feasibility study of using frac-packed wells to produce natural gas from subsea gas hydrate resources. *Energy Sci. Eng.* **2020**, *8*, 1247–1259. [CrossRef]
32. Ye, J.; Qin, X.; Xie, W.; Lu, H.; Ma, B.; Qiu, H.; Liang, J.; Lu, J.; Kuang, Z.; Lu, C.; et al. Main progress of the second gas hydrate trial production in the South China Sea. *Geol. China* **2020**, *47*, 557–568.
33. Too, J.L.; Cheng, A.; Khoo, B.C.; Palmer, A.; Linga, P. Hydraulic fracturing in a penny-shaped crack. Part II: Testing the frackability of methane hydrate-bearing sand. *J. Nat. Gas Sci. Eng.* **2018**, *52*, 619–628. [CrossRef]
34. Liu, X.; Zhang, W.; Qu, Z.; Guo, T.; Sun, Y.; Rabiei, M.; Cao, Q. Feasibility evaluation of hydraulic fracturing in hydrate-bearing sediments based on analytic hierarchy process-entropy method (AHP-EM). *J. Nat. Gas Sci. Eng.* **2020**, *81*, 103434. [CrossRef]
35. Ito, T.; Igarashi, A.; Suzuki, K.; Nagakubo, S.; Matsuzawa, M.; Yamamoto, K. Laboratory Study of Hydraulic Fracturing Behavior in Unconsolidated Sands for Methane Hydrate Production. In Proceedings of the Offshore Technology Conference, Houston, TX, USA, 5–8 May 2008.
36. Konno, Y.; Jin, Y.; Yoneda, J.; Uchiumi, T.; Shinjou, K.; Nagao, J. Hydraulic fracturing in methane-hydrate-bearing sand. *Rsc Adv.* **2016**, *6*, 73148–73155. [CrossRef]
37. Zhang, W.; Shi, X.; Jiang, S.; Cao, Q.; Wang, F.; Wang, Z.; Ge, Y.; Du, Y. Experimental study of hydraulic fracture initiation and propagation in highly saturated methane-hydrate-bearing sands. *J. Nat. Gas Sci. Eng.* **2020**, *79*, 73148–73155. [CrossRef]

38. Yao, Y.; Guo, Z.; Zeng, J.; Li, D.; Lu, J.; Liang, D.; Jiang, M. Discrete Element Analysis of Hydraulic Fracturing of Methane Hydrate-Bearing Sediments. *Energy Fuels* **2021**, *35*, 6644–6657. [CrossRef]
39. Yu, Y.; Liu, J.; Ma, X.; Yang, G.; Sun, Y.; Sun, W.; Shi, W. Mechanism Analysis of Multi-Cluster Fracture Interference in Horizontal Wells of Hydrate Reservoirs in the South China Sea. *Energy Fuels* **2022**, *36*, 3580–3595. [CrossRef]
40. Lu, C.; Qin, X.; Mao, W.; Ma, C.; Geng, L.; Yu, L.; Bian, H.; Meng, F.; Qi, R. Experimental Study on the Propagation Characteristics of Hydraulic Fracture in Clayey-Silt Sediments. *Geofluids* **2021**, *2021*, 6698649. [CrossRef]
41. Sun, Y.; Li, S.; Ma, X.; Jiang, D. Experimental study on hydraulic fracturing behavior of frozen clayey silt and hydrate-bearing clayey silt. *Fuel* **2022**, *322*, 124366.
42. Zhang, R.; Hou, B.; Tan, P.; Muhadasi, Y.; Fu, W.; Dong, X.; Chen, M. Hydraulic fracture propagation behavior and diversion characteristic in shale formation by temporary plugging fracturing. *J. Pet. Sci. Eng.* **2020**, *190*, 107063. [CrossRef]
43. Tan, P.; Jin, Y.; Han, K.; Hou, B.; Chen, M.; Guo, X.; Gao, J.J.F. Analysis of hydraulic fracture initiation and vertical propagation behavior in laminated shale formation. *Fuel* **2017**, *206*, 482–493. [CrossRef]
44. Zhao, H.; Wang, X.; Liu, Z. Experimental investigation of hydraulic sand fracturing on fracture propagation under the influence of coal macrolithotypes in Hancheng block, China. *J. Pet. Sci. Eng.* **2019**, *175*, 60–71. [CrossRef]
45. Lee, J.; Lee, D.; Seo, Y. Experimental investigation of the exact role of large-molecule guest substances (LMGSs) in determining phase equilibria and structures of natural gas hydrates. *Energy* **2021**, *215*, 119219. [CrossRef]
46. Iida, T.; Mori, H.; Mochizuki, T.; Mori, Y.H. Formation and dissociation of clathrate hydrate in stoichiometric tetrahydrofuran-water mixture subjected to one-dimensional cooling or heating. *Chem. Eng. Sci.* **2001**, *56*, 4747–4758. [CrossRef]
47. Nie, S.; Zhong, X.; Song, J.; Tu, G.; Chen, C. Engineering, Experimental study on hydraulic fracturing in clayey-silty hydrate-bearing sediments and fracability evaluation based on multilayer perceptron-analytic hierarchy process. *J. Nat. Gas Sci. Eng.* **2022**, *106*, 104735. [CrossRef]
48. Waite, W.F.; Santamarina, J.C.; Cortes, D.D.; Dugan, B.; Espinoza, D.N.; Germaine, J.; Jang, J.; Jung, J.W.; Kneafsey, T.J.; Shin, H.; et al. Physical Properties of Hydrate-Bearing Sediments. *Rev. Geophys.* **2009**, *47*. [CrossRef]
49. Zhang, H.Q.; Yang, S.X.; Wu, N.Y.; Su, X.; Gary, H. Successful and surprising results for China's first gas hydrate drilling expedition. *Fire Ice Fall* **2007**, *7*, 6–9.
50. Nengyou, W.U.; Haiqi, Z.; Shengxiong, Y.; Jinqiang, L.; Hongbin, W. Preliminary Discussion on Natural Gas Hydrate (NGH) Reservoir System of Shenhu Area, North Slope of South China Sea. *Nat. Gas Ind.* **2007**, *27*, 1–6.
51. Su, Z.; Moridis, G.J.; Zhang, K.; Wu, N. A huff-and-puff production of gas hydrate deposits in Shenhu area of South China Sea through a vertical well. *J. Pet. Sci. Eng.* **2012**, *86–87*, 54–61. [CrossRef]
52. Su, M.; Luo, K.; Fang, Y.; Kuang, Z.; Yang, C.; Liang, J.; Liang, C.; Chen, H.; Lin, Z.; Wang, C.; et al. Grain-size characteristics of fine-grained sediments and association with gas hydrate saturation in Shenhu Area, northern South China Sea. *Ore Geol. Rev.* **2021**, *129*, 103889. [CrossRef]
53. Lu, H.; Chen, H.; Chen, F.; Liao, Z.J. Mineralogy of the sediments from gas-hydrate drilling sites, Shenhu area, South China Sea. *Geol. Res. S. China Sea* **2009**, 28–39.
54. Yun, T.S.; Santamarina, J.C.; Ruppel, C. Mechanical properties of sand, silt, and clay containing tetrahydrofuran hydrate. *J. Geophys. Res.* **2007**, *112*. [CrossRef]
55. Ning, F.; Wu, N.; Li, S.; Zhang, K.; Yu, Y.; Liu, L.; Sun, J.; Guosheng, J.; Sun, C.; Chen, G. Estimation of in-situ mechanical properties of gas hydrate-bearing sediments from well logging. *Pet. Explor. Dev.* **2013**, *40*, 542–547. [CrossRef]
56. Hou, B.; Chen, M.; Li, Z.; Wang, Y.; Diao, C. Propagation area evaluation of hydraulic fracture networks in shale gas reservoirs. *Pet. Explor. Dev.* **2014**, *41*, 833–838. [CrossRef]
57. Zhao, H.; Xiong, Y.; Zhen, H.; Liu, C.; Li, X. Experimental investigation on the fracture propagation of three-stage acid fracturing of tight sandstone gas reservoir. *J. Pet. Sci. Eng.* **2022**, *211*, 110143. [CrossRef]

energies

MDPI

Article

Development of Virtual Flow-Meter Concept Techniques for Ground Infrastructure Management

Ruslan Vylegzhanin [1], Alexander Cheremisin [2,*], Boris Kolchanov [1], Pavel Lykhin [1], Rustam Kurmangaliev [1], Mikhail Kozlov [1,3], Eduard Usov [3,4] and Vladimir Ulyanov [1,3]

1 Novosibirsk R&D Center LLC, 630090 Novosibirsk, Russia
2 Skolkovo Institute of Science and Technologies, 121205 Moscow, Russia
3 Physical Department, Novosibirsk State University, 630090 Novosibirsk, Russia
4 Novosibirsk Branch of the Nuclear Safety Institute of the Russian Academy of Sciences, 630090 Novosibirsk, Russia
* Correspondence: a.chermeisin2@skoltech.ru

Abstract: This paper describes the further development of the virtual flow meter concept based on the author's simulator of an unsteady gas–liquid flow in wells. The results of comparison with commercial simulators based on real well data are given as practical applications. The results of the comparison of the simulators demonstrated high correspondence (<10% error) for a number of target parameters. The description of the architecture and results of testing the algorithm for automatic settings of the model parameters are given. Operating speed was the key criterion in the architecture development. According to the test results, it became possible to achieve the adaptation accuracy of 5% specified.

Keywords: virtual flow measurement; multiphase flow; reservoir fluid; simulator; numerical methods

Citation: Vylegzhanin, R.; Cheremisin, A.; Kolchanov, B.; Lykhin, P.; Kurmangaliev, R.; Kozlov, M.; Usov, E.; Ulyanov, V. Development of Virtual Flow-Meter Concept Techniques for Ground Infrastructure Management. *Energies* **2023**, *16*, 400. https://doi.org/10.3390/en16010400

Academic Editor: Hossein Hamidi

Received: 12 September 2022
Revised: 9 November 2022
Accepted: 14 November 2022
Published: 29 December 2022

1. Introduction

The so-called digital-twin technologies designed to offer a commercial solution to a number of process problems are being actively implemented in many industrial areas.

A simulator of an unsteady flow of a multiphase fluid in a reservoir, a well, and surface gathering lines, which takes into account all the key physical processes occurring during hydrocarbon production, is actually used as a twin for oil and gas fields. This paper contains a brief description of the key features of the unsteady multiphase flow simulator developed by the authors taking into account the choke and electric centrifugal pump models.

Correlating the data obtained during the simulation with the actually measured data is a necessary step when using a simulator in a digital twin. This process is called adaptation or adjustment to an actual value.

The simulator parameters can be grouped according to two criteria: impact on the result, i.e., sensitivity, and accuracy of their measurement in a real field.

The figure (Figure 1) shows some parameters for the black oil model. Adaptation consists of selecting highly sensitive and weakly measured parameters so that the input parameters sent to the simulator provide the output ones corresponding to the measured ones with sufficient accuracy for a wide set of input data. Gas viscosity, gas compressibility, and pipe roughness were used as adjustment parameters.

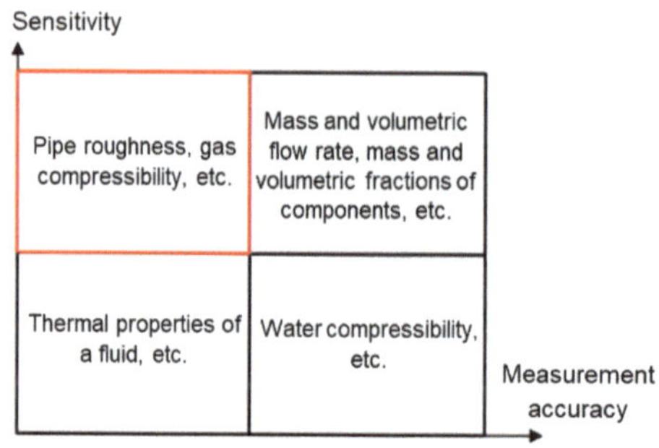

Figure 1. Simulator parameters grouped by the two criteria.

It should be noted that it is incorrect to equate this process with the calibration of an instrument because the development of an instrument concept consists in minimizing the number of highly sensitive and weakly measured parameters, while calibration consists of selecting highly sensitive and accurately measured parameters.

Two neural networks were used as the adaptation algorithm of the simulator. The first was used to make a forecast of the simulator output parameters based on the input parameters.

The second was used to assess the optimal input parameters for the given output parameters.

The neural network implementations were taken from the TensorFlow software library.

2. Methodological Approaches

2.1. Mathematical Model of an Unsteady Flow of a Multiphase Fluid

The proposed simulator uses the systems of equations of continuity, conservation of mass, energy, and momentum (for the gas and liquid phases) [1–4], as well as the equations for the total enthalpy of a mixture in the allocated volume and for the mass of an individual component taking account of friction, specific enthalpy, and heat exchange (gas/liquid, wall/gas, liquid/gas, wall/liquid) in order to simulate the gas–liquid flows in a well and a gas-gathering network.

The Peng–Robinson equation of state (modified Van der Waals equation) was used as the equation of state, where p is pressure, T is temperature, V is molar volume, R is the gas constant, and a and b parameters are the coefficients responsible for the forces of attraction between molecules and the finite volume of the molecules calculated through the parameters at the critical point:

$$p = \frac{RT}{V-b} - \frac{a\alpha}{V(V+b)+b(V-b)} \quad (1)$$

.

It is necessary to take into account the temperature dependence of the a and b coefficients. If compressibility factor Z is introduced, then the Peng–Robinson equation of state can be represented as a cubic equation of Z. After solving this equation, the maximum root was selected for the gas phase, while the minimum root was selected for the liquid phase. Additional ratios are required to determine a and b coefficients and the molar mass for a mixture of several components, which are called blend recipes (Soave ratios).

The model for calculating the phase state of the mixture allows obtaining the quantities (in moles) of the hydrocarbon liquid and gas coexisting in a tank or vessel at a given pressure and temperature. These calculations also make it possible to define the composition of the existing liquid and gas hydrocarbon phases.

In a multicomponent system, the separation of the components between the liquid and gas phases is described by the equilibrium ratio (coefficient) for the given component.

The equilibrium coefficient is defined as the ratio of the molar fraction of the component in the gas phase to the molar fraction of the component in the liquid phase. At low pressures, Dalton's law is applicable to a mixture of gases, and Raoult's law is applicable to regular solutions. Raoult's law allows the expression of the partial pressure of a component in a solution through the pressure of saturated vapors of the given component. Dalton's law expresses the partial pressure of a component in a gas through the pressure in the system.

This model assumes that the value of the equilibrium coefficients for any component does not depend on the total composition of the mixture, and since the saturation pressure depends only on temperature, the equilibrium coefficients depend only on pressure and temperature in the system.

The temperature of the mixture can be determined if the total enthalpy is known, and the empirical Lohrenz–Bray–Clark formula was used to calculate the viscosities.

The Lee–Gonzalez–Eakin method was used to calculate the gas phase viscosity. This method is a semi-empirical correlation in which the gas viscosity is expressed through temperature, gas phase density, and molar mass. The proposed correlation predicts the viscosity value with a standard deviation of 2.7% and a maximum deviation of 8.99%, which is why this method cannot be used for sulfurous gases.

Calculation of friction with the channel walls for a Newtonian fluid is made using the standard approaches as closing relations of the system.

For a single-phase liquid or gas flow, the friction pressure loss due to the wall friction included in the momentum conservation equations is usually expressed through the specific mass flow rate of the mixture with a density of the medium, while the friction coefficient is a function of the Reynolds number at the fixed channel geometry. Various correlations of the friction coefficient (in laminar or turbulent mode) are used to calculate the pressure loss. Transition flow uses linear interpolation between the expressions for the laminar and turbulent modes.

The situation is more complicated in the case of a two-phase flow. Here, the pressure loss depends not only on the flow rates and Reynolds numbers of the individual components but also on the structure of the two-phase flow. The Lockhart–Martinelli approach is common for calculating friction in two-phase modes: according to it, a pressure loss in a two-phase flow can be calculated as the product of the pressure loss of a single-phase flow of any phase with the flow rate of the two-phase mixture and the two-phase friction multiplier. Simpler, Hagedorn–Brown [2], Duns–Ros [3], and Orkiszewski [4] models can be used to calculate the friction in a two-phase flow. There are also transient two-phase flow models [5,6], but they are not covered in this article.

Heat fluxes coming from the wall into the liquid and gas phases were calculated in order to determine the heat exchange with the wall in the proposed model (heat transfer coefficients for each phase are calculated through the dimensionless Nusselt number).

The proposed mathematical models contribute to the development of the theoretical aspects of time simulation of multiphase media, as well as the research into important practical problems in the industry. In order to make a computational algorithm, all the previous conservation equations were rewritten in a finite-difference form.

2.2. Choke Model

In the basic model, a choke is a local abrupt taper of a pipeline and then its abrupt enhancer (Figure 2).

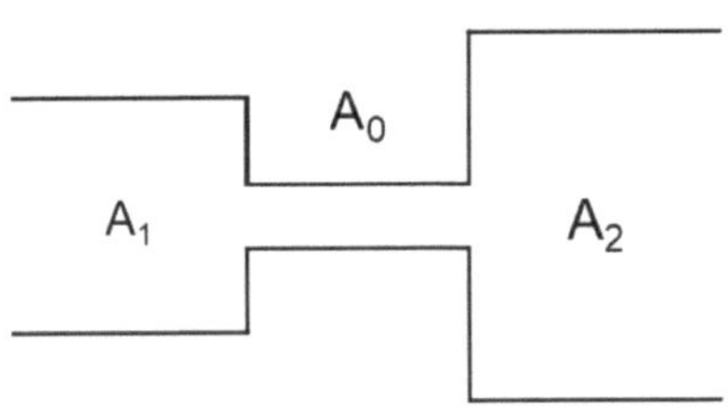

Figure 2. Choke model geometry. A_0, A_1, and A_2 are flow areas.

In this case, the pressure loss at the choke is the sum of irreversible losses in case of the abrupt taper of the pipeline from section A_1 to section A_0, irreversible losses in case of the abrupt enhancer of the channel from section A_0 to section A_2, and local friction losses in the narrow section of the channel:

$$\Delta p = \xi \frac{\rho u_0^2}{2} = \left(\xi_{taper} + \xi_{enhancer} + \xi_{friction} \frac{l_0}{D_0} \right) \frac{\rho u_0^2}{2} \tag{2}$$

where ρ is the fluid density, u_0 is the fluid velocity, ξ is the loss factor, l_0 is the choke length, and D_0 is its inner diameter.

The friction loss coefficient is calculated using the friction model [7]. The taper and enhancer loss coefficients are calculated according to the ratios in [8].

The taper loss coefficient is calculated as follows:

- at $Re < 10$: $\xi_{taper} = \frac{30}{Re}$
- at $10 \leq Re < 10^4$: $\xi_{taper} = A \cdot B\left(1 - \frac{A_0}{A_1}\right)$

where $A = \sum\limits_{i=0}^{7} a_i (\lg Re)^i$, $a_0 = -25.12458$, $a_1 = 118.5076$, $a_2 = -170.4147$, $a_3 = 118.1949$, $a_4 = -44.42141$, $a_5 = 9.09524$, $a_6 = -0.9244027$, $a_7 = 0.03408265$.

$B = \sum\limits_{i=0}^{2} \left\{ \left[\sum\limits_{j=0}^{2} a_{ij} \left(\frac{A_0}{A_1} \right)^j \right] (\lg Re)^i \right\}$, a_{ij} coefficients are presented in Table 1.

Table 1. Values a_{ij}.

i/j	$10 \leq$ Re $\leq 2 \times 10^3$			2E3 $\leq$ Re $\leq 4 \times 10^3$		
	0	**1**	**2**	**0**	**1**	**2**
0	1.07	1.22	2.9333	0.5443	−17.298	−40.715
1	0.05	−0.51668	0.8333	−0.06518	8.7616	22.782
2	0	0	0	0.05239	−1.1093	−3.1509

- at $Re > 10^4$: $\xi_{taper} = \frac{1}{2}\left(1 - \frac{A_0}{A_1}\right)^{3/4}$

The enhancer loss coefficient is as follows:

- at $Re < 10$: $\xi_{taper} = \frac{30}{Re}$
- at $10 \leq Re \leq 500$:

$$\begin{aligned} \xi_{taper} = {} & 3.62536 + 10.744a - 4.41041a^2 + \\ & + b\left(-18.13 + 56.77855a + 33.40344a^2\right) + \\ & + b^2\left(30.8558 + 99.9542a - 62.78a^2\right) + \\ & + b^3\left(-13.217 - 53.9555a + 33.8053a^2\right) \end{aligned}$$

where $a = \left(1 - \frac{A_0}{A_2}\right)^2$, $b = \frac{1}{\lg Re}$;

- at $500 < Re < 3.3 \cdot 10^3$:

$$\begin{aligned} \xi_{taper} = {} & -8.44556 - 26.163a - 5.38086a^2 + c\left(6.007 + 18.5372a + 3.9978a^2\right) + \\ & + c^2\left(-1.02318 - 3.0916a - 0.680943a^2\right) \end{aligned}$$

where $a = \left(1 - \frac{A_0}{A_2}\right)^2$, $c = \lg Re$;

- at $Re \geq 3.3 \cdot 10^3$: $\xi_{taper} = \left(1 - \frac{A_0}{A_2}\right)^2$.

2.3. ESP Model

The ESP model allows for defining the head created by the pump depending on the flow rate of the mixture. Therefore, the momentum conservation equation is replaced by a pressure differential equation:

$$L_c \frac{\partial p}{\partial l} = g\rho_m \Delta H$$

where L_c is the length of the pump stage, ΔH is the pump head, and $\langle \rho \rangle$ is density of the mixture.

$$\Delta H = \Delta H_B^* K_H (2 - q - A(1-q)^2),\ K_H = K_Q = 1 - \frac{150}{Re},\ K_\eta = 1.08 - 12Re^{-0.45}$$

$$\eta = \frac{g\langle \rho \rangle \Delta H Q}{\Delta N},\ q = \frac{Q}{Q^* = \frac{Q}{(K_q Q_B{}^*)}},\ A = 0.66 + 16Re^{-0.68},\ Re = \frac{4.3 + 0.816 n_s{}^{0.274}}{n_s{}^{0,575}} Q_B^* \frac{\langle \rho \rangle}{\langle \mu \rangle} \left(\frac{\omega}{Q_B^*}\right)^{1/3},$$

$$n_s = 193\omega (Q_B^*)^{0.5} (g\Delta H_B^*)^{-0.75}$$

where K_Q, K_H, and K_η are the supply, head, and efficiency coefficients; ω is the angular speed of the shaft rotation; $\langle \mu \rangle$ is the effective viscosity of the mixture; and n_s is the speed coefficient.

$$Q_B^* = Q_{BH}^* \frac{\omega}{\omega_H},\ \Delta H_B^* = \Delta H_{BH}^* \left(\frac{\omega}{\omega_H}\right)^2,\ \Delta N_B^* = \Delta N_{BH}^* \left(\frac{\omega}{\omega_H}\right)^3$$

where Q_{BH}^*, ΔH_{BH}^*, and ΔN_{BH}^* are the motor stage characteristics at the rated angular velocity ω_H = 50 Hz.

3. Method of Automatic Adjustment to the Actual Data

The number of measured parameters of wells in the existing fields is less than necessary for a comprehensive description of the physical system. In order to adjust the model intended for forecasting and calculation, the unknown parameters should be as close as possible to the actual ones. An automatic adaptation algorithm based on neural networks was used to identify any hidden dependencies between the measured and unmeasured values.

The adaptation algorithm consists of two main modules: a decision-making module and an assessment module. The assessment module is used to approximate the aggregate data specified by the objective function based on the data calculated by the hydrodynamic simulator (Figure 3), whose computational kernel performs a deterministic calculation of the multiphase fluid movement in the reservoir and the well [9]. The objective function can be set arbitrarily based on the measured parameters and the simulator output so that all values are within the [−1, 1] range, where higher values correspond to higher accuracy. The decision-making module was used to determine the values of the adjusted parameters of the simulator based on the approximation performed by the assessment module.

Since direct training of the network requires knowledge of the optimal set of parameters for each set of the input values, the network was trained by end-to-end gradient propagation through the assessment module (Figure 4) by the chain rule formula:

$$\frac{\delta v}{\delta w} = \frac{\delta v}{\delta a} \cdot \frac{\delta a}{\delta w} \tag{3}$$

where v is the output value of the assessment module, a is the vector of the adjusted parameter values, and w is the internal variables of the decision-making model.

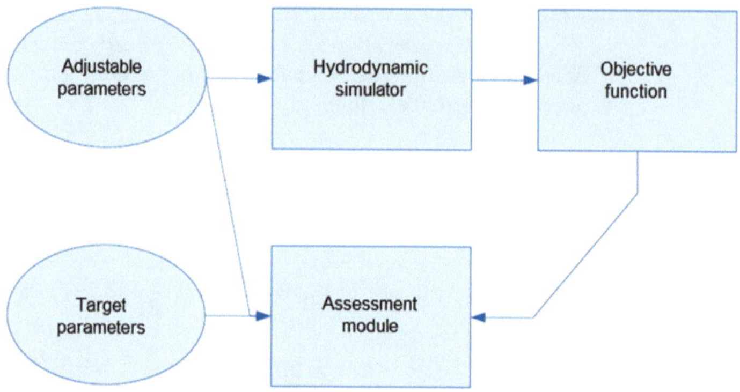

Figure 3. Assessment module training.

The architecture of a fully connected two-layer neural network (Figure 5) with RectifiedLinearUnit as an activation function was chosen for the decision-making module. The output layer uses the hyperbolic tangent activation function. The network receives measurement data from a real well as input. The network provides the values of the parameters sent to the simulator as its output.

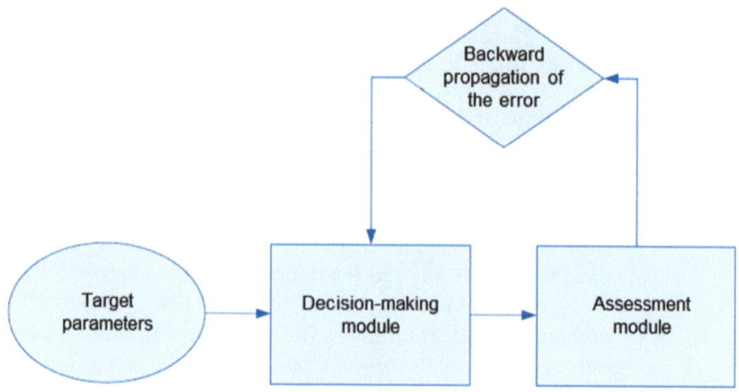

Figure 4. Decision-making module training.

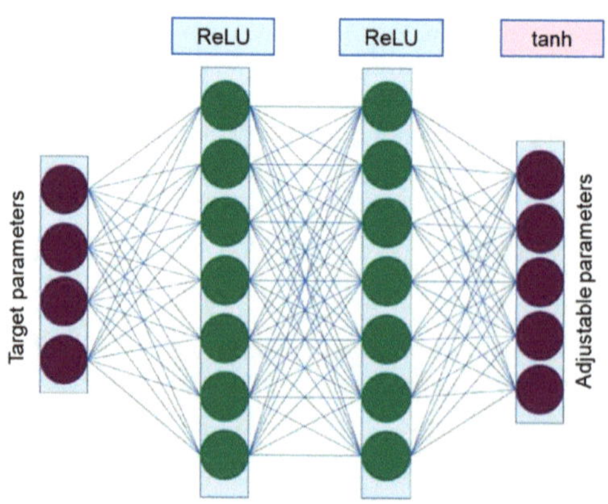

Figure 5. Architecture of the neural network for the decision-making module.

This calculation can be performed directly since both modules were implemented on the basis of the TensorFlow [10] computation graph. Adam from the TensorFlow package was used as an optimizer that implements the backpropagation of the error algorithm based on the calculated gradient value.

The architecture of a fully connected three-layer neural network (Figure 6) with RectifiedLinearUnit as an activation function was chosen for the assessment module. The output layer uses a single neuron with the hyperbolic tangent activation function. The network receives measurement data from a real well and the vector of the simulator parameters as input. The network provides the predicted value of the objective function as its output. RectifiedAdam from TensorFlow Addons package with the mean-square deviation as a loss function was used as an optimizer.

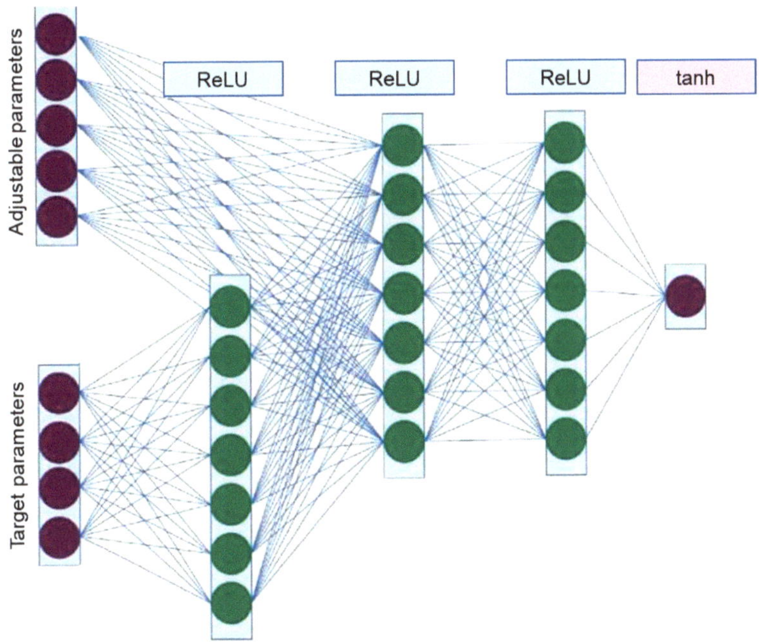

Figure 6. Architecture of the neural network for the assessment module.

3.1. Problem Statement

Two tasks were considered in this paper: validation of the correctness of the developed simulator and assessing the quality of the automatic adaptation algorithms.

3.2. Validation of Correctness

In order to validate correctness, it is necessary to compare with a reference, with the PIPESIM simulator used as one.

A deviated well (Table 2) with a maximum deflection angle of about 32° was adopted as the input parameters for the model in PIPESIM software and the methodology used.

Table 2. Well geometry.

Depth along the Wellbore, m	Azimuthal Angle, °
0	1.9
270	2.0
400	0.9
820	0.7
950	0.9
2070	3.2
2220	11.9
2340	24.4
2470	29.0
2620	32.1
2770	28.5
2910	26.2
3060	28.9
3200	28.4
3350	23.8
3460	11.6
3580	1.3
3880	0.0

Fluid properties table (Table 3):

Table 3. Fluid properties.

Properties	Density, kg/m^3
Gas	0.8087
Liquid	765.8

The following table (Table 4) descibe boundary conditions were used for calculation:

Table 4. Reservoir conditions.

Pressure, Bar	Mixture Temperature, K	OGR	Watercut, %
422	378.5	0.000715	0

Beggs and Brill's methods were used to calculate the parameters in the well in the author's simulator and in PIPESIM software.

Two PVT models were used to compare the calculation results: the black oil model and the compositional model. Compositional model was: N2, CO2, C1, C2, C3, C5+, C7+, C11+, C19+, and C36+.

There was no surface or bottom hole equipment such as choke or ESP.

The graphs of the distribution of the parameters of pressure, temperature, gas mass rate, and gas volume gate (Figures 7–10) along the wellbore are shown below.

As mentioned above, the proposed model of the gas–liquid flow is unsteady. Therefore, the steady flow (stationary mode) (Figure 11) is adopted as a comparison of the method developed with PIPESIM.

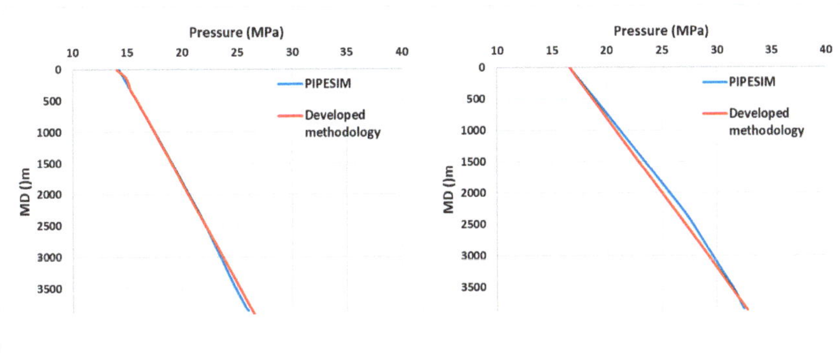

(a) (b)

Figure 7. Pressure distribution along the wellbore. The blue line represents the result of the PIPESIM calculation, while the red line represents the result of the calculation of the proposed model. The black oil model is (**a**), while the compositional model is (**b**).

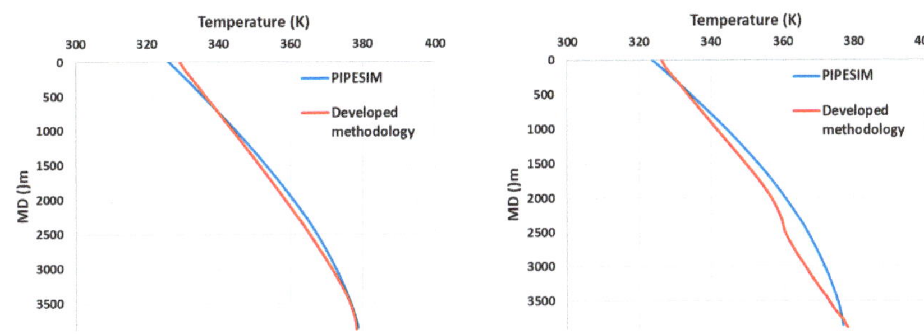

(a) (b)

Figure 8. Borehole wall temperature distribution along the wellbore. The blue line represents the result of PIPESIM calculation, while the red line represents the result of calculation of the proposed model. The black oil model is (**a**), while the compositional model is (**b**).

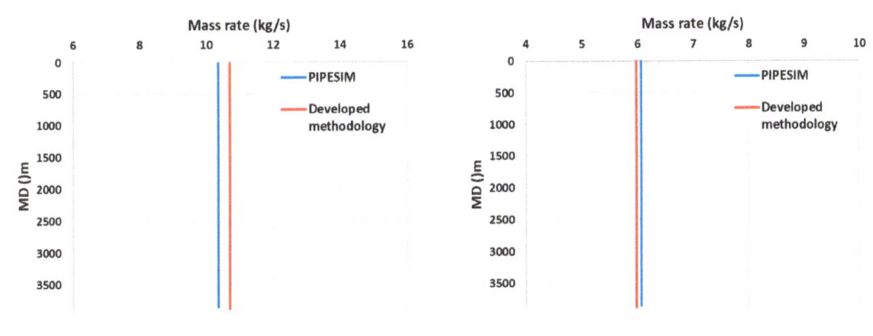

(a) (b)

Figure 9. Mass gas flow rate distribution along the wellbore. The blue line represents the result of PIPESIM calculation, while the red line represents the result of calculation of the proposed model. The black oil model is (**a**), while the compositional model is (**b**).

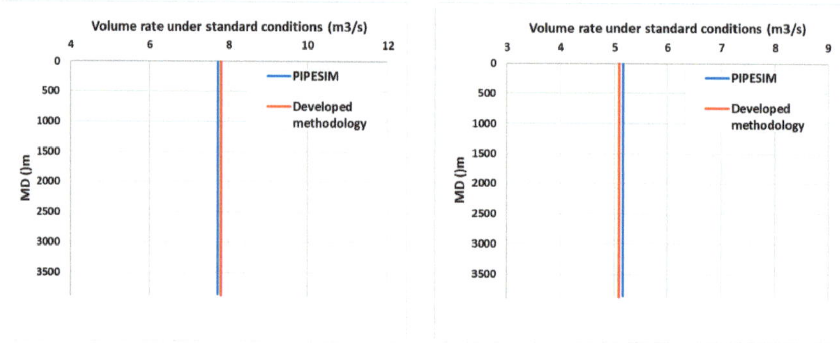

(a) (b)

Figure 10. Volume gas flow rate distribution (standard condition) along the wellbore. The blue line represents the result of PIPESIM calculation, while the red line represents the result of calculation of the proposed model. The black oil model is (**a**), while the compositional model is (**b**).

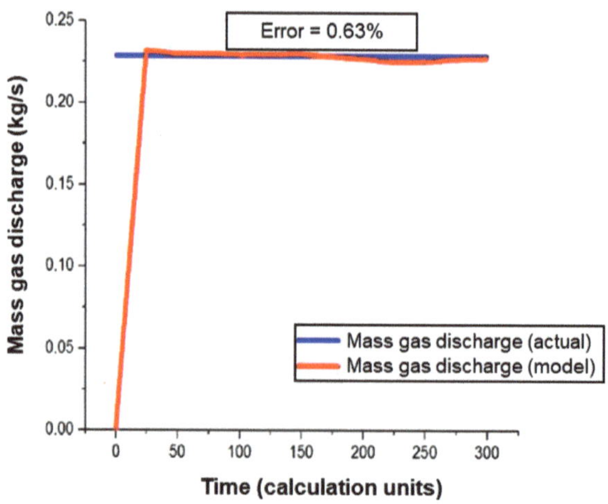

Figure 11. Adjustment of the model to actual values in the virtual flow meter mode.

In addition to the above parameters, the proposed method allows calculating the parameters of density, velocity, volume fraction, and viscosity of each phase (liquid, water, gas), as well as the mass, density, and inner radius of the layer of paraffin formed at any time.

3.3. *Automatic Adjustment Quality*

Quality was assessed based on two criteria: velocity and accuracy.

The velocity of the algorithm was determined by the speed of training and the speed of picking. In the tests carried out on a PC with the following configuration: Intel Core i5-7600 CPU at 3.5 GHz and 12 GB memory, the training took 5–7 min of real time, while the time of the calculation itself was about 3 s.

In order to assess the accuracy of the adaptation algorithm, it is necessary to define the accuracy of determining the measured parameters. In case of adjustment to actual values, there is no need to achieve a mistie less than the measurement accuracy. A 5% accuracy for the target parameters was used in this paper.

A set of input and output data divided into the training and test sets was used to assess the quality of the automatic adjustment. In the case of the training set, the algorithm was adjusted to the target parameters with an accuracy of 5%, then the adjusted model parameters were sent to the test set, and the target values were compared. The quality of the adjustment was considered acceptable if the mistie in the test set did not exceed 5%.

A table (Table 5) comparing the target values on the training and test sets is shown below.

Table 5. Comparison of the target parameters on the training and test sets.

Parameter	Training Set			Test Set		
	Simulator	Experiment	Error, %	Simulator	Experiment	Error, %
Volume oil fraction	6.5 (%)	6.3 (%)	3.3	5.9 (%)	6.2 (%)	4.8
Volume gas fraction	93.5 (%)	93.7 (%)	0.2	94.1 (%)	93.8 (%)	0.3
Mass oil flow rate	0.87 (kg/s)	0.91 (kg/s)	4.4	0.85 (kg/s)	0.82 (kg/s)	3.5
Mass gas flow rate	20.12 (kg/s)	19.67 (kg/s)	2.23	19.12 (kg/s)	19.93 (kg/s)	4.0

4. Discussion

By using the example of modeling a part of an oil-gathering network, which includes a number of sensors and measurement systems, this paper presented the concept of a virtual measurement system based on a hydrodynamic simulator and a deterministic model of the measurement system. The concept was tested on real experimental data and applied in real time. Acceptable accuracy of predictions of gas and oil discharge rates was obtained. Differences in calculated parameters are related to the difference in viscosity models [9,11].

It was found, based on the analysis, that the relations used to calculate friction of the multiphase flow with the borehole walls are one of the main sources of uncertainty. An exhaustive search of the typical friction models (in addition to the Beggs and Brill model used in the calculations) with the remaining parameters fixed showed a spread of the mass flow rate values most likely related to the use of the models outside the scope of their application. The effect of the pipe wall roughness and the gas compressibility parameter on the flow rate—due to its significant volume fraction of more than 89%—was also shown. Thus, it is necessary to measure the properties of the multiphase mixture and to know the process and design features of the well, as well as to understand the ranges within which the measured parameters might vary in order to use the developed methods in real situations to improve the prediction accuracy.

Further development of the methods will allow for fundamentally new opportunities to improve the measurement systems created on the principle of multiple run metering systems to appear, whose virtualization multiplies the density and reliability of data, to practically implement the principles described in the above-mentioned patents, and to significantly improve the quality of the information base for further development of the smart field technologies.

5. Results and Conclusions

The results of the operation of the previous proprietary unsteady influx simulator were compared with the results of the PIPESIM simulator based on actual data. The comparison showed a satisfactory fit for a number of target parameters, such as pressure, mass flow rate, and temperature.

An algorithm for the automatic adaptation of the model to actual data was developed and tested. The selected algorithm architecture allows quick fine-tuning and retraining neural networks to adjust to certain data types: GOR and WOR parameters of a well, typical flow rates, borehole geometry, etc.

Assessment of the quality of adjustment showed that it is possible to achieve the required level of reliability subject to a sufficient amount of data available in the training set. However, further research using actual data is needed to determine the limits of the applicability of this algorithm.

Author Contributions: R.V.: investigation and writing—original draft preparation; A.C.: writing—original draft preparation, writing—review and editing, and conceptualization; B.K.: visualization, software, and writing—review and editing; P.L.: conceptualization and writing—review and editing; R.K.: formal analysis and writing—original draft preparation; M.K.: writing—review and editing; E.U.: methodology and resources; V.U.: resources and supervision. All authors have read and agreed to the published version of the manuscript.

Funding: This research received no external funding.

Informed Consent Statement: Not applicable.

Data Availability Statement: Not applicable.

Conflicts of Interest: The authors declare no conflict of interest.

References

1. Klapchuk, O.V. Hydraulic characteristics of gas-liquid flows in wells. *Gas Ind.* **1981**, *2*, 32–35.
2. Hagedorn, A.R.; Brown, K.E. Experimental study of pressure gradients occurring during continuous two-phase flow in small-diameter vertical conduits. *J. Pet. Technol.* **1965**, *17*, 475–484. [CrossRef]
3. Duns, H., Jr.; Ros, N.C.J. Vertical flow of gas and liquid mixtures in wells. In Proceedings of the 6th World Petroleum Congress, Frankfurt, Germany, 19–26 June 1963.
4. Orkiszewski, J. Predicting two-phase pressure drops in vertical pipe. *J. Pet. Technol.* **1963**, *19*, 829–838. [CrossRef]
5. Oloruntoba, O.; Kara, F. Simplified Transient Two-Phase Model for Pipe Flow. *Int. J. Mod. Eng.* **2017**, *17*, 2.
6. Taitel, Y.; Barnea, D. Simplified transient simulation of two phase flow using quasi-equilibrium momentum balances. *Int. J. Multiph. Flow* **1997**, *23*, 493–501. [CrossRef]
7. Lockhart, R.W.; Martinelli, R.C. Proposed correlation of data for isothermal two-phase, two-component flow in pipes. *Chem. Eng. Prog.* **1949**, *45*, 39–48.
8. Idelchik, I.E. *Reference Book on Hydraulic Resistance*, 3rd ed.; Steinberg, M.O., Ed.; Mashinostroyeniye: Moscow, Russia, 1992.
9. Lykhin, P.A.; Usov, E.V.; Chukhno, V.I.; Kurmangaliyev, R.Z.; Ulyanov, V.N. Simulation of gas-liquid flows in a deviated well. *Autom. Telemech. Commun. Oil Ind.* **2019**, *10*, 22–27. [CrossRef]
10. Available online: https://www.tensorflow.org/ (accessed on 8 November 2022).
11. Usov, E.V.; Tokarev, D.N.; Ulyanov, V.N.; Vylegzhanin, R.I. Problems of integration of monitoring and simulation tasks in distributed hydraulic systems of an oil-gathering network. *Autom. Telemech. Commun. Oil Ind.* **2019**, *11*, 9–14. [CrossRef]

energies MDPI

Article

Testing the INSIM-FT Proxy Simulation Method

Mkhitar Ovsepian [1], Egor Lys [2], Alexander Cheremisin [1,*], Stanislav Frolov [2], Rustam Kurmangaliev [2], Eduard Usov [3], Vladimir Ulyanov [2], Dmitry Tailakov [3] and Nikita Kayurov [2]

1 Petroleum Engineering, Skolkovo Institute of Science and Technologies, 121205 Moscow, Russia
2 Novosibirsk R&D Center LLC, 630090 Novosibirsk, Russia
3 Department of Physics, Novosibirsk State University, 630090 Novosibirsk, Russia
* Correspondence: a.cheremisin2@skoltech.ru

Abstract: This paper describes testing of the INSIM-FT proxy simulation method (interwell-numerical-simulation model improved with front-tracking method) to assess the dependencies between production and injection wells, as well as to assess the forecast of oil/liquid production by wells depending on their operation parameters. The paper proposes the approach of taking into account the influence of various production enhancement operations. The method was tested on a synthetic hydrodynamic model and on a sector of a real field. The results show a good match between historical data and simulation results and indicate significant computational efficiency compared to classical reservoir simulators.

Keywords: proxy modeling; ES-MDA; history matching; reservoir optimization

Citation: Ovsepian, M.; Lys, E.; Cheremisin, A.; Frolov, S.; Kurmangaliev, R.; Usov, E.; Ulyanov, V.; Tailakov, D.; Kayurov, N. Testing the INSIM-FT Proxy Simulation Method. *Energies* **2023**, *16*, 1648. https://doi.org/10.3390/en16041648

Academic Editor: Reza Rezaee

Received: 26 December 2022
Revised: 19 January 2023
Accepted: 30 January 2023
Published: 7 February 2023

1. Introduction

Waterflooding of oil fields is used to displace oil with water from the formations and maintain the formation pressure at a given level by injecting water. Water injection through injection wells is the main and most common way to maintain reservoir pressure. In Russia over 90% of oil fields are developed using this technology [1].

Mature oil and gas fields are characterized by a rapid increase in the water cut of the produced liquid and a decrease in the oil flow rate. This stage is associated with complications of the production process. The sweep efficiency and the oil recovery factor can be improved by application of efficient waterflood systems, primarily, focal ones as well as by changing the filtration flows [2]. Moreover, with the increase in the share of hard-to-recover reserves under development, the task of increasing oil production rates becomes relevant.

With the development of the oil industry, geological and hydrodynamic simulation of hydrocarbon recovery has become one of the main tools for both predicting future performance of the reservoir in either the long-term or short-term period and selecting a reasonable development strategy for oil and gas fields. Moreover, it is an essential tool for the decision-making process in field development and petroleum production optimization where different operating conditions and actions are tested to increase productivity from existing wells or well patterns [3].

Full-scale hydrodynamic models are usually built on poor, averaged, and roughened geological models. Therefore, the insufficient accuracy of calculations performed on such models is associated with a high degree of uncertainty of initial information, especially of reservoir properties in the interwell space [4]. Moreover, to assess the efficiency of development methods, it is necessary to be able to carry out operational multivariate calculations. The use of three-dimensional hydrodynamic models in such cases is inefficient due to the following reasons [5]:

1. Long duration of the simulation;
2. Need to use large computational resources;

3. Complexity of modifying the model when making adjustments or changing various parameters.

The use of machine-learning algorithms or simplified models (proxy models) based on material balance methods and various analytical dependencies is practical, making it possible to account for the most important factors affecting the calculated properties. Such models are less time-consuming, adapt faster to actual data, and allow for quick calculations while maintaining the required accuracy [6]. The advantage of proxy models is historical data are used as the input, namely, production and injection data.

It is obvious that in the development of hydrocarbon fields, interconnection and mutual influence of all wells must be taken into account. Under reservoir pressure maintenance conditions, understanding the interaction between production and injection wells, assessing the degree of injection impact on oil production, and plotting the dependencies of recovery on water injection plays a key role in selecting the most optimal waterflood strategy. The use of proxy modeling makes it possible to solve the problems of accounting for the relationship and mutual influence of all wells.

Thus, a promising area of simulation of hydrocarbon recovery is the improvement of approaches to the creation and implementation of proxy models, as well as the development of methods for automated optimization of the oil field development system.

One of the important stages in proxy simulation is history matching, which solves the inverse task of hydrodynamics, i.e., the main reservoir properties (porosity, permeability, net pay thickness, heterogeneity, etc.) are adjusted to fit the actual field data with certain accuracy [7]. These algorithms are used in predicting well performance and in redistributing the volume of injected fluid between the injection wells with minimal costs and high payback.

Currently, one of the most common proxy modeling approaches are based on reduced-order models, data-driven models, physics-based models, and flow-network models. These models provide for solving problems of waterflooding optimization, identifying the efficiency of injection and production, assessing the interaction of wells, forecast fluid, and oil production in a specific period of time [8].

Reduced-order models are used to replace large reservoir models by reducing high-fidelity simulator models or by creating new reduced-order, data-driven models [9]. There are different model reduction techniques in the literature [10–14]. The main disadvantage of such models is that geologic data is required.

As opposed to reduced-order models, data-driven models require no prior knowledge of petrophysical properties or other specific geological information. Recurrent neural networks are common data-driven models used in the petroleum industry [15,16]. Because these are data-driven models, the accuracy of these models solely depends on the quality of the data used for training and can be affected by data noise. Moreover, these models are difficult to train and have pitfalls, such as overtraining, extrapolation, or lack of validation.

One example of the physics-based approach is the capacitance-resistance model (CRM), which uses production and injection rate data and bottomhole pressure to match the model against a particular reservoir [17–19]. CRM is analogous to electrical circuits where the compressibility and transmissibility, respectively, are analogous to capacitance and resistance. It is based on a model of nonlinear signal processing in which injection rates are treated as input signals and production rates are output signals [20]. The disadvantages of the CRM method are the productivity of each production well is considered constant throughout the entire history of production, which is not quite correct physically in the conditions of multiphase filtration. Moreover, the saturation distribution in this model is not calculated, and the oil production rate is adjusted indirectly through the displacement characteristics [17].

Flow-network models represent reservoir flow by a coupled-network model in which each pair of wells is connected with a one-dimensional (1D) finite-difference reservoir-simulation model. Each reservoir is defined by two parameters: absolute permeability and pore volume. The advantages of this approach are the construction of this model does not require a high-fidelity geological data and that it follows the physical laws of the realistic multiphase system. In this approach, relative permeabilities are assumed to be known. This model uses the true relative permeability curves, which is physically correct but requires a priori knowledge of the relative permeability curves [21].

This paper focuses on the interwell-numerical-simulation model (INSIM-FT), which is somewhat similar to the CRM and flow-network model. In the INSIM assumptions, a reservoir is viewed as a series of units connecting well pairs, but instead of discretizing those connections as in the flow-network model, INSIM only defines a pair of parameters for each connection. That is a significant reduction in the number of parameters compared to a set of 1D finite-difference reservoir models. The model uses the correct front-tracking procedure to calculate water saturation [22]. A more detailed description and methods used in the work are described in the next chapters.

This article presents software implementation of the method [4] and modification of the method to account for the influence of geological and technical measures, and on this basis, conducts a series of numerical experiments to assess the accuracy, convergence, and sensitivity of the method to parameterization of the historical performance of the hydrocarbon field.

2. Model Description

2.1. INSIM-FT: Theoretical Background

The interwell-numerical-simulation model (INSIM) is used as a tool to calculate the approximation of the well production rate under waterflooding [23]. In INSIM, the reservoir is characterized as a rough model consisting of several interwell control units (flow tubes), where each unit has two specific properties: conductivity, T_{ij}, and control pore volume, V_{pij} (Figure 1).

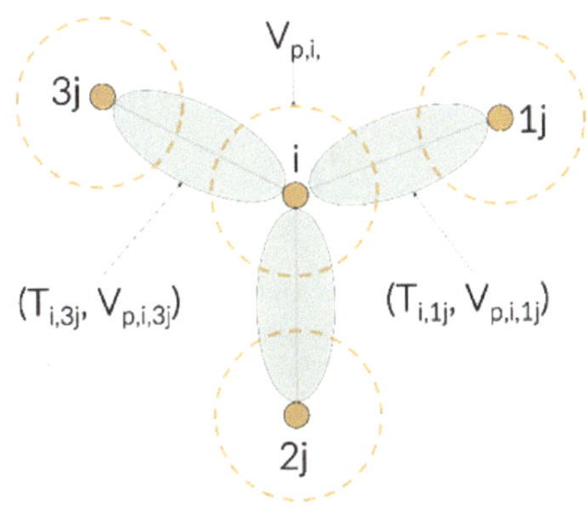

Figure 1. Schematic illustration of the pore channels network model in the INSIM model.

By solving the mass balance equation and front motion equations for each of the units, it is possible to obtain the velocities and saturation of interwell fluids for further prediction of the production rate. INSIM is used to adapt the model based on the available data to estimate parameters and to determine interwell correlation and geological characteristics. INSIM has the following advantages:

- The model parameters estimated based on historical data provide a relative characteristic of the reservoir properties between the wells. The model can handle changes in flow direction caused by well flow rate changes, including well shut-in or conversion of production wells to injection wells;
- INSIM is able to calculate the oil and water flow rate and the adapted water cut data;
- It can be used to optimize waterflooding but with considerably less computational effort.

In the described model, only a two-phase flow rate (liquid oil and water) is considered, and the conductivity, T_{ij}, is set as the average total permeability between the i-th and j-th wells. INSIM-FT solves the material balance equation for the j-th well (without accounting for capillary pressure and gravity):

$$\sum_{j=1}^{n_w} T_{i,j}(t)\cdot\left(P_j(t)-P_i(t)\right)-q_i(t)=c_i(t)\cdot V_{p,i}\cdot\frac{dP_i}{dt} \quad (1)$$

where n_w is the total number of wells; $q_{i(t)}$ is the flow rate of the i-th well at time, t, and is positive for injection and negative for production; $c_i(t)$ is full compressibility, $V_{p,i}$. If well j is not interlinked with well i, then $T_{i,j} = 0$ in the equation.

The physical meaning of the above-mentioned equation is that the change in the control pore volume due to compressibility is equal to the difference between the injection into the volume, ij, and the production from the volume, i. The equation is a combined pressure equation. Approximation of the equation by the implicit finite difference scheme used in reservoir simulation gives the following expression:

$$P_i^n-P_i^{n-1}=\frac{\Delta t_n}{c_{t,i}^n V_{p,i}^n}\left(-\left(\sum_{j=1}^{n_w} T_{i,j}^n\right)P_i^n+\sum_{j=1}^{n_w}\left(T_{i,j}^n P_j^n\right)+q_i^n\right) \quad (2)$$

for $i = 1, 2, \ldots, n_w$; where $\Delta t_n = t_n - t_{n-1}$ and $t_0 = 0$. Throughput (capacity) ($T_{i,j}^n$) well drainage pore volumes ($V_{p,i}^n$-th) and compressibility ($c_{t,i}^n$-th) may change with time.

In INSIM, similar to the pressure equation in reservoir simulation, non-linear terms are estimated at the previous time level, that is, terms that depend on the pressure and water saturation estimated at the moment in time, t_{n-1}, instead of t_n. Accordingly, in Equation (2) it is necessary to use:

$$T_{i,j}^{n-1}=\frac{\alpha k_{i,j}A_{i,j}\lambda_{t,i,j}^{n-1}}{L_{ij}}=T_{i,j}^0\frac{\lambda_{t,i,j}^{n-1}}{\lambda_{t,i,j}^0} \quad (3)$$

$$V_{p,i}^{n-1}=V_{p,i}^0\left(1+c_r\left(p_i^{n-1}-p_i^0\right)\right) \quad (4)$$

$$c_{t,i}^{n-1}=S_{o,i}^{n-1}c_o+S_{w,i}^{n-1}c_w+c_r \quad (5)$$

instead of $T_{i,j}^n$; $V_{p,i}^n$; $c_{t,i}^n$, respectively. Here, $S_{o,i}$ and $S_{w,i}$ are the corresponding volumes of oil and water saturation of the well i; c_o,c_w, and c_r, respectively, represent the compressibility of oil, water, and rock; $\lambda_{t,i,j}$ is the total mobility, which is calculated by upstream weighting. If $p_i^n > p_i^{n-1}$, then $\lambda_{t,i,j}$ is replaced by the overall fluidity of the well i, $\lambda_{t,i}$. It is assumed that the viscosity of oil and water is constant. Weighing with an upward flow means $p_i^n < p_i^{n-1}$:

$$\lambda_{t,i,j}^{n-1}=\lambda_{t,i}^{n-1}=\frac{k_{ro}\left(S_{w,i}^{n-1}\right)}{\mu_o}+\frac{k_{rw}\left(S_{w,i}^{n-1}\right)}{\mu_w} \quad (6)$$

Otherwise:

$$\lambda_{t,i,j}^{n-1}=\lambda_{t,i}^{n-1}=\frac{k_{ro}\left(S_{w,i}^{n-1}\right)}{\mu_o}+\frac{k_{rw}\left(S_{w,i}^{n-1}\right)}{\mu_w} \quad (7)$$

where μ_o and μ_w are the viscosities of oil and water, respectively. As $V_{p,i,j}^{n-1}$ changes, we use the following equation:

$$V_{p,i,j}^{n-1}=V_{p,i,j}^0\left[1+c_r\left(0.5\left(p_i^{n-1}+p_j^{n-1}\right)-p_i^0\right)\right] \quad (8)$$

We designate:

$$E_i^n=\frac{\Delta t_n}{c_{t,i}^{n-1}V_{p,i}^{n-1}} \quad (9)$$

$$G_i^n = E_i^n \sum_{j=1}^{n_w} T_{i,j}^{n-1} \tag{10}$$

$$M_i^n = \frac{\Delta t_n q_i^{n-1}}{c_{t,i}^{n-1} V_{p,i}^{n-1}} \tag{11}$$

Then, at each step t, the following system of equations is solved:

$$\begin{pmatrix} G_1^n+1 & -E_1^n T_{1,2}^n & \cdots & -E_1^n T_{1,n_w}^n \\ -E_2^n T_{2,n_w}^n & G_2^n+1 & \cdots & -E_1^n T_{1,n_w}^n \\ \vdots & \vdots & \vdots & \vdots \\ -E_{n_w}^n T_{n_w,1}^n & -E_1^n T_{1,n_w}^n & \cdots & G_{n_w}^n+1 \end{pmatrix} \begin{pmatrix} p_1^n \\ p_2^n \\ \vdots \\ p_{n_w}^n \end{pmatrix} = \begin{pmatrix} p_1^{n-1} \\ p_2^{n-1} \\ \vdots \\ p_{n_w}^{n-1} \end{pmatrix} + \begin{pmatrix} M_1^n \\ M_2^n \\ \vdots \\ M_{n_w}^n \end{pmatrix} \tag{12}$$

where M_i^n is positive for the injection well and negative for the production well.

The system of Equation (12) can be solved for p_i^{n-1} on the assumption the necessary saturations at t_{n-1} are known. After finding p_i^n, it is necessary to solve the 'fractional flow' equations to obtain the water saturation at the moment in time, t_n.

In Formula (12), the number of pressure equations is equal to the number of downhole assemblies, while in traditional reservoir modeling, the number of pressure equations is equal to the number of grid blocks. Therefore, the INSIM method requires considerably less computational resources than 3D direct simulation of dynamic processes in the reservoir. After the pressures are calculated, the saturation values are estimated. Further, it is easy to calculate the flow rate between the pairs of wells, which can be used for further diagnostics of downhole dynamics. Since the number of downhole assemblies is limited, the methodology may only give an approximate distribution of the formation fluid. However, the results of the calculations to date show the saturation or pressure distributions calculated by the INSIM model compared with historical data are sufficient to obtain a successful production forecast.

The INSIM model was improved to become the INSIM-FT model—an interwell simulation model with fluid front tracking. The key difference is the calculation of the water cut distribution between the wells. Unlike INSIM, INSIM-FT also includes parameters that define relative permeability power curves.

The distribution of water saturation between two wells is described by the following form of the one-dimensional Buckley–Leverett equation [24]:

$$\frac{\partial S_w(x,t)}{\partial t} + \frac{q_{t,i,j}(t)}{\phi_{i,j} A_{i,j}} \frac{\partial f_w(x,t)}{\partial x} = 0,\ 0 \le x \le L_{i,j}, t^{n-1} \le t \le t^n \tag{13}$$

$$f(S_w) = \frac{\frac{k_{rw}}{\mu_w}}{\frac{k_{rw}}{\mu_w} + \frac{k_{ro}}{\mu_o}} \tag{14}$$

The curves of relative phase permeabilities are set by analytical relations. In this paper, we used the Corey correlation:

$$k_{rw}(S_{wn}) = \alpha \cdot S_{wn}^{n_w} \tag{15}$$

$$k_{ro}(S_{wn}) = S_{wn}^{n_o} \tag{16}$$

$$S_{wn} = \frac{S_w - S_{iw}}{1 - S_{iw} - S_{or}} \tag{17}$$

To solve Equation (13), the front-tracking method is used, which is an adaptation of the method [25]. This method is stable and has a relatively low variance.

The general idea of the solution is as follows:

- Each pair of wells is represented as a quasi-one-dimensional model. This model is divided into cells in which the water cut value is set. The water cut function is approximated as a sequence of constant values (Figure 2).

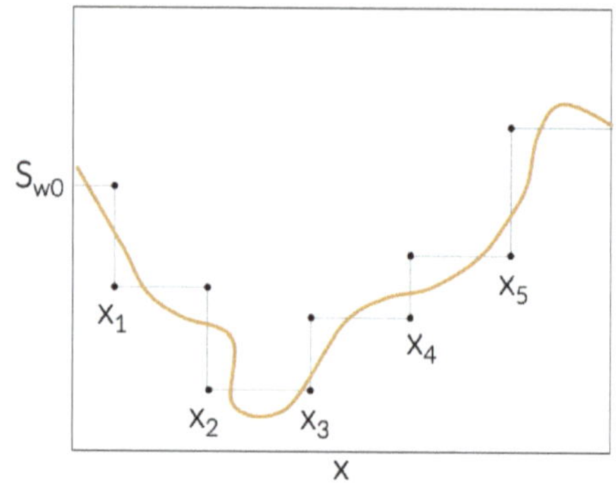

Figure 2. Partitioning of the quasi-one-dimensional model.

- In each interval, the task of calculating the velocity of the shock wave cluster is solved.

Determination of the right and left boundary values of the water cut, as well as the accuracy of the water cut determination is $S_{wl}, S_{wr}, \delta S_w$

Calculation of the Rankine–Hugoniot condition is $\sigma_{trial} = \frac{f_w(S_{wl}) - f_w(S_{wr})}{S_{wl} - S_{wr}}$

If $f'_w(S_{wl}) > \sigma_{trial} > f'_w(S_{wr})$, then the result is asingle shock wave.

If $f'_w(S_{wl}) < f'_w(S_{wr})$ and $f''_w(S_{wl}) f''_w(S_{wr}) > 0$, then the result is a depression wave. It is simulated as a sequence of shock waves connecting values S_{wl}, S_{wr}.

In other cases, a sequence of shock waves to approximate a depression wave connecting values S_{wl}, S_{w*} that follows the shock wave connecting the values S_{w*}, S_{wr} is used.

The value calculation, S_{w*}, is based on the equation $f'_w(S_{w*}) = \frac{f_w(S_{w*}) - f_w(S_{wr})}{S_{w*} - S_{wr}}$.

For the calculation of the first intersections, in the x-t diagram, the calculated velocities are straight lines with different slopes. As seen on Figure 3, the first intersections appear at the boundaries of the primary areas.

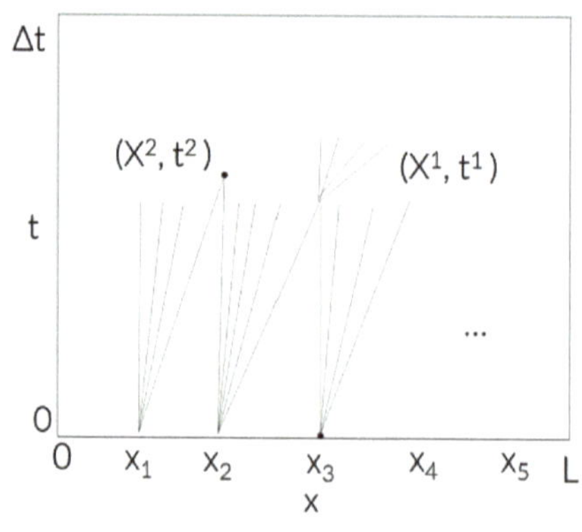

Figure 3. Intersections of shock wave clusters.

- The point of intersection becomes a new boundary, and it is necessary to calculate new velocities of the shock waves in new intervals. Then, the process continues until the time value, Δt, is reached.
- Go to the calculation in the next time interval.

2.2. Model Adaptation

As the model is adapted to history, interwell conductivity and interwell volumes are selected, as well as parameters that determine the power functions of relative phase permeabilities used in the calculation of saturation and other parameters required for solving the previously mentioned equations. The complete list of parameters used in this paper is presented in Appendix A. The INSIM-FT model itself consists of at least 14 adaptation parameters.

Adaptation parameters are distribution functions having the following parameters: minimum value, maximum value, mean value, distribution type, and value of dispersion parameter, σ (Figure 4). During the adaptation period, the parameters variation intervals may change. For correct calibration, it is necessary to determine the intervals of these parameters' variations with high accuracy.

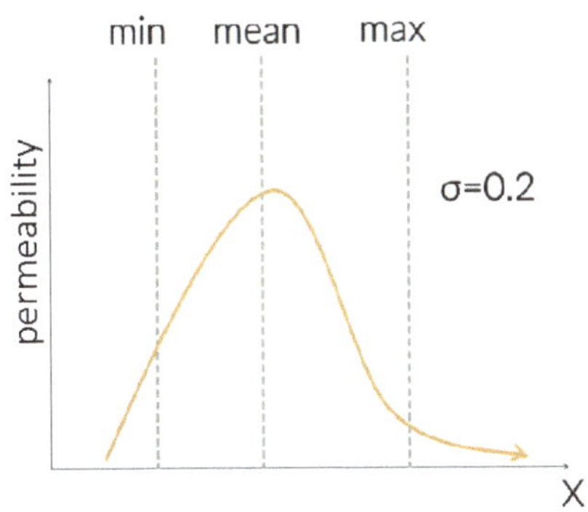

Figure 4. Schematic distribution of the parameter in the model.

In adaptation, we use the ES-MDA method [26] (smoothing assembly with multiple data assimilation). The task of adapting historical data in matrix form looks like:

$$O(m) = \frac{1}{2}[g(m) - d_{obs}]^T C_D^{-1}[g(m) - d_{obs}] \tag{18}$$

where C_D is the covariance matrix that adjusts the sensitivity of the model. The general view of the matrix is as follows:

$$C_D = \begin{pmatrix} cov\left(\left[d^1_{calc} - d^1_{obs}\right]\left[d^1_{calc} - d^1_{obs}\right]\right) & \cdots & 0 \\ \vdots & \ddots & \vdots \\ 0 & \cdots & cov\left(\left[d^n_{calc} - d^n_{obs}\right]\left[d^n_{calc} - d^n_{obs}\right]\right) \end{pmatrix} \tag{19}$$

Volumetric flow rates and water cut values are used as calculated and observed parameters.

In the task of historical data adaptation, it is necessary to minimize the discrepancy between the observed and calculated data. The calculated data are calculated as a direct hydrodynamic task based on the model parameters (capacity, drainage pores volumes, compressibility, etc.). At the same time, it is necessary to periodically check and adjust the model parameters for compliance with historical information. The model parameters are restored from the inverse optimization problem, reducing the error with the observed data, according to the following formula:

$$m_j^u = m_j^p + \left[C_{MD}(C_{DD} + \alpha_i C_D)^{-1}\right]\left(d_{uc} - d_j^p\right) \tag{20}$$

$$\sum_{i=1}^{N_i} \frac{1}{\alpha_i} = 1,\ i = 1 \ldots N \tag{21}$$

The time interval of using the historical and calculated data is called the data assimilation step N.

2.3. Forecast of Oil and Liquid Flow Rate

The flow rate forecast is based on the Dupuis formula:

$$Q_r = \frac{2\pi kh(P_k - P_c)}{\mu ln\frac{R_k}{R_c}} \quad (22)$$

where k is the permeability coefficient; h—reservoir capacity; P_k and P_c—pressure on the feed loop and in the well, respectively; R_k and R_c—the radii of the feed loop and the well, respectively; μ—the fluid viscosity; Q_r—well flow rate. Assume all parameters are constants except for the pressure difference. Then, the Dupuis formula can be rewritten as follows:

$$\eta = \frac{Q}{\Delta P} \quad (23)$$

where ΔP is pressure differential, Q—the flow rate, η—productivity index.

Then, the task of forecasting the flow rate is reduced to finding the productivity index based on the known values of the flow rate and pressure difference. To solve the problem, it is proposed to use the linear regression method.

Knowing the pressure, it is possible to predict the flow rate. Figure 5 shows the forecast results: the blue curve is the initial flow rate; yellow is the initial pressure; the red section of the pressure curve is the time interval, the points of which were used to plot the regression; the green curve is the forecasted flow rate for the entire interval.

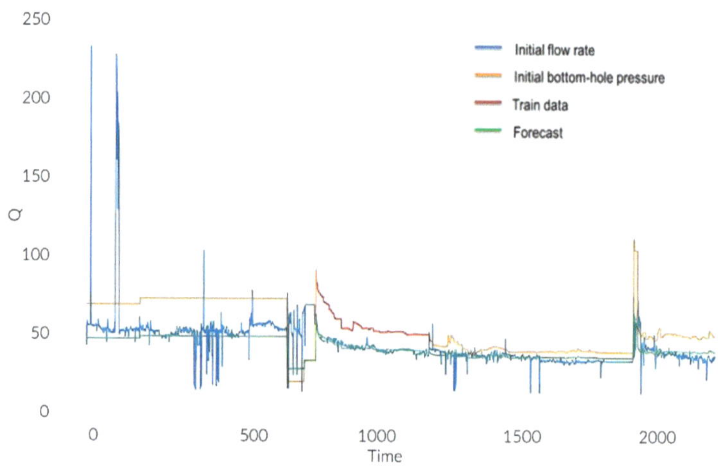

Figure 5. Forecast results.

3. Testing the Method on Synthetic Test Data

The main purpose of this section is to verify the correctness of the model adaptation on synthetic data. In this paper, a hydrodynamic model in tNavigator format was used. It represents a sector with geomechanical and filtration-capacitive properties. To assess the quality of adaptation on synthetic data, four production and three injection wells were added to the model, and calculations were performed with a period of 25 years.

Figure 6 shows the distribution of water saturation in the field, as well as the layout of production and injection wells.

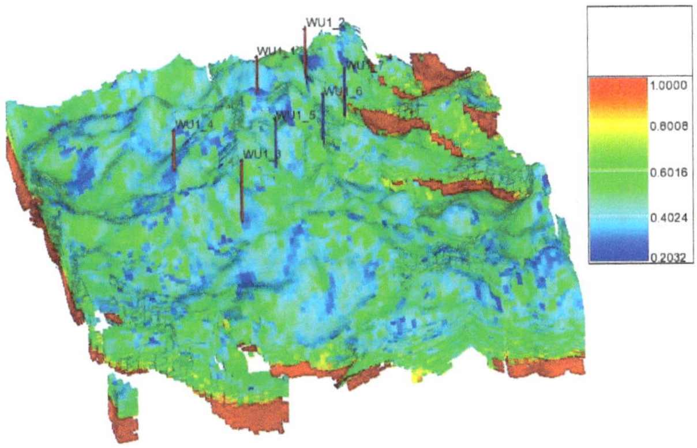

Figure 6. Hydrodynamic model of the field (red vertical lines—production wells, blue—injection wells).

Below are the results of the developed INSIM-FT prototype (Figures 7 and 8).

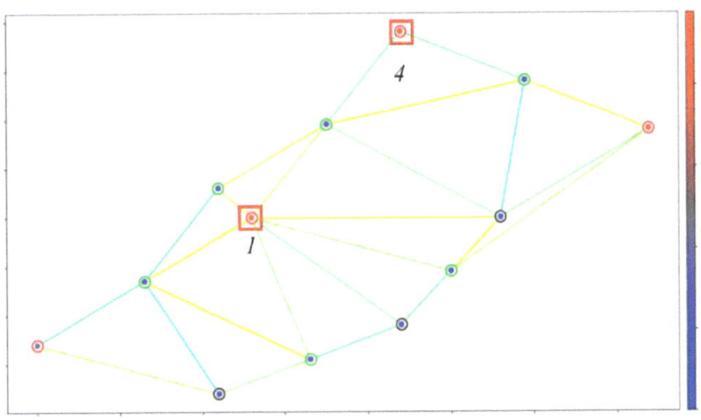

Figure 7. Displaying the field map in the INSIM-FT prototype. The studied production wells 1 and 4 are marked with a red rectangle.

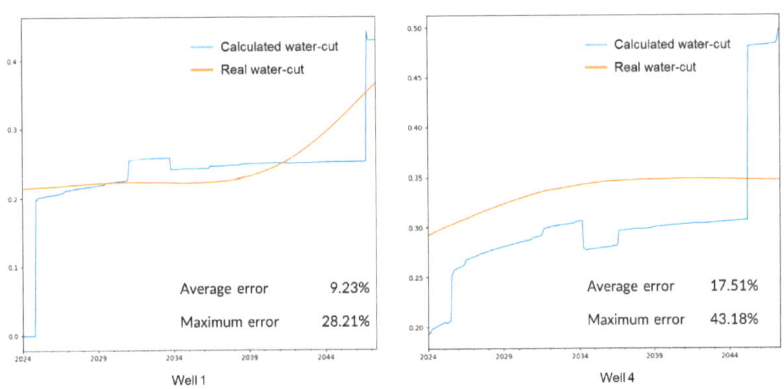

Figure 8. Approbation of the model on synthetic data.

The fluid flow rate and water cut were adapted using synthetic data for the time period from 2020 to 2045. Based on the results of the adaptation, it can be concluded the fluid flow rate is reproduced well, and the water cut is reproduced satisfactorily.

To improve the results of water cut adaptation at well-4, sensitivity analysis was carried out to the number of assimilation steps (Figure 9).

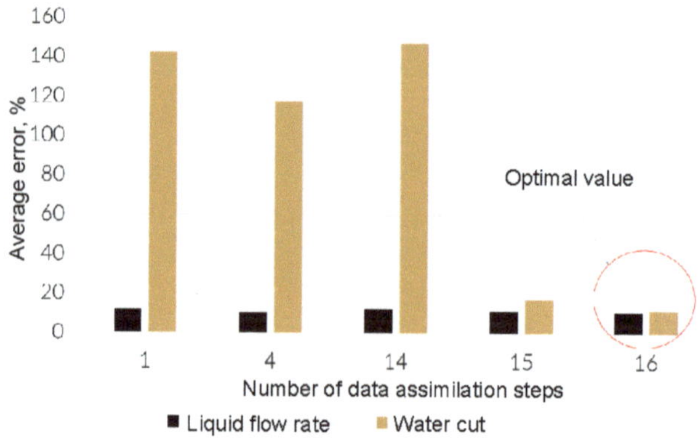

Figure 9. Analysis of the forecast sensitivity to the number of assimilation steps.

It was determined that the number of assimilation steps significantly affects the final result. Increasing the number of steps to 16 reduced the average error from 17.51% to 11.65%.

It should also be noted that using the approach described in the article [27], the adaptation rate directly depends on the number of assimilation steps. The increase in speed is due to the decrease in the number of data assimilation steps since it is multiplied by the sum of the reciprocal values of the MDA coefficients. Thus, it is necessary to select such a number of assimilation steps that will maintain the quality of adaptation at the required level but with minimal adaptation time. These algorithms require a more detailed study.

4. Methods of Accounting for Geological and Technical Measures

4.1. Hydraulic Fracturing

For the INSIM-FT model, we proposed the following solution: according to the source [27], during the injection of water into the injection well, as well as during the formation and propagation of cracks, hydraulic auto-fracturing takes place due to several factors:

- Equality of pressure in the injection well and horizontal stress, which can cause the formation of a developed pattern of cracks causing a high rate of pressure drop of the injection fluid;
- Change in effective horizontal stress due to temperature change at the bottomhole;
- Change in effective horizontal stresses due to changes in pore pressure;
- Stress contrast in the rock between different geological layers (clays and sandstones);
- Difference between vertical and horizontal stresses (the stress anisotropy coefficient is higher than 1.15).

It is assumed that during the hydraulic fracturing procedure or the occurrence of auto-fracturing, there is a sharp increase in the absolute permeability parameter k, i, j of the flow tube. However, over time, the parameter should return to the "shelf" (due to natural contamination of the bottom hole zone). That is, it is necessary to multiply it by a time-dependent monotonically decreasing function. The possible form of the additional function will be exponential, and the following parameters may be present in the exponent index:

- Borehole radius across the bit;
- Effective well radius;
- Reservoir permeability;
- Fluid viscosity;
- Geometric characteristics of the producing reservoir;

The following type of additional function is proposed as the first approximation:

$$g(s,t) = a \cdot e^{-stb} + c \tag{24}$$

where a, b, c are adaptable parameters that depend on the type of geological and technical measures performed, s is skin factor, which is expressed by dependence $s = \ln\left(\frac{R}{r}\right)$, r is the wellbore radius across the bit, and R is the effective radius of the wellbore.

An illustration of this dependence is shown in Figure 10.

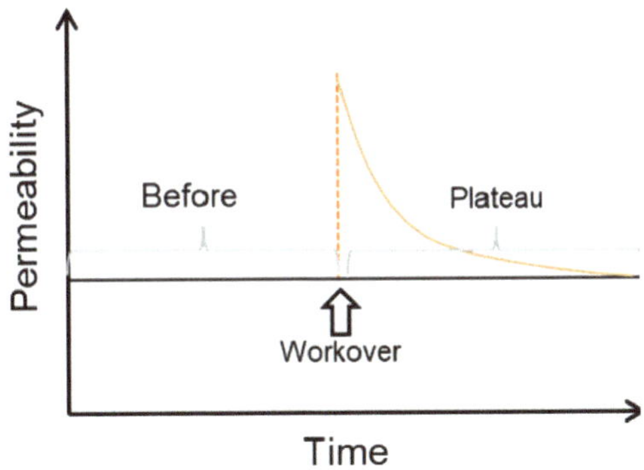

Figure 10. Effect of geological and technical measures on the hydraulic conductivity of the flow tube.

Thus, it is necessary to create a correspondence table "type of geological and technical measures"—adaptable coefficients a, b, c.

In the case of hydraulic fracturing and hydraulic auto-fracturing, it is possible to consider the formula for the skin factor, which takes into account the dimensionless conductivity of the fracture, C_{FD}, the value depending on the difference in permeability of the proppant and the formation:

$$C_{FD} = \frac{k_f \cdot w}{k \cdot x_f} \tag{25}$$

where k_f is proppant permeability (mD), k is formation permeability (mD), w is fracture width (m), x_f is the crack half-length (m).

The skin factor is calculated according to the following algorithm.

- Option 1:

$$C_{FD} = \frac{k_f \cdot w}{k \cdot x_f} \tag{26}$$

$$r_{ef} = r_w + \frac{x_f}{2\left(1 + \left(\frac{C_{FD}}{1.7}\right)^{-1.01}\right)} \tag{27}$$

$$S = -ln\left(\frac{r_{ef}}{r_w}\right) \tag{28}$$

- Option 2:

$$C_{FD} = \frac{k_f \cdot w}{k \cdot x_f} \tag{29}$$

$$u = ln(C_{FD}) \tag{30}$$

$$f = \frac{1.65 - 0.328 \cdot u + 0.116 \cdot u^2}{1 + 0.18 \cdot u + 0.064 \cdot u^2 + 0.005 \cdot u^3} \quad (31)$$

$$S = f + ln\left(\frac{r_w}{x_f}\right) \quad (32)$$

4.2. Hydraulic Auto-Fracturing

Assume that auto-fracturing occurs when the injection pressure exceeds the sum of the formation pressure and horizontal stress of the formation:

$$P_{hfr} > P_{form} + \sigma_{hor} \quad (33)$$

The horizontal rock stress can be expressed through the vertical stress as follows:

$$\sigma_{hor} = \left(\frac{1}{3} \div \frac{1}{2}\right)\sigma_{vert} \quad (34)$$

In turn, vertical intensity is related to geostatic pressure:

$$P_{geost} = P_{form} + \sigma_{vert} \quad (35)$$

Therefore:

$$\sigma_{vert} = P_{geost} - P_{form} \quad (36)$$

Geostatic pressure is calculated according to the following relationship:

$$P_{geost} = \rho_{prock} g H \quad (37)$$

The correction introduced during hydraulic auto-fracturing has the same structure as in the case of hydraulic fracturing, but the point in time when this multiplicative additive is introduced is determined by the condition recorded above.

The next step was to regulate the multiplicative additive: setting the exponent degree, amplitude, and shift. As the first approximation, it is proposed to determine these parameters manually for each type of geological and technical measures. In the INSIM-FT code, these values were used as averages, after which they are edited and adapted.

5. Testing the Model on Real Data While Accounting for the Geological and Technical Measures

Initially, calculations in the INSIM-FT model are carried out without accounting for geological and technical measures. As mentioned in the previous paragraph, it is necessary to introduce a multiplicative exponential additive, which, in fact, shows the pollution of the bottomhole zone and decrease in water saturation. Figure 11 shows an illustration of the real field sector. The wells for which the adjustment for geological and technical measures have been introduced are shown in red.

Let us compare the influence of geological and technical measures on the calculation of water cut. Let us consider two cases: neutral and positive effect of the amendment. The results are presented below (Figures 12 and 13).

Despite a correction introduced on well 1 (hydraulic fracturing was performed on it), the algorithm could not significantly reduce the error. This is probably due to the mutual influence of various types of geological and technical measures. It may be necessary to introduce additional corrections. Adaptation of water cut accounting for the geological and technical measures at well 2 yielded a positive effect. It is clear to the naked eye (Figure 13) that the calculated curve significantly approached the actual one and repeated its shape, accounting for the features.

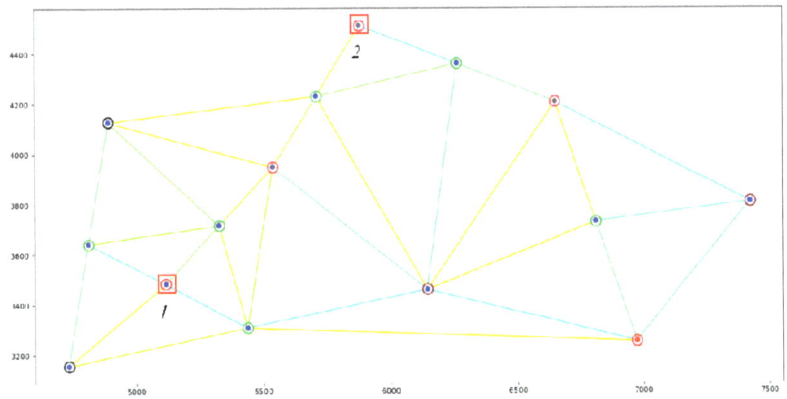

Figure 11. Illustration of the field sector where the geological and technical measures effect was tested. Production wells 1 and 2 are in red rectangles.

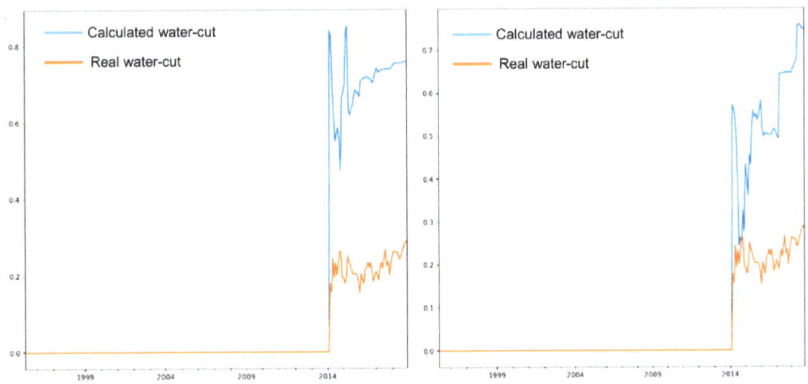

Figure 12. Comparison of water saturation adaptation for well 1 (on the **left**—without accounting for geological and technical measures, on the **right**—accounting for geological and technical measures).

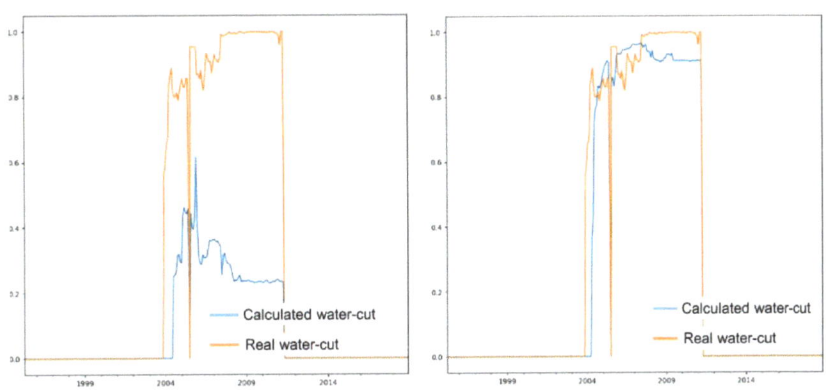

Figure 13. Comparison of water saturation adaptation for well 2 (on the **left**—without accounting for geological and technical measures, on the **right**—accounting for geological and technical measures).

Calculations have demonstrated the efficiency of this approach. Further, a software realization of the algorithm is required, which will take into account all types of geological and technical measures according to the input parameters. It is also necessary to create a library of multiplicative correction parameters for various types of geological and technical measures or to select coefficients in automatic mode.

6. Discussion

Our results showed the introduction of the geological and technical measures into the approach described in the paper allows in several cases to significantly increase the design parameters approximation accuracy. We associate the partial low sensitivity of the solution to corrections for geological and technical measures with insufficient knowledge and elaboration of the parameterization of certain types of geological and technical measures. It is necessary to carry out a complex analysis of the mutual influence of various types of geological and technical measures and analysis of certain geological and technical measure zones of influence.

The choice of assembly and assimilation parameters, such as: expectation, dispersion, number of iterations of the ES-MDA algorithm, and assembly size, plays a key role for fine settings the model and predictive calculations. The search for balance between computational costs and procedure for approximating historical indicators makes it possible to find sufficiently accurate and computationally efficient approaches to assessing the operating parameters of the hydrocarbon field.

The results of the experiments along with the high computational efficiency of the method showed good potential for parallel implementations since the operations of multithreaded computations on the CPU were used in the paper; it is reasonable to further implement the method for GpGPU platforms using low-level languages, for example, C ++. It is also necessary to carefully consider the choice of a framework for the implementation of calculations on graphic cards; the largest modern vendors of graphic equipment created software solutions only for their "hardware" components without the support of competitors' hardware and software.

The authors of the paper are convinced that the presented approach has rich opportunities for "tuning", for example, building proxy connections of wells based on more advanced algorithms, Voronoi diagrams, PEBI grids, etc., using optimized solvers or more efficiently solving differential equations and SLAEs computational approaches.

7. Conclusions

The method of the field hydrodynamic model based on the data-driven (setting to historical data) approach and the proxy model of fluid migration dynamic processes in hydrocarbon reservoir was implemented. Based on efficient averaging of reservoir hydrodynamic parameters for a finite set of production and injection wells, the method shows significant computational efficiency compared to classical reservoir simulators (Eclipse, tNavigator). The proposed additional adaptation of the model to the geological and technical measures parameters has shown its efficiency in predicting the operating conditions of the hydrocarbon field.

The use of multithreaded calculation technologies made it possible to significantly reduce the simulation time for a particular case; in the future, it is planned to transfer the algorithm to the GpGPU platform, which, according to preliminary experiments, will increase the performance by at least an order of magnitude.

For the correct settings of water saturation and reservoir pressures, it is necessary to fine-tune the model parameters: selection of the optimal interval of parameter variation, selection of the parameter distribution function, selection of the initial geological and technical measures adaptation coefficients while considering the interference of different types of geological and technical measures.

Author Contributions: M.O.—investigation, writing—original draft preparation; E.L.—investigation, writing—original draft preparation; A.C.—writing—original draft preparation, writing—review and editing, conceptualization; S.F.— writing—review and editing; R.K.—formal analysis, writing—original draft preparation; E.U.—methodology, resources; V.U.—resources, supervision; D.T.—conceptualization, writing—review and editing; N.K.—writing—review and editing. All authors have read and agreed to the published version of the manuscript.

Funding: The research was carried out in accordance with the FSUS-2022-0020 project.

Data Availability Statement: Not applicable.

Conflicts of Interest: The authors declare no conflict of interest.

Appendix A

Table A1. List of adaptation parameters.

Parameter	Symbol	Units
Initial reservoir pressure	P	Pa
Oil viscosity	μ_o	Pa·s
Water viscosity	μ_w	Pa·s
Rock compressibility	c_r	Pa^{-1}
Water compressibility	c_w	Pa^{-1}
Oil compressibility	c_o	Pa^{-1}
Residual water saturation	S_{iw}	fraction
Residual oil saturation	S_{ro}	fraction
Parameter in Corey correlation	n_w	
Parameter in Corey correlation	n_o	
Parameter in Corey correlation	α	
Porosity of flow-tube	$\varphi_{i,j}$	fraction
Permeability of flow-tube	$k_{i,j}$	mD
Cross-section area of flow-tube	$A_{i,j}$	m^2

References

1. Yaskin, S.A.; Mukhametshin, V.V.; Andreev, V.E.; Dubinskiy, G.S. Forecasting the Parameters of Non-Stationary Water Flooding of Oil Deposits. *IOP Conf. Ser. Earth Environ. Sci.* **2018**, *194*, 062037. [CrossRef]
2. Mukhametshin, V.S.; Tyncherov, K.T.; Rakhimov, N. Geological and Technological Substantiation of Waterflooding Systems in Deposits with Hard-to-Recover Reserves. *IOP Conf. Ser. Mater. Sci. Eng.* **2021**, *1064*, 012068. [CrossRef]
3. Aziz, K.; Settari, A. *Petroleum Reservoir Simulation*; Applied Science Publishers Ltd.: London, UK, 1979. [CrossRef]
4. Khalimov, E.M. Detailed Geological Models and Three-Dimensional Simulation. *Neft. Geologiâ. Teor. I Pract.* **2012**, *7*, 17–27.
5. Chudin, Y.S.; Chumakov, G.N.; Fedorov, I.A.; Pyatakova, O.A.; Gilev, D.V.; Petrukhin, V.A.; Tarasov, G.V.; Arkhipov, Y.A. Using Proxy Models of Gas Deposits for Production Optimization. *Gazov. Promishlennost* **2020**, *4*, 30–36.
6. Bahrami, P.; Sahari, M.F.; James, L.A. A Review of Proxy Modeling Highlighting Applications for Reservoir Engineering. *Energies* **2022**, *15*, 5247. [CrossRef]
7. Nehoroshkova, A.A.; Danko, M.Y.; Zavyalov, A.S.; Elisheva, A.O. Critical Analysis of the INSIM-FT Proxy Modeling Method (Interwell Numerical Simulation Front Tracking Models) at Synthetic Models and Real Field. *Neft. Gas. Novacii* **2019**, *12*, 49–55.
8. Udy, J.; Hansen, B.; Maddux, S.; Petersen, D.; Heilner, S.; Stevens, K.; Lignell, D.; Hedengren, J. Review of Field Development Optimization of Waterflooding, EOR, and Well Placement Focusing on History Matching and Optimization Algorithms. *Processes* **2017**, *5*, 34. [CrossRef]
9. Jafroodi, N.; Zhang, D. New method for reservoir characterization and optimization using CRM–EnOpt approach. *J. Pet. Sci. Eng.* **2011**, *77*, 155–171. [CrossRef]
10. Jansen, J.D.; Durlofsky, L.J. Use of Reduced-Order Models in Well Control Optimization. *Optim. Eng.* **2017**, *18*, 105–132. [CrossRef]
11. Cardoso, M.A.; Durlofsky, L.J.; Sarma, P. Development and Application of Reduced-Order Modeling Procedures for Subsurface Flow Simulation. *Int. J. Numer. Meth. Eng.* **2009**, *77*, 1322–1350. [CrossRef]

12. Gildin, E.; Klie, H.; Rodriguez, A.; Wheeler, M.F.; Bishop, R.H. Projection-Based Approximation Methods for the Optimal Control of Smart Oil Fields. In Proceedings of the ECMOR X—10th European Conference on the Mathematics of Oil Recovery, Amsterdam, The Netherlands, 4–7 September 2006; European Association of Geoscientists & Engineers: Amsterdam, The Netherlands, 2006. [CrossRef]
13. Gildin, E.; Klie, H.; Rodriguez, A.; Wheeler, M.F.; Bishop, R.H. Development of Low-Order Controllers for High-Order Reservoir Models and Smart Wells. In *All Days*; SPE: San Antonio, TX, USA, 2006; p. SPE-102214-MS. [CrossRef]
14. Liu, F.; Mendel, J.M. Forecasting Injector-Producer Relationships From Production and Injection Rates Using an Extended Kalman Filter. In *All Days*; SPE: Anaheim, CA, USA, 2007; p. SPE-110520-MS. [CrossRef]
15. Jansen, F.E.; Kelkar, M.G. Non-Stationary Estimation of Reservoir Properties Using Production Data. In *All Days*; SPE: San Antonio, TX, USA, 1997; p. SPE-38729-MS. [CrossRef]
16. Bailer-Jones, C.A.L.; MacKay, D.J.C.; Withers, P.J. A Recurrent Neural Network for Modelling Dynamical Systems. *Netw. Comput. Neural Syst.* **1998**, *9*, 531–547. [CrossRef]
17. Sayarpour, M. Development and Application of Capacitance-Resistive Models to Water/CO_2 Floods. Ph.D. Thesis, The University of Texas at Austin, Austin, TX, USA, August 2008. [CrossRef]
18. Yousef, A.A.; Gentil, P.; Jensen, J.L.; Lake, L.W. A Capacitance Model to Infer Interwell Connectivity from Production and Injection Rate Fluctuations. In *All Days*; SPE: Dallas, TX, USA, 2005; p. SPE-95322-MS. [CrossRef]
19. Cao, F.; Luo, H.; Lake, L.W. Oil-Rate Forecast by Inferring Fractional-Flow Models From Field Data With Koval Method Combined With the Capacitance/Resistance Model. *SPE Reserv. Eval. Eng.* **2015**, *18*, 534–553. [CrossRef]
20. Sayarpour, M.; Zuluaga, E.; Kabir, C.S.; Lake, L.W. The Use of Capacitance–Resistance Models for Rapid Estimation of Waterflood Performance and Optimization. *J. Pet. Sci. Eng.* **2009**, *69*, 227–238. [CrossRef]
21. Lerlertpakdee, P.; Jafarpour, B.; Gildin, E. Efficient Production Optimization With Flow-Network Models. *SPE J.* **2014**, *19*, 1083–1095. [CrossRef]
22. Guo, Z.; Reynolds, A.C.; Zhao, H. Waterflooding Optimization with the INSIM-FT Data-Driven Model. *Comput. Geosci.* **2018**, *22*, 745–761. [CrossRef]
23. Guo, Z.; Reynolds, A.C.; Zhao, H. A Physics-Based Data-Driven Model for History Matching, Prediction, and Characterization of Waterflooding Performance. *SPE J.* **2017**, *23*, 367–395. [CrossRef]
24. Buckley, S.E.; Leverett, M.C. Mechanism of Fluid Displacement in Sands. *Trans. AIME* **1942**, *146*, 107–116. [CrossRef]
25. Godunov, S.K.; Bohachevsky, I. Finite Difference Method for Numerical Computation of Discontinuous Solutions of the Equations of Fluid Dynamics. *Mat. Sb.* **1959**, *47*, 271–306.
26. Emerick, A.A.; Reynolds, A.C. History-Matching Production and Seismic Data in a Real Field Case Using the Ensemble Smoother With Multiple Data Assimilation. In *All Days*; SPE: The Woodlands, TX, USA, 2013; p. SPE-163675-MS. [CrossRef]
27. Baykov, V.A.; Burakov, I.M.; Latypov, I.D.; Yakovlev, A.; Asmandiyarov, R.N. Waterflood Induced Hydraulic Fracturing Control under Reservoir Pressure Maintenance Conditions on RN-Yuganskneftegas Oilfields. *Oil Ind.* **2012**, *11*, 30–33.

energies

MDPI

Article

A Physics-Informed Neural Network Approach for Surrogating a Numerical Simulation of Fractured Horizontal Well Production Prediction

Taiyu Jin [1], Yang Xia [1,*] and Haolin Jiang [2,*]

[1] National Key Laboratory of Petroleum Resources and Engineering, China University of Petroleum (Beijing), Beijing 102249, China

[2] School of Integrated Circuit Science and Engineering, University of Electronic Science and Technology of China (UESTC), Chengdu 611731, China

* Correspondence: xiayang@vip.126.com (Y.X.); orrinnj@gmail.com (H.J.)

Abstract: With the popularity of deep learning (DL), more and more studies are focusing on replacing time-consuming numerical simulations with efficient surrogate models to predict the production of multi-stage fractured horizontal wells. Previous studies on constructing surrogate models for the prediction of the production of fractured horizontal wells often relied on directly applying existing deep learning architectures without incorporating physical constraints into the model. When dealing with the large number of variables necessary for characterizing the properties of fractures, the input variables of proxy models are often oversimplified; meanwhile, lots of physical information is lost. Consequently, predictions are sometimes physically inconsistent with the underlying principles of the domain. In this study, by modifying the traditional Seq2Seq (LSTM–LSTM) deep learning architecture, a physics-informed encoder–decoder (PIED) architecture was developed to surrogate the numerical simulation codes for predicting the production of horizontal wells with unequal-length intersecting hydraulic fractures on a 2D plane. The encoder is a LSTM network, and the decoder consists of LSTM and fully connected layers. The attention algorithm is also applied in the Seq2Seq architecture. The PIED model's encoder is capable of extracting the physical information related to fractures. And the attention module effectively passes on the most relevant physical information related to production to the decoder during the training process. By modifying Seq2Seq architecture, the decoder of the PIED incorporates the intermediate input, which is the constant production time, along with the extracted physical information to predict production values. The PIED model excels in extracting sufficient physical information from high-dimensional inputs while ensuring the integrity of the production time information. By considering the physical constraints, the model predicts production values with improved accuracy and generalization capabilities. In addition, a multi-layer perceptron (MLP) which is broadly used as a proxy model; a regular Seq2Seq model (LSTM–Attention–LSTM); and the PIED were compared via a case study, and their MAE values were shown to be 241.76, 184.07, 168.81, respectively. Therefore, the proposed model has higher accuracy and better generalization ability. In the case study, a comparative experiment was conducted by comparing LSTM–MLP (with an MAE of 221.50) and LSTM–LSTM to demonstrate that using LSTM as the decoder structure is better for predicting production series. Moreover, in the task of predicting production sequences, LSTM outperforms MLP. The Seq2Seq architecture demonstrated excellent performance in this problem, and it achieved a 48.4% reduction in MSE compared to MLP. Meanwhile, the time cost for build datasets was considered, and the proposed model was found to be capable of training in a small dataset (e.g., in the case study, 3 days were used to generate 450 samples for training.); thus, the proposed model has a certain degree of practicality.

Keywords: fractured well; production; proxy model; physics-informed neural network; deep learning; machine learning

Citation: Jin, T.; Xia, Y.; Jiang, H. A Physics-Informed Neural Network Approach for Surrogating a Numerical Simulation of Fractured Horizontal Well Production Prediction. *Energies* **2023**, *16*, 7948. https://doi.org/10.3390/en16247948

Academic Editor: Reza Rezaee

Received: 23 June 2023
Revised: 19 July 2023
Accepted: 2 August 2023
Published: 7 December 2023

1. Introduction

1.1. Traditional Numerical Simulation of Fractured Horizontal Well Production

Hydraulic fracturing is a treatment widely applied to enhance oil and gas recovery, especially in low-permeability reservoirs. The artificial fractures created during hydraulic fracturing substantially improve the seepage resistance of fluid flow in a reservoir. The success of hydraulic fracturing in increasing production depends largely on the geometry of the hydraulic fractures, so before fracturing is implemented, numerical simulation methods are usually required to predict the geometry of the hydraulic fractures and the fracture conductivity, as well as other parameters, and then calculate important indices such as well production to evaluate or optimize the fracturing schemes.

Currently, the main method for fractured horizontal well production prediction is the numerical simulation method. Numerical simulation methods usually include the finite difference method (FDM), the finite element method (FEM), the boundary element method (BEM), and the discrete element method (DEM). Due to the multiple parameters involved (e.g., fracture permeability, fracture inclination, fracture length, etc.), it is time-consuming to perform high-accuracy production prediction using numerical simulation methods. To optimize hydraulic fracturing parameters and thus maximize production, traditional trial-and-error methods tend to be very time-consuming and expensive. Therefore, there is a need to develop some proxy models for numerical simulation methods for hydraulic fracturing production prediction. Proxy models are approximations of numerical simulation models, usually based on machine learning algorithms, which map the inputs and outputs of the numerical simulation model. Proxy models for fractured well production prediction can be efficiently performed to significantly reduce computational costs and speed up the optimization process in large-scale parameter space search and optimization.

1.2. Review of Research

In recent years, with the popularity of machine learning, many studies have been performed on data-driven machine learning models to surrogate physically driven numerical simulations for fractured well production. These studies simplify the input data or output data of numerical simulators and then use them to build datasets and train machine learning models. Gaussian process regression (GPR), convolutional neural networks (CNNs), and support vector machines (SVMs) have been used as proxy models for numerical simulation models of fractured horizontal wells with parallel, equal-length hydraulic fractures. These proxy models have been used to predict gas production and to obtain optimal fracture half-lengths and horizontal well lengths (Wang et al., 2021) [1]. Multi-layer perceptron (MLP) can be used to construct surrogate models to predict the production of gas wells with parallel, equal-length hydraulic fractures on a two-dimensional plane (Wang et al., 2021) [2]. The tree-based ensemble method was used as a proxy model of a numerical simulator of 2D discrete fracture networks. This proxy model can predict 5-year oil cumulative production (Xue et al., 2019) [3]. The regular neural network (NN), random forests (RFs), adaptive boosting (AdaBoost), and support vector machines (SVMs) were used to build several proxy models of numerical simulation codes which can predict the production and the net present value (NPV) of a horizontal well with parallel and unequal-length hydraulic fractures (Li et al., 2022) [4]. Transformer, a Seq2Seq DL model, was used to predict the time series of production data based on the integration of geology and engineering time series production data (Wang et al., 2023) [5].

However, the input variables of the proxy models mentioned in these studies are often oversimplified. For example, among the previously mentioned proxy models, CNN, GPR, and SVM (Wang et al., 2021) [1] require that the length of the hydraulically fractured fractures in the dataset take only a fixed number of four values; the training set of the MLP network (Wang et al., 2021) [2] contains only four variables, and these four features can only take a fixed number of four values; the tree-based ensemble model (Xue et al., 2019) [3] requires that the dataset can only have these four variables, which can only take a fixed number of three values; and the input data of the transformer (Wang et al. [5]) need to be preprocessed by PCA (principal component analysis) to decrease its dimension, but the variables generated by PCA often pose challenges in terms of interpretation. Although reducing the dimensions of the input data or simplifying the input data can reduce the complexity of machine learning models and make training easier, this also decreases the performance of the proxy models.

Incorporating prior physical knowledge into neural networks empowers the proxy model to be easily trained in the presence of numerous input variables, endowing it with enhanced accuracy and generalization capabilities [6–8]. Physics-informed neural networks have the advantages of increased learning efficiency [9], improved generalization [10], enhanced interpretation, and improved robustness in solving physical problems [11], and this is supported by relevant studies.

Therefore, the machine learning architecture of the proxy models for fractured well production prediction needs to be adapted to the specific engineering problem by introducing physical information, rather than simply applying machine learning algorithms. For example, a physics-constrained Bi-GRU-DHNN method was used to predict the production of hydraulic fractured wells (Li et al., 2022) [12]. The model consists of two parts. The first part consists of MLP, CNN, and Bi-GRU. Different network structures in the first part can extract different kinds of features (i.e., ecological properties, logging curves, well information, fracturing parameters, and field operations) and generate a preliminary prediction value as well as two feature vectors of physical information. In the second part, a MLP can establish a nonlinear mapping between the physical constraints and the production target. The model can input very high-dimension features which include six numerical variables and ten sequence variables. A physics-informed MLP (Li et al., 2021) [4] was developed as a proxy model of numerical simulation codes which could predict the production of a horizontal well with parallel- and unequal-length hydraulic fractures. The proxy model was introduced to prior physical knowledge by pre-training it with a large number of variables of all the fractures and then training the model with the small number of input variables that remained. The authors of this study demonstrate in the paper that the MLP integrated with prior physical knowledge has a stronger generalization ability and higher accuracy than a normal MLP. Guevara et al. [13] used generalized additive models (GAMs) to predict the cumulative production of fractured wells. By using GAMs, they could modify the smooth functions of each input variable to align the relationship between these input variables and the output with prior knowledge, thereby improving the accuracy of the results. Qu et al. [14] proposed a physics-informed MLP which could combine information regarding the static parameters of a fractured well with the input variables to predict fracture parameters. Yang et al. [15] developed a hybrid NN called GRU–MLP. They used MLP to combine the static parameters of hydraulic fractures with the production predicted by GRU, so that the production predicted by MLP was physically constrained and had a lower bias. Wang et al. [16] proposed a theoretical guide to CNN to surrogate the numerical simulation of two-phase flow in fields by using the coupled and discretized equations as loss functions in the training process. Cornelio et al. [17] used an auto encoder (AE) to extract the physical information of the hydraulic fractures and the reservoir properties. Then, they combined the physical information outputted from AE with the production predicted by a numerical simulator into an MLP to predict the production of fractured wells. Razak et al. [18] leveraged neural networks to rectify the errors in production calculated by parsing formulas, thus constraining the model's predictive values with physical knowledge. These physics-informed models can exhibit higher accuracy as well as learning significantly more features than ML models without physical constraints.

1.3. A Novel Proxy Model

In these previous studies, the numerical simulations surrogated by neural networks did not consider the influence of the geometry and permeability of each hydraulic fracture on production. Therefore, their proxy models cannot be used for the optimization of the geometric parameters of hydraulic fractures. In this study, we developed a physics-informed neural network proxy model for a numerical simulator that can calculate production based on the geometric parameters and permeability of each hydraulic fracture. The proxy model, called a physics-informed encoder–decoder (PIED), was developed for the numerical simulation of the production prediction of horizontal wells with non-equal-length intersecting hydraulic fractures. The proxy model has a modified Seq2Seq architecture and consists of an encoder and a decoder.

To develop this proxy model, the Seq2Seq structure was modified in this study so that the decoder could have the intermediate input of a static time series.

The proxy model utilizes the encoder structure and the attention module to integrate physical constraints into the decoder structure for predicting production, and it is robust enough to handle fracture intersections, so that it can predict production accurately in complex physical scenarios.

2. Data Preparation from Numerical Simulations

Before performing the machine learning model, the input and output data should be carefully prepared, and a large dataset is required to mine the relationship between the inputs and the outputs. In this section, the characterization of a fracture, a coupled reservoir flow model, and the numerical simulation method are proposed to prepare the required inputs and outputs.

2.1. Fracture Characterization

In the discrete fracture model, the fracture is modelled as a zero-thickness plate or a slit [19]. The schematic diagram of the fractured horizontal well is shown in Figure 1. The morphology of each hydraulic fracture is different, varying in two geometric parameters: the half-length and dip angle, shown in Figure 1. During the fracturing process, the amount of sand injected into each fracture is different, so the permeability of each hydraulic fracture is different. Therefore, three parameters were selected to characterize the hydraulic fracture and serve the inputs into the network. The dip angle, half-length and permeability of the fracture were evaluated by a uniform variation of (60–120)°, (5–75) m, (1–1000) D.

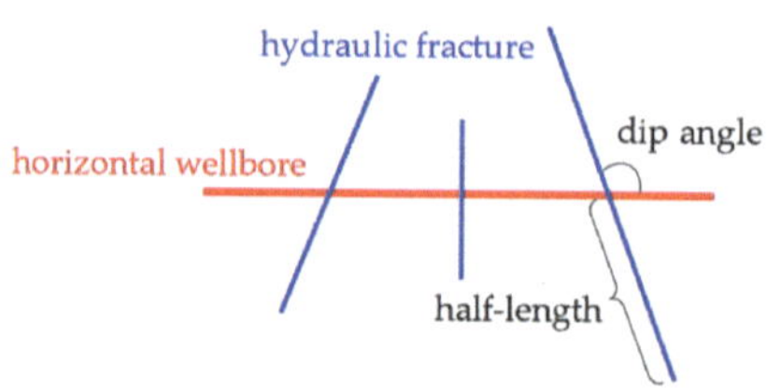

Figure 1. A schematic diagram of the fractured horizontal well.

2.2. Reservoir Flow Model

The basic governing equations for fluid flow in a fractured reservoir are [20]:

$$\frac{\partial(\rho\phi_m)}{\partial t} + \nabla \cdot (\rho v_m) = 0 \quad (1)$$

$$\frac{\partial\left(\rho\phi_f\right)}{\partial t} + \nabla \cdot \left(\rho v_f\right) = 0 \quad (2)$$

where ρ, ϕ, and v are the pore fluid density, porosity, and Darcy velocity, respectively. The subscripts m and f represent the physical parameters of the matrix and fracture, respectively. The Darcy velocity is given by:

$$v_{m,f} = -\frac{\kappa_{m,f}}{\mu} \nabla p_{m,f} \quad (3)$$

where κ and μ are the reservoir permeability and fluid viscosity. The left sides of Equations (1) and (2) can be rewritten as [21]

$$\frac{\partial\left(\rho\phi_{m,f}\right)}{\partial t} = c_{tm,f}\rho_0\frac{\partial p_{m,f}}{\partial t} \tag{4}$$

with the total compressibility $c_{tm,f}$ considered to be a lumped, constant parameter which contains both the compressibility of the solid and the fluid. After substituting Equations (3) and (4) into Equations (1) and (2), the governing equations for coupled flow in a fractured reservoir can be obtained:

$$\frac{\partial p_{m,f}}{\partial t} = \gamma_{m,f}\nabla^2 p_{m,f} \tag{5}$$

where $\gamma_{m,f} = \kappa_{m,f}/\mu c_{tm,f}$. The total compressibility can be related to the rock modulus and fluid modulus:

$$c_{tm,f} = \frac{\phi}{K_{ff}} + \frac{1}{K_{m,f}} \tag{6}$$

where K_{ff} is the fluid modulus and $K_{m,f}$ is the modulus of the matrix or fracture. When the reservoir is considered fully saturated, the total mass of the drained fluid, i.e., the cumulative production, can be expressed by:

$$M_q = \int(\rho_0\phi_{m0} - \rho\phi_m)d\Omega \tag{7}$$

where the subscript 0 represents the initial value. Using the state equations, the cumulative mass production can be calculated by:

$$M_q = \rho_0 c_{tm}\int(p_{m0} - p_m)d\Omega \tag{8}$$

2.3. Weak Formulations

The Gauss divergence theorem is used for discrete formulations [22]:

$$\int_\Omega \mathrm{div}F d\Omega = \int_\Gamma F\cdot n_\Gamma d\Gamma - \sum_{i=1}^{N}\int_{f_{\Gamma_i}} F^{\pm}\cdot n_{f_{\Gamma_i}} d\Gamma \tag{9}$$

where F represents a discontinuous function, and n_Γ and n_{f_Γ} are the outward normal directions of the external boundary and fracture surface, respectively. The jump generated by the discontinuous function at the fracture surface is denoted by $F^{\pm}$, $F^{\pm} = F^+ - F^-$. F^+ and F^- represent the function values on the positive and negative sides of the fracture, respectively. To derive the weak form of the matrix flow equation, we multiply the strong form by a test function δp and construct an equivalent integral form. Then, we utilize the Gauss divergence theorem to derive the weak form.

$$\int_\Omega \delta p\frac{\partial p_m}{\partial t}d\Omega + \sum_{i=1}^{N}\int_{f_{\Gamma_i}}\gamma_m\delta p\left(\nabla p_m\cdot n_{f_{\Gamma_i}}\right)^{\pm}d\Gamma + \int_\Omega \gamma_m\nabla(\delta p)\cdot\nabla p_m d\Omega + \int_{\Gamma_q}\frac{\delta p\bar{q}}{c_{tm}}d\Gamma = 0 \tag{10}$$

$d\Omega$ represents the reservoir domain, while Γ_q is a constant flow rate boundary. On Γ_q, the flow rate $q = \bar{q}$. The third term on the left side of the equation is the discontinuous term. By combining with Darcy's law, we can obtain the following:

$$\int_{f_{\Gamma_i}}\gamma_m\delta p\left(\nabla p_m\cdot n_{f_{\Gamma_i}}\right)^{\pm}d\Gamma = \int_{f_{\Gamma_i}}\frac{\delta p}{c_{tm}}\left(\frac{\kappa_m}{\mu}\nabla p_m\right)^{\pm}\cdot n_{f_{\Gamma_i}}d\Gamma = -\int_{f_{\Gamma_i}}\frac{1}{c_{tm}}\delta p(n_{f_{\Gamma_i}}\cdot q_f)d\Gamma \tag{11}$$

where q_f is used to characterize the fluid exchange between fractures and the matrix. The exchange term q_f can be obtained by solving the flow governing equation within the fractures. To derive the weak form of the fracture flow equation, we multiply the strong form by a test function δp and construct an equivalent integral form. Then, we utilize the Gauss divergence theorem to derive the weak form.

$$\int_{\Omega_{fi}} \delta p \frac{\partial p_f}{\partial t} d\Omega + \int_{\Omega_{fi}} \gamma_f \nabla(\delta p) \cdot \nabla p_f d\Omega - \int_{f\Gamma_i} \gamma_f \delta p \nabla p_f \cdot n_{f\Gamma_i} d\Gamma = 0 \tag{12}$$

Considering that the fracture's width is significantly smaller than its length, the pressure variation in the width direction can be disregarded. By combining this with Darcy's law, the expression for the exchange term q_f can be derived as follows:

$$\int_{f\Gamma_i} \delta p (n_{f\Gamma_i} \cdot q_f) d\Gamma = -c_{tf} (\int_{f\Gamma_i} d_{f_i} \delta p \frac{\partial p_f}{\partial t} d\Gamma + \int_{f\Gamma_i} d_{f_i} \gamma_f \frac{\partial \delta p}{\partial x} \frac{\partial p_f}{\partial x} d\Gamma) \tag{13}$$

By substituting the exchange term between the matrix and fractures (Equation (13)) into the weak form of matrix flow (Equation (10)), we can obtain the weak form of the matrix fracture coupled flow equation.

$$\begin{aligned} &\int_\Omega \frac{\delta p}{\gamma_m} \frac{\partial p_m}{\partial t} d\Omega + \int_\Omega \nabla(\delta p) \nabla p_m d\Omega + \sum_{i=1}^{n} \int_{f\Gamma i} \left(\frac{d^i_f k^i_f}{\kappa_m} \frac{\partial(\delta p)}{\partial x^i} \frac{\partial p_m}{\partial x^i} + \frac{d^i_f c_{tf}}{\gamma_m c_{tm}} \delta p \frac{\partial p_m}{\partial t} \right) d\Gamma \\ &+ \int_{\Gamma_q} \frac{\delta p \bar{q}}{\gamma_m c_{tm}} d\Gamma = 0 \end{aligned} \tag{14}$$

In Equation (14), a line integral along the fracture segment is performed (the third item on the left of the equal sign); this action is related to the fracture length and fracture dip.

2.4. XFEM Solver

The discrete fracture model is commonly solved using the finite element method (FEM), finite volume method (FVM), boundary element method (BEM) and extended finite element method (XFEM). In the FEM and FVM, unstructured grids are used to discrete the reservoir domain, and nodal points are set on the fracture segment to maintain the flow consistency between the porous matrix and the fractures [23,24]. These methods can discretize a geometrically complex reservoir when using mesh points optimally and exactly represent the flow in the matrix and fractures; however, the fracture segment is restricted to the inter-element boundaries and the problem of mesh dependency limits their application [25,26]. The BEM can reduce the dimension of the problem, but the flow distribution around the fracture junction always results in computational difficulties [27]. To eliminate the requirement for the mesh topology to conform with that of the fracture network, an XFEM was developed that relies on adding enrichment functions to enhance the conventional finite element interpolation [28]. In the XFEM, a structured mesh can be used for explicit simulation on complex fracture networks; thus, the XFEM can overcome the constraint of mesh conformity between the reservoir and the fractures [26,29]. During the data preparation from numerical simulations, the fracture geometry changes in each simulation example, it becomes convenient to run multiple simulations with different fracture configurations without meshing the domain each time. Therefore, XFEM was used in this study for the numerical simulation.

In our XFEM solver, a structured grid is employed to discretize the matrix. The fractures in the computational domain are cut off by the element edges and the fracture segments can be classified as belonging to two groups: (I) fracture segments that completely cut an element and (II) fracture segments that partly cut an element, as shown in Figure 2.

Based on the partition of unit method, additional terms are added to the standard finite element approximation of the pressure:

$$p = \sum_{i \in N_s} N_i p_i + \Psi(x) = \sum_{i \in N_s} N_i p_i + \sum_{j \in N_{enr}} \varphi_j(x) \widetilde{p}_j \tag{15}$$

where N_i is the standard FEM shape function; p_i represents the standard pressure degrees of freedom for nodes N_S. $\tilde{p}_j$ is the added enriched degrees of freedom for nodes N_{enr}. $\varphi_j(x)$ is the superimposed enrichment function:

$$\varphi_j(x) = N_j(x)\left[\Phi^n(x) - \Phi^n(x_j)\right], n = I, II \tag{16}$$

where the corresponding enrichment functions for the two types of fracture segments are:

$$\Phi^I(x) = \sum_i |f(x)|N_i(x) - \left|\sum_i f(x)N_i(x)\right| \tag{17}$$

$$\Phi^{II}(x) = \sqrt{r}cos\frac{\theta}{2} \tag{18}$$

where f is the level set function, and r and θ are the local polar coordinates near the fracture tip. By substituting Equation (15) into the weak form (14), we obtain the discretized governing equation:

$$\begin{bmatrix} K^{pp} & K^{p\tilde{p}} \\ K^{\tilde{p}p} & K^{\tilde{p}\tilde{p}} \end{bmatrix}\begin{pmatrix} p \\ \tilde{p} \end{pmatrix} + \begin{bmatrix} M^{pp} & M^{p\tilde{p}} \\ M^{\tilde{p}p} & M^{\tilde{p}\tilde{p}} \end{bmatrix}\begin{pmatrix} \dot{p} \\ \dot{\tilde{p}} \end{pmatrix} = \begin{pmatrix} F \\ \tilde{F} \end{pmatrix} \tag{19}$$

where

$$K_{IJ}^{mn} = \int_{\Omega^e} B_I^m (B_J^n)^T d\Omega + \sum_{i=1}^{N} \int_{\Gamma_i} \frac{d_f^i k_f^i}{\kappa_m} B_I^{mi}(B_J^{ni})^T d\Gamma \,(m, n = p, \tilde{p})$$

$$\begin{cases} M_{IJ}^{pp} = \int_{\Omega^e} \frac{1}{\gamma_m} N_I N_J^T d\Omega + \sum_{i=1}^{N} \int_{\Gamma_i} \frac{d_f^i c_{tf}}{\gamma_m c_{tm}} N_I N_J^T d\Gamma \\ M_{IJ}^{p\tilde{p}} = (M_{IJ}^{\tilde{p}p})^T = \int_{\Omega^e} \frac{1}{\gamma_m} N_I \varphi_J^T d\Omega + \sum_{i=1}^{N} \int_{\Gamma_i} \frac{d_f^i c_{tf}}{\gamma_m c_{tm}} N_I \varphi_J^T d\Gamma \\ M_{IJ}^{\tilde{p}\tilde{p}} = \int_{\Omega^e} \frac{1}{\gamma_m} \varphi_I \varphi_J^T d\Omega + \sum_{i=1}^{N} \int_{\Gamma_i} \frac{d_f^i c_{tf}}{\gamma_m c_{tm}} \varphi_I \varphi_J^T d\Gamma \end{cases}$$

$$\begin{cases} F_I = \int_{\Gamma_q} N_I \bar{q} d\Gamma \\ \tilde{F}_I = \int_{\Gamma_q} \varphi_I \bar{q} d\Gamma \end{cases}$$

where $B = \nabla N$.

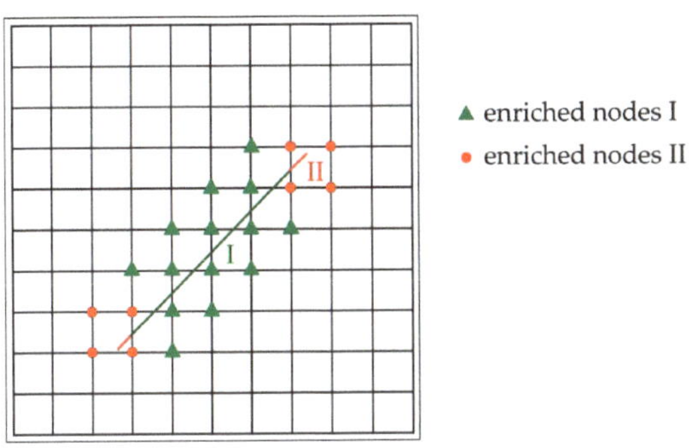

Figure 2. Two types of fracture segments and the enriched nodes (the green segment completely crosses the element, and the red segment partly crosses the element).

2.5. Inputs and Outputs

Ten hydraulic fractures are simulated. During the simulations, the fracture geometry and permeability change while the matrix and fluid properties are fixed. For all the simulation cases, the fixed parameters are listed in Table 1, and the inputs and outputs of a simulation example are listed in Table 2. Presentations of mesh discretization, the calculated pressure distributions at different times, and the cumulative production curve are shown in Figure 3.

Table 1. Fixed parameters across all simulations.

Parameters	Value	Unit
Matrix permeability	5×10^{-18}	m^2
Fluid modulus	3×10^{9}	Pa
Matrix modulus	20×10^{9}	Pa
Fracture modulus	1.05×10^{9}	Pa
Fluid viscosity	3×10^{-3}	Pa·s
Initial fluid density	1000	kg/m^3
Porosity	0.15	/
Wellbore pressure	15	MPa
Reservoir pressure	20	MPa
Reservoir thickness	1	m

Table 2. Inputs and outputs of a simulation example.

Inputs			Outputs			
Fracture Half-Length (m)	**Fracture Dip**	**Fracture Permeability (m^2)**	**Cumulative Production (kg)**			
62.03	69.45°	6.65×10^{-10}	Day 1	1164.32	Day 120	7984.58
68.41	118.23°	1.25×10^{-10}	Day 3	1905.52	Day 150	8567.65
13.89	117.43°	3.22×10^{-12}	Day 6	2591.89	Day 180	9074.56
68.94	89.12°	1.37×10^{-12}	Day 9	3086.58	Day 210	9527.55
49.27	108.01°	1.78×10^{-9}	Day 15	3814.71	Day 240	9939.82
11.83	68.51°	7.59×10^{-10}	Day 21	4359.01	Day 270	10,319.74
24.49	85.30°	6.56×10^{-10}	Day 30	4990.75	Day 300	10,672.95
43.28	114.94°	3.54×10^{-11}	Day 45	5775.01	Day 330	11,003.43
72.03	107.53°	8.60×10^{-12}	Day 60	6374.02	Day 360	11,314.08
72.54	117.56°	1.28×10^{-12}	Day 90	7283.71		

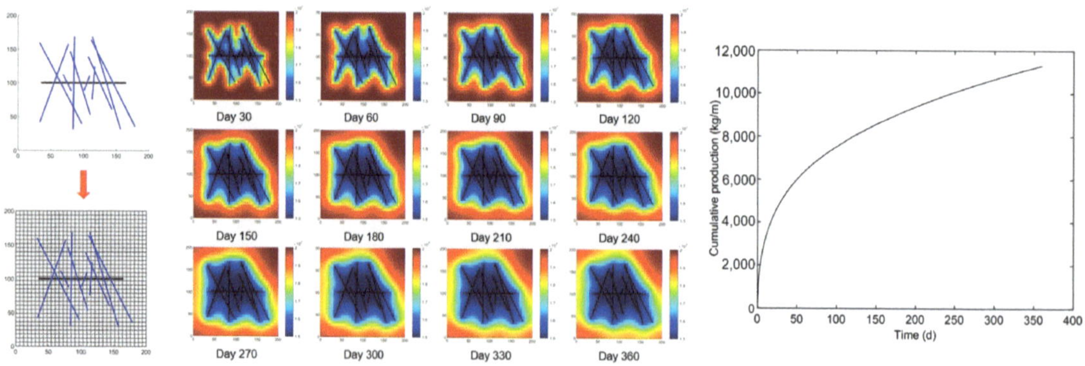

Figure 3. Presentations of mesh discretization, the calculated pressure distributions at different times, and the cumulative production curve.

3. AI Methodology

In this study, three deep learning algorithms were used to construct three proxy models. These three models included MLP, regular Seq2Seq (LSTM–Attention–LSTM), and the PIED proposed in this study. MLP is widely used to construct proxy models, but MLP is not the structure specific to the sequence problem. Meanwhile, LSTM–Attention–LSTM is a common sequence-to-sequence model but does not have physical constraints integrated. In this section, the functions, structures, and principles of these three models are introduced in detail. And the training workflow of PIED is also elaborated.

3.1. MLP

Many studies have utilized multi-layer perceptron (MLP) to predict the production of hydraulic fracturing wells (Wang et al., 2019; Luo et al., 2018; Panja et al., 2018; Li et al., 2022) [4,30–32]. In this study, we also adopted MLP as one of our alternative proxy models. MLP consists of an input layer, hidden layers, and an output layer. Each hidden layer is a fully connected layer. A fully connected layer, also known as a dense layer, is a type of layer commonly used in neural networks. It connects every neuron from the previous layer to every neuron in the current layer, creating a fully connected network structure. The fully connected layers of the MLP are illustrated in Figure 4.

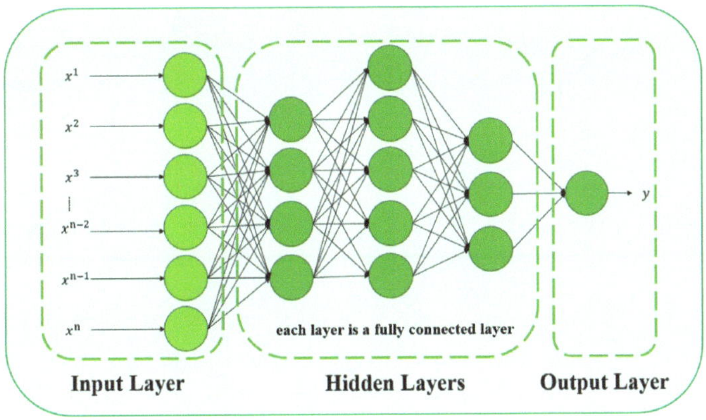

Figure 4. Illustration of the MLP.

The operation of the MLP is based on the forward propagation process. Input data are passed through the input layer and transmitted to the hidden layers. Upon receiving the input data, each neuron in the hidden layers performs a weighted sum and undergoes a non-linear transformation through an activation function. The transformed results are then propagated to the neurons in the next layer. The neurons in the output layer generate the final prediction.

3.2. Seq2Seq Architecture

3.2.1. LSTM

Long short-term memory (LSTM) is a variant of recurrent neural networks (RNNs) that addresses the issue of long-term dependencies in traditional RNNs (Hochreiter et al., 1997) [33]. It provides powerful modeling capabilities for various sequence tasks. LSTM utilizes memory cells to store and update information and employs gate mechanisms to control the flow of information, effectively capturing long-term dependencies in sequences. LSTM consists of input gates, forget gates, output gates, and candidate memory cells. The input gate determines the relevance of the current input, the forget gate determines what is retained or forgotten from the previous memory, and the output gate controls the output of the memory. The candidate memory cell computes the candidate value for the current time step. The structure of a LSTM cell is depicted in Figure 5. The computation process of LSTM is as follows.

Input sequence: $X = [x_1, x_2, \ldots, x_m]$
Initial hidden state: h_1
Initial cell state: c_1
For each time step t,
$y_t,\ h_t,\ c_t = LSTM(x_t,\ h_{t-1},\ c_{r-1})$
Final hidden state at the last time step: h_n
Final cell state at the last time step: c_n

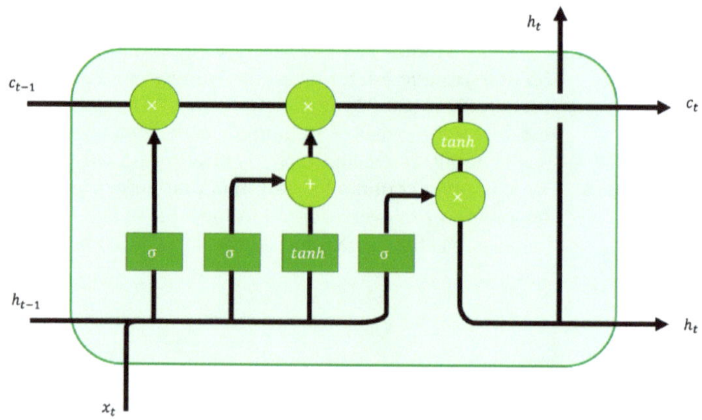

Figure 5. Illustration of the computational process of an LSTM cell at a single time step.

The LSTM (long short-term memory) uses the following formulas:

Input gate: $i_t = \sigma(W_i[h_{t-1}, x_t] + b_i)$
Forget gate: $f_t = \sigma(W_f[h_{t-1}, x_t] + b_f)$
Output gate: $o_t = \sigma(W_o[h_{t-1}, x_t] + b_o)$
Candidate memory cell: $\tilde{c}_t = tanh\ (W_c[h_{t-1}, x_t] + b_c)$
Memory cell: $c_t = f_t \odot c_{t-1} + i_t \odot \tilde{c}_t$
Hidden state (output): $h_t = o_t \odot tanh(c_t)$

In the above formulas:
σ denotes the sigmoid activation function.
tanh denotes the hyperbolic tangent activation function.
The symbol $\odot$ represents element-wise multiplication (also known as the Hadamard product).
W_i, W_f, W_o, W_c are weight matrices associated with input, forget, output, and candidate memory cells, respectively.

b_i, b_f, b_o, b_c are bias vectors associated with input, forget, output, and candidate cell states, respectively.

3.2.2. Regular Encoder–Decoder Architecture and Attention Mechanism

(1) Regular Encoder–Decoder architecture

The Seq2Seq (Sequence-to-Sequence) structure is a neural network model used to convert one sequence into another (Sutskever et al., 2014) [34]. It is a special type of encoder–decoder architecture. The encoder is used to map the input sequence into a fixed-dimension vector, often referred to as a context vector. The decoder then uses the context vector to generate the target sequence. When the Seq2Seq structure is used for regression problems, it can learn the non-linear mapping relationship between the input and output sequences.

In general, the encoder utilizes LSTM as the encoder model. The encoder processes the input sequence step by step and performs recurrent computations to integrate the information from each time step into the hidden state. The hidden states at each time step are also passed through an attention module to obtain the most relevant information for the prediction. These pieces of information are then fed into the decoder structure. The decoder consists of an LSTM and a fully connected layer. It receives the feature vector ((h_n, c_n) in Figure 6) generated by the encoder and the context vector (α in Figure 6) attention module, and then computes the output sequence (h' in Figure 6). At each time step, the output sequence of the LSTM of the decoder is inputted into a fully connected layer of the decoder, and the fully connected layer can calculate a production value. The Seq2Seq structure is illustrated in Figure 6.

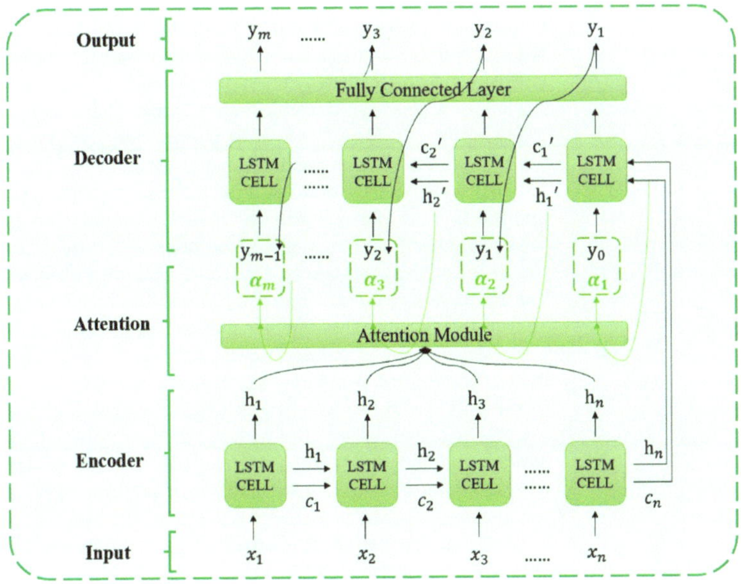

Figure 6. Illustration of a regular Seq2Seq structure.

(2) Attention mechanism

When it comes to the sequence-to-sequence (Seq2Seq) architecture, the attention mechanism is a crucial component that enables the model to selectively focus on different parts of the input sequence (in this study, the fracture information at each time step) while predicting each element of the target sequence (in this study, the predicted production at each time step).

To achieve this, the attention mechanism introduces a learnable attention model that calculates the relevance scores between the hidden states of the encoder at each time step and the hidden state of the decoder at the current time step. These scores reflect the importance of the fracture information processed at various time steps in the encoder for the current decoding step.

Specifically, the computation process of the attention mechanism is as follows:

Given the decoder's hidden state h'_{t-1} at time step $t-1$ and the hidden state vectors of the encoder $H = (h_1, h_2, \ldots, h_n)$,

$$Q = h'_{t-1} W_q$$
$$K = HW_k$$
$$V = H = (h_1, h_2, \ldots, h_n)$$
$$scores = softmax(QK^T / \sqrt{n})$$
$$\alpha_t = scores\ V^T$$

Here, W_q and W_k are the learnable weight matrices used for linear transformations; 'softmax' is the softmax function commonly used in machine learning; and α_t is the context vector.

As illustrated in Figure 6, the context vector α_t is concatenated with the output of the decoder at time step $t-1$ and then serves as the input to the decoder at time step t. This allows the decoder to receive the most relevant fracture information that needs attention at time step t.

3.2.3. Physics-Informed Encoder–Decoder Architecture

For horizontal wells with non-equal intersecting hydraulic fractures, the numerical simulator considers many variables in predicting production. For example, geological factors, fracture properties and distribution, production time, etc. For a horizontal well which requires hydraulic fracturing, the geological information and information regarding entrance hole have already been determined, and the factors influencing the production are left to be considered, including fracture properties and production time. Therefore, the PIED inputs include all hydraulic fracture lengths, permeability values, and dip angles, while production times are used as intermediate inputs.

Since we have assumed that the fractures have different inclinations and lengths, there are intersections between the fractures. Different orders of fractures will lead to different intersections of fractures and thus, necessarily, different production rates. Therefore, the order of the fractures is physically meaningful to the model, and we need to treat the fractures as ordered sequential data. The predicted production is also a time series. Therefore, we chose the Seq2Seq architecture to construct the proxy model.

However, we found that the regular Seq2Seq (LSTM–Attention–LSTM) has shortcomings in solving the problem—the input of the decoder is automatically set to the output of the previous time step. This leads to the problem that the production time has to be input to the encoder together with the fracture parameters. However, in fact, in numerical simulations, the properties of the fractures were not related to the production time, so it is not reasonable to couple the time information with the physical information of the fractures in the encoder. Since the encoder will extract the information of the input, if the production time is input into the encoder structure, then the time information received by the decoder will not be complete.

Based on the considerations mentioned above, PIED adopts a modified Seq2Seq structure, which causes the decoder to have a constant intermediate input by manually defining the input for each time step when constructing the decoder. The intermediate input is the production time used in the numerical simulator. The PIED consists of an encoder and a decoder. The encoder consists of an LSTM combined with an attention module, while the decoder consists of an LSTM and a fully connected layer. The input of the encoder is a sequence with a shape of $3 \times n$ consisting of the length, permeability, and dip angle of n fractures (10 in this study). The intermediate input of the decoder is a sequence consisting of 15 numerical type variables. The structure of the PIED is shown in Figure 7.

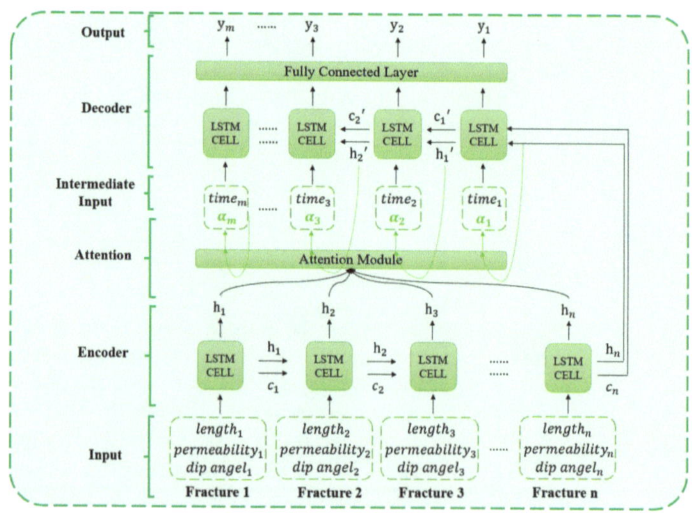

Figure 7. Illustration of the computational process of the PIED.

The encoder is able to extract the input (i.e., the sequence of $3 \times n$ fractures) to feature vectors (i.e., the hidden states and cell states of the encoder at the last time step), and by introducing the attention module, the proxy model can automatically extract the fracture information, which has an impact on the decoder from all the hidden states of the encoder, thus reducing the risk of overfitting. The LSTM of the decoder combines the feature vector output by the encoder, the production time series (i.e., the intermediate input), and the vectors output by the attention module to output a vector in each time step. And after each time step, the fully connected layer inputs the vector to predict a production. In detail, the hidden state and cell state of the encoder at the last time step will be used as the initial hidden state and initial cell state of the decoder's LSTM. And at each time step of the decoder's LSTM, the vector output from the attention module is combined with the production time and then input into the LSTM. Thus, the decoder will be constrained by physical information regarding the fractures in predicting the production. We also compare the PIED with the regular LSTM–Attention–LSTM in Section 4 to demonstrate the improvement in performance as a result of this modification.

In addition, we chose LSTM as the decoder because the output production is a time series and LSTM, as a model specialized in sequential problems, can fully consider the effect of the production in the earlier period on the production in the later period. The plot of correlation coefficients among the variables in one case is shown in Figure 8, showing a high correlation between the productions at different production times. We also show the superiority of LSTM in this problem by comparing it with the widely used MLP proxy model in Section 4.

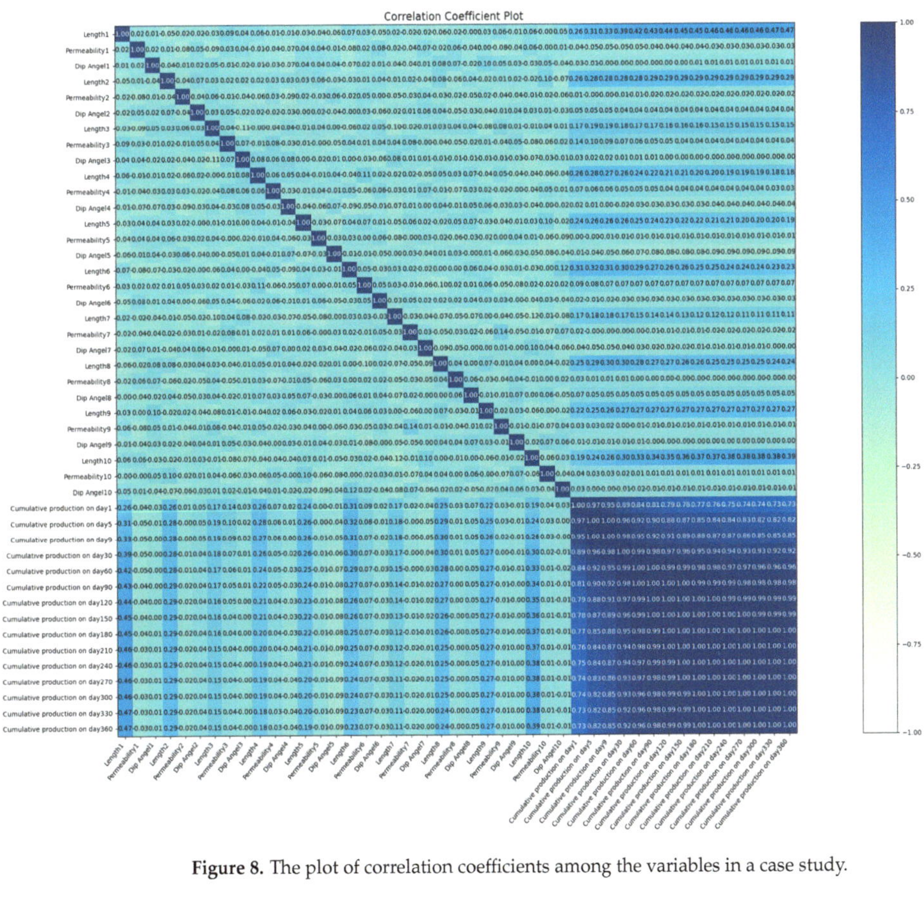

Figure 8. The plot of correlation coefficients among the variables in a case study.

3.3. The Training Workflow of PIED

A schematic diagram of the training process of the PIED is shown in Figure 9.

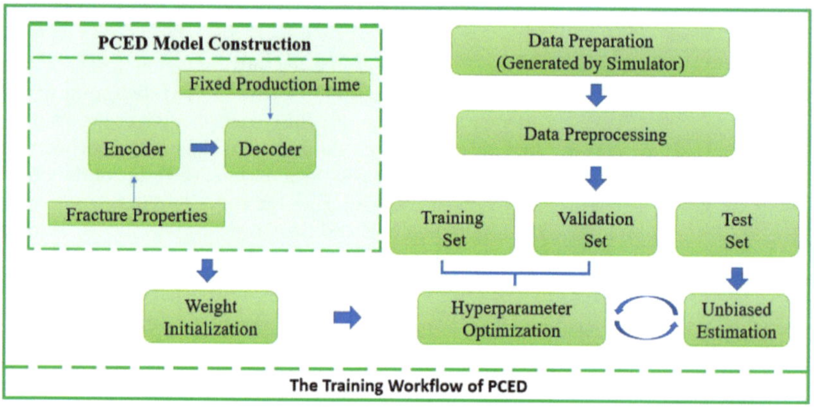

Figure 9. The training workflow of the PIED.

3.3.1. Data Preparation

The dataset used to train the proxy models is generated by the numerical simulator mentioned in Section 2. For the horizontal wells that are required for production prediction, some static parameters such as the geological parameters (e.g., permeability and porosity of the matrix) and parameters of well completion (e.g., perforated interval) need to be determined at first. Then, the number of hydraulic fractures as well as the ranges of values of fracture parameters is set according to the engineering background and a series of time points at which the production need to be predicted are determined. After that, the fracture length, permeability, and dip angle are randomly generated using a uniform distribution. Finally, the generated fracture parameters and the static parameters determined previously are input into the numerical simulator. Finally, the generated fracture parameters and the previously determined static parameters are input into the numerical simulator. Finally, the numerical simulator predicts the production at each time point. As an example, a sample of 10 fractures was generated with this method and is presented in Table 3.

Table 3. A sample generated by the numerical simulation.

A Sample Generated by the Numerical Simulator				
	Length	Permeability	Dip angel (radian)	
Fracture1	9.173320731	3.33587×10^{-13}	1.737934557	Generated randomly according to the uniform distribution
Fracture2	52.73803329	4.45321×10^{-13}	1.590268905	
Fracture3	7.970179625	1.1239×10^{-8}	2.066094122	
Fracture4	10.00118252	8.87251×10^{-12}	1.726819853	
Fracture5	41.51548897	2.33735×10^{-9}	1.88530177	
Fracture6	11.7711018	1.84973×10^{-10}	1.522413401	
Fracture7	62.27039877	3.97928×10^{-11}	1.499996875	
Fracture8	62.22829645	2.58495×10^{-13}	1.911464137	
Fracture9	55.57077147	4.9175×10^{-10}	1.134606937	
Fracture10	15.49058097	3.71595×10^{-13}	1.186653904	
PRODUCTION Series	[767.3, 1580.8, 2030.5, 3271.0, 4266.5, 4974.3, 5537.9, 6013.7, 6430.8, 6805.8, 7149.1, 7467.8, 7766.5, 8048.6, 8316.7]			Generated by the numerical simulator (in Section 2)

After the dataset is generated by the numerical simulator, the dataset will be randomly split into a training set, a validation set, and a test set at a ratio of 8:1:1.

3.3.2. Data Preprocessing

Normalization is a common method of data preprocessing. Normalization scales the data so that different features have similar scales, thus removing the impact of scales on the data and helping to speed up the convergence process of the model. This reduces the sensitivity of the model to noise or outliers and improves the robustness and generalization ability of the model.

Min–Max normalization has shown favorable results in data preprocessing. This technique employs a linear transformation on the original data, mapping the minimum value to 0 and the maximum value to 1, while scaling the remaining data proportionally based on their relative positions. The advantages of Min–Max normalization are preserving the original distribution shape of the data and ensuring a linear mapping of the data within the specified range. Min–Max normalization is calculated as follows:

$$x\_normalized = (x - min(x))/(max(x) - min(x))$$

In this representation, "$x\_normalized$ " represents the normalized value, "x " represents the original value, "$min(x)$" represents the minimum value in the dataset, and "$max(x)$" represents the maximum value in the dataset.

3.3.3. Model Construction and Weight Initialization

The coding of the PIED and its forward propagation steps can be done using PyTorch 1.13.0 based on Python 3.9.

Before the forward propagation, Xavier initialization (Glorot et al., 2010) [35] is used to ensure that the variances of the activation values and gradients remain roughly the same across different layers, which helps effectively propagate gradients during the training process. For the weight matrix of an LSTM unit, assuming the input dimension is n_{in} and the output dimension is n_{out}, during Xavier initialization, the elements of the weight matrix are still randomly sampled from a uniform distribution, and the range can be calculated using the following formula:

$$a = sqrt\left(\frac{6}{(n_{in}+n_{out})}\right)$$
$$w = uniform(-a,\ a)$$

Here, n_{in} is the input dimension and n_{out} is the output dimension.

For LSTM, since it has complex structures such as input gates, forget gates, and output gates, each gate has its own weight matrix. During Xavier initialization, the weight matrices of each gate should be initialized based on the corresponding input and output dimensions. This ensures that the weight initialization range for each gate is appropriate and facilitates gradient propagation within the LSTM unit.

During the backward propagation process, the Adam optimization algorithm (Kingma et al., 2014) [36] is Used. The Adam algorithm combines the characteristics of adaptive learning rate and momentum, enabling effective learning rate adjustment and fast convergence.

3.3.4. Hyperparameter Optimization

Grid Search is a commonly used hyperparameter optimization method that aims to determine the best combination of hyperparameters to optimize model performance. The PIED hyperparameter space consists of the learning rate, the hidden size of LSTM, and the batch size. For a more comprehensive search, additional parameters such as weight decay, the number of fully connected layers, and the number of neurons can also be considered.

Once a promising combination of hyperparameters is found, learning curves can be plotted for further fine-tuning. In this study, the training curve vs. validation curve method was employed to evaluate model overfitting by simultaneously observing the learning curves of the training and testing datasets. Typically, the x-axis represents the number of iterations or epochs of training, while the y-axis represents the model's loss or accuracy.

When the training loss continues to decrease and the testing loss starts to increase, this indicates the model is starting to overfit.

3.3.5. Model Evaluation

When solving regression problems, three commonly used model evaluation metrics include mean absolute error (MAE), mean squared error (MSE), and root mean squared error (RMSE). These model evaluation metrics can be calculated as follows:

$$MAE = \frac{1}{n}\sum_{i=1}^{n} |y_i - \hat{y}_i|$$
$$RMSE = \sqrt{\frac{1}{n}\sum_{i=1}^{n} (y_i - \hat{y}_i)^2}$$
$$MSE = \frac{1}{n}\sum_{i=1}^{n} (y_i - \hat{y}_i)^2$$

In the above formula, y_i represents the true values (i.e., the production series calculated by the numerical simulator mentioned in Section 2), $\hat{y}_i$ represents the predicted values by the DL proxy model, and n represents the number of samples.

4. Case Study

4.1. Data Preparation and Preprocessing

Based on the data generation method proposed in Section 3 and the numerical simulator mentioned in Section 2, a dataset was created for predicting the cumulative production of horizontal wells with unequal-length and intersecting hydraulic fractures on a two-dimensional plane during depletion. Under the given geological static parameters and perforation spacing conditions, we randomly generated 10 hydraulic fractures with different lengths, permeabilities, and dip angles, as shown in Figure 10. The production was then computed using the numerical simulator for the following time intervals: 1, 5, 9, 30, 60, 90, 120, 150, 180, 210, 240, 270, 300, 330, and 360 days. The lengths, permeabilities, and dip angles of these hydraulic fractures are uniformly distributed within specified ranges. In the end, we generated 500 such samples as the dataset, and a sample is presented as an example in Table 3. The distributions of each feature in the dataset are visualized in Figure 11. Specifically, the range of fracture lengths is 5~75 m, the range of permeabilities is 1×10^{-13}~1×10^{-8} D, and the range of inclinations is $\pi/6 \sim 2\pi/3$. After applying Min–Max normalization, the dataset was split into training, validation, and testing sets at a ratio of 8:1:1.

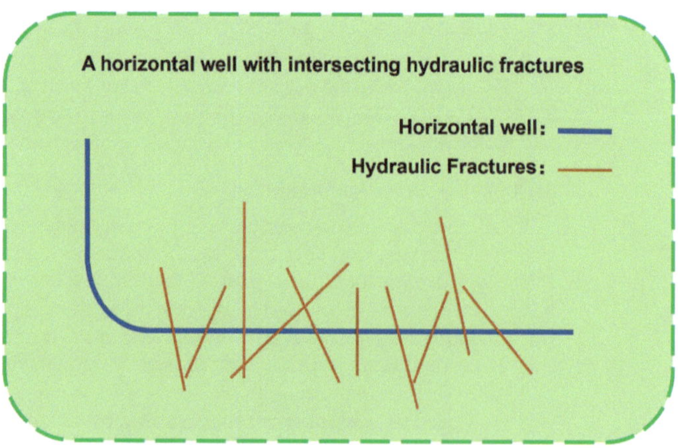

Figure 10. Illustration of a sample of 10 fractures from the case study.

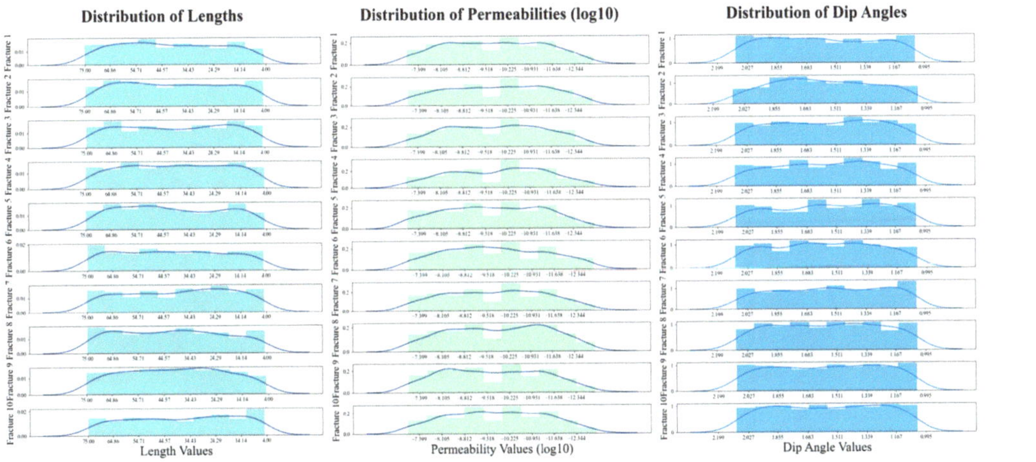

Figure 11. The distribution of the lengths, permeabilities, and dip angles of all the fractures in the dataset.

4.2. Configuration of Proxy Models and Hyperparameter Optimization

Based on the "training workflow" described in Section 3, MLP, LSTM–LSTM, and the PIED were constructed and optimized with hyperparameters. The hyperparameters used for training the three surrogate models are presented in Table 4.

Table 4. The hyperparameters of the proxy models we trained in the case study.

Proxy Models	Hidden Size of LSTM	Fully Connected Layers	Dropout	Activation Function	BatchNorm
PIED	Encoder: 13	Encoder: 0	—	—	—
	Decoder: 13	Decoder: 1 (neurons: 13)	—	—	Not Applied
Regular LSTM-Attention-LSTM	Encoder: 12	Encoder: 0	—	—	—
	Decoder: 12	Decoder: 1 (neurons: 12)	—	—	Not Applied
MLP	—	4 (neurons: 28, 24, 24, 16)	0.2, 0.1, 0, 0	Leaky Relu	Applied

4.3. Evaluation of the Proxy Models

The performance of the three proxy models were evaluated based on the metrics mentioned in the 'model evaluation' part of Section 3. By predicting the data from the test set, the MSE, MAE, and RMSE values of these three models are shown in Figure 12. The PIED achieved the best results regarding MSE, MAE, and RMSE, while regular LSTM–Attention–LSTM outperformed MLP in all three metrics. The predictions of the three proxy models for the test set are shown in Figure 13.

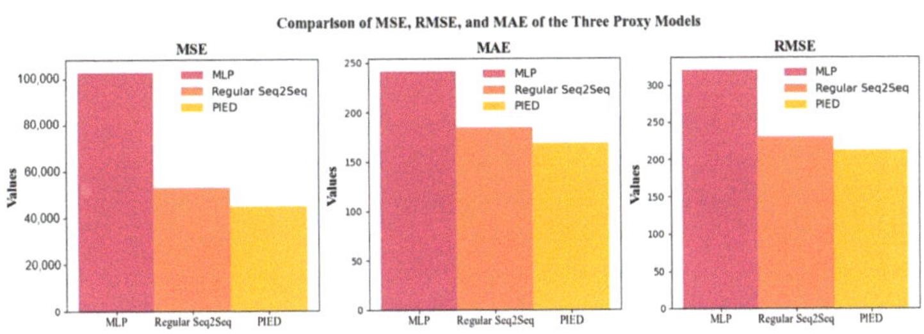

Figure 12. The comparison of MSE, RMSE and MAE of the three proxy models.

Model / MAE	PIED	Regular Seq2Seq	MLP
Mean	168.81	184.07	241.76
25%	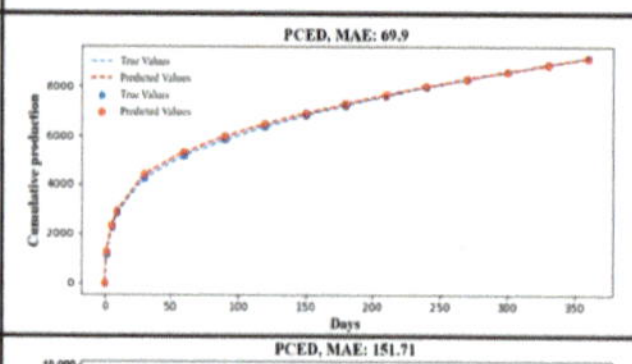	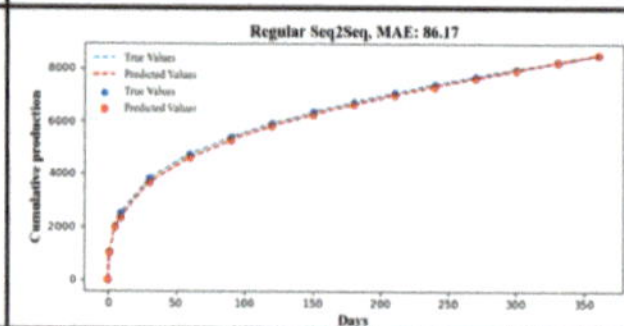	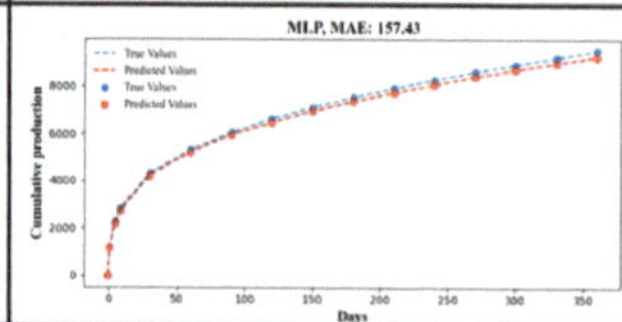
50%	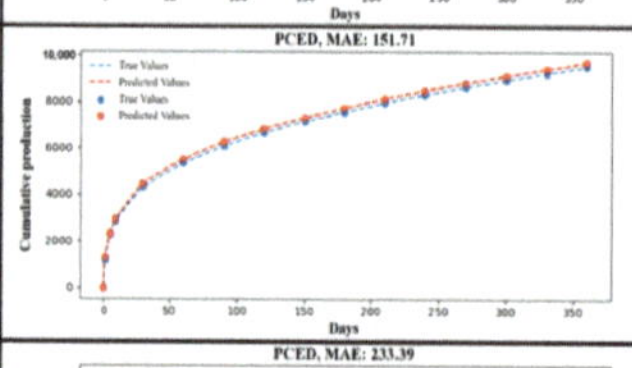	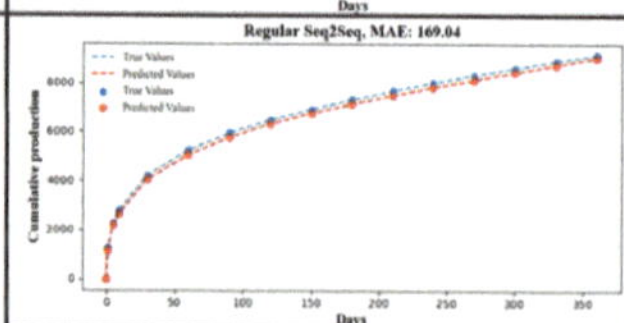	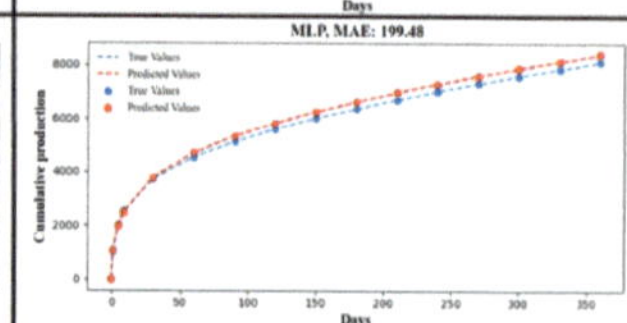
75%	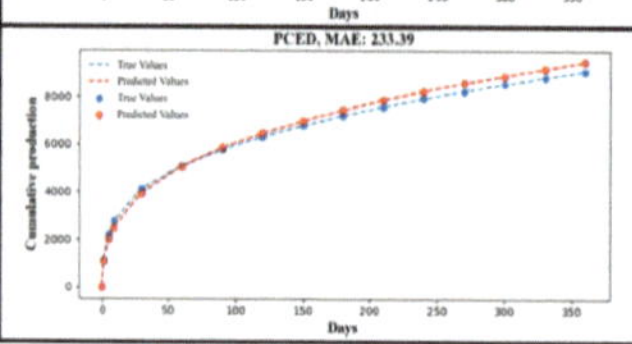	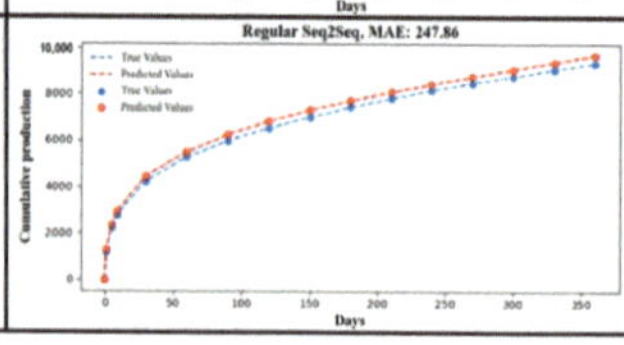	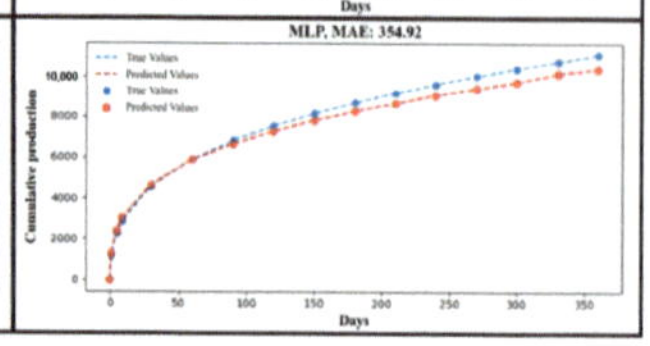

Figure 13. Illustration of the predictive performance of the three agent models on the test set.

4.4. Contribution of Physics Information and Seq2Seq Structure

(1) Superiority of the Seq2Seq Structure:

In terms of model performance, compared to MLP's MSE, LSTM–Attention–LSTM's MSE decreased by 48.4%. Seq2Seq architecture outperformed MLP in predicting the production sequence.

When analyzing the model structures, we first employed a multi-layer perceptron (MLP) as the decoder and constructed an LSTM–MLP architecture for comparative experimentation in the test set. The comparative results of MLP, LSTM–MLP, and LSTM–Attention–LSTM are illustrated in Figure 14. The superior performance of LSTM–MLP over MLP demonstrates that the LSTM structure is better equipped to handle fracture information. Given that fractures intersect and their positions and order affect production, leveraging LSTM to process fissures as sequential features enhances its performance. The fact that the performance of LSTM–Attention–LSTM surpassed that of LSTM–MLP further validates its stronger performance in predicting production sequences using LSTM.

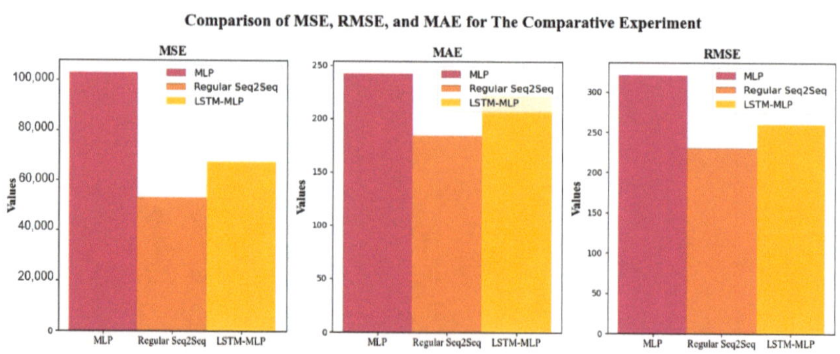

Figure 14. The comparison of the MSE, RMSE, and MAE values of MLP, regular Seq2Seq, and the LSTM–MLP.

From the perspective of the number of internal parameters (i.e., weights and biases), in this case, MLP has 2557 internal parameters, while LSTM–Attention–LSTM has 3020 internal parameters. This is one of the reasons why MLP exhibits more severe overfitting compared to LSTM–Attention–LSTM. Since MLP treats the input fractures as discrete input features, while LSTM–Attention–LSTM treats them as a sequence, the internal parameters of MLP increases with the number of fractures, whereas that of LSTM–Attention–LSTM does not. In this case, where 10 fractures have 30 fracture variables and 15 time variables, the input dimension is high. This results in MLP having more internal parameters compared to LSTM–Attention–LSTM. Consequently, training MLP requires more data compared to training LSTM–Attention–LSTM, and it also implies that constructing the MLP model incurs more costs to generate additional samples.

(2) Improvement of Seq2Seq Performance using Physical Information:

Compared to that of LSTM–Attention–LSTM, the PIED's MSE decreased by 11.18%. It is evident that by separately inputting the production time series into the decoder, the PIED exhibits superior performance.

From the perspective of production calculation, the geometry attributes of fractures and the set production time are independent in the XFEM numerical simulator when it is calculating the production. the geometric parameters of the fractures remain unchanged, and therefore, PIED has an encoder structure that can independently process all the fracture information. Meanwhile, when it comes to calculating production, the time information will be coupled with fracture information in the decoder. However, in the regular Seq2Seq model, time information and fracture information are input together from the beginning.

From the viewpoint of model structure, the regular LSTM exhibits error accumulation, while the PIED's structure mitigates error accumulation. Specifically, for the regular LSTM, it takes the output of the previous time step as the input for the current time step. As the output of the previous time step contains errors, each LSTM at every time step is provided with information containing accumulated errors. Therefore, the more time steps, the more error accumulation in LSTM. However, in the PIED, at each time step, the input of the decoder's LSTM are the predetermined static values form the production time series and the vectors outputted by the attention module, so that the accumulated error in the PIED is reduced.

5. Conclusions

In this study, the PIED, a new physics-informed proxy model of numerical simulator for fractured horizontal well production prediction, is proposed, and the Seq2Seq architecture is initially used to do this. The encoder structure and attention algorithm successfully extract physical information regarding intersecting hydraulic fractures, and the modified decoder structure succeeds in predicting production series within the constraints of production time and the fracture information outputted by the encoder and the attention module. We also proved that the intersecting hydraulic fractures processed as the series type data by the LSTM can enhance the performance of the proxy model. Finally, according to the case study, the performance of the Seq2Seq architecture significantly surpassed the commonly used MLP, highlighting its superiority. Furthermore, the performance of the PIED exhibited further improvement upon the foundation of Seq2Seq.

In the case study, where the feature dimensionality was high, with 30 variables forming a sequence of size 3×10 as the input and 15 fixed time values as the intermediate input for the decoder, the PIED exhibited satisfactory performance in handling high-dimensional features. The MAE of the PIED was 168.8, which is equal to 2.7% bias of the true cumulative production values. Compared to the MSE values of MLP and LSTM–Attention–LSTM, the MSE of PIED was 56.0% and 11.8% less, respectively. The PIED model demonstrates its stronger generalization ability and higher accuracy. Only 500 samples need to be generated over 3 days to build datasets, so that PIED exhibits practicality as it can be trained and fine-tuned on small datasets, achieving satisfactory performance.

6. Discussion

The physics-informed encoder–decoder (PIED) model, while capable of achieving strong performance on smaller datasets, still exhibits some limitations. Firstly, the PIED currently incorporates only two types of physical information, namely, the geometric features of fractures and production time. However, incorporating more physical information, such as the geological conditions of the matrix, should be considered. Secondly, while the PIED performed well in the case of 10 hydraulic fractures, its encoder's ability to handle a much larger number of input variables, say a hundred fractures, needs to be validated.

Fortunately, the aforementioned limitations of the PIED can be addressed. To incorporate additional types of physical information, other network structures such as fully connected layers can be employed to map the information into vectors, which can then be used as input or hidden states in the PIED. Moreover, when dealing with a higher number of fractures, the encoder can be equipped with a self-attention mechanism to further mitigate overfitting.

Author Contributions: Data preparation and numerical simulations, Y.X.; proxy models, AI methodology, and case Study, T.J.; AI methodology and case study, H.J. All authors have read and agreed to the published version of the manuscript.

Funding: This work was financially supported by the National Natural Science Foundation of China (Grant NO. U19B6003-05).

Data Availability Statement: Not applicable.

Acknowledgments: The financial support provided by the National Natural Science Foundation of China is highly appreciated.

Conflicts of Interest: The authors declare no conflict of interest.

References

1. Wang, H.; Qiao, L.; Lu, S.; Chen, F.; Fang, Z.; He, X.; Zhang, J.; He, T. A Novel Shale Gas Production Prediction Model Based on Machine Learning and Its Application in Optimization of Multistage Fractured Horizontal Wells. *Front. Earth Sci.* **2021**, *9*, 675–689. [CrossRef]
2. Wang, T.; Wang, Q.; Shi, J.; Zhang, W.; Ren, W.; Wang, H.; Tian, S. Productivity Prediction of Fractured Horizontal Well in Shale Gas Reservoirs with Machine Learning Algorithms. *Appl. Sci.* **2021**, *11*, 12064. [CrossRef]
3. Xue, H.; Malpani, R.; Agrawal, S.; Bukovac, T.; Mahesh, A.L.; Judd, T. Fast-Track Completion Decision Through Ensemble-Based Machine Learning. In Proceedings of the SPE Reservoir Characterisation and Simulation Conference and Exhibition, Abu Dhabi, United Arab Emirates, 17–19 September 2019. [CrossRef]
4. Lizhe, L.; Fujian, Z.; You, Z. The prediction and optimization of Hydraulic fracturing by integrating the numerical simulation and the machine learning methods. *Energy Rep.* **2022**, *8*, 15338–15349. [CrossRef]
5. Wang, B.; Han, G.; Lu, X.; Ma, H.; Zhu, Z.; Liang, X. The integrated geosciences and engineering production prediction in tight reservoir based on deep learning. *Geoenergy Sci. Eng.* **2023**, *223*, 211571. [CrossRef]
6. Krishnapriyan, A.; Gholami, A.; Zhe, S.; Kirby, R.; Mahoney, M.W. Characterizing possible failure modes in physics-informed neural networks. *Adv. Neural Inf. Process. Syst.* **2021**, *34*, 26548–26560.
7. Cai, S.; Mao, Z.; Wang, Z.; Yin, M.; Karniadakis, G.E. Physics-informed neural networks (PINNs) for fluid mechanics: A review. *Acta Mech. Sin.* **2021**, *37*, 1727–1738. [CrossRef]
8. Cursi, J.S.D.; Koscianski, A. Physically constrained neural network models for simulation. In Proceedings of the Advances and Innovations in Systems, Computing Sciences and Software Engineering, Bridgeport, CT, USA, 3–12 December 2007; Springer: Amsterdam, The Netherlands, 2007.
9. Cuomo, S.; Di Cola, V.S.; Giampaolo, F.; Rozza, G.; Raissi, M.; Piccialli, F. Scientific machine learning through physics–informed neural networks: Where we are and what's next. *J. Sci. Comput.* **2022**, *92*, 88. [CrossRef]
10. Raissi, M.; Perdikaris, P.; Karniadakis, G.E. Physics-informed neural networks: A deep learning framework for solving forward and inverse problems involving nonlinear partial differential equations. *J. Comput. Phys.* **2019**, *378*, 686–707. [CrossRef]
11. Zhu, Y.; Zabaras, N.; Koutsourelakis, P.S.; Perdikaris, P. Physics-constrained deep learning for high-dimensional surrogate modeling and uncertainty quantification without labeled data. *J. Comput. Phys.* **2019**, *394*, 56–81. [CrossRef]
12. Li, X.; Ma, X.; Xiao, F.; Xiao, C.; Wang, F.; Zhang, S. A physics-constrained long-term production prediction method for multiple fractured wells using deep learning. *J. Petrol. Sci. Eng.* **2022**, *217*, 110844. [CrossRef]

13. Guevara, J.; Zadrozny, B.; Buoro, A.; Lu, L.; Tolle, J.; Limbeck, J.W.; Hohl, D. A machine-learning methodology using domain-knowledge constraints for well-data integration and well-production prediction. *SPE Reserv. Eval. Eng.* **2019**, *22*, 1185–1200. [CrossRef]
14. Qu, H.Y.; Zhang, J.L.; Zhou, F.J.; Peng, Y.; Pan, Z.J.; Wu, X.Y. Evaluation of hydraulic fracturing of horizontal wells in tight reservoirs based on the deep neural network with physical constraints. *Pet. Sci.* **2023**, *20*, 1129–1141. [CrossRef]
15. Yang, R.; Qin, X.; Liu, W.; Huang, Z.; Shi, Y.; Pang, Z.; Wang, T. A physics-constrained data-driven workflow for predicting Coalbed methane well production using artificial neural network. *SPE J.* **2022**, *27*, 1531–1552. [CrossRef]
16. Wang, N.; Chang, H.; Zhang, D. Surrogate and inverse modeling for two-phase flow in porous media via theory-guided convolutional neural network. *J. Comput. Phys.* **2022**, *466*, 111419. [CrossRef]
17. Cornelio, J.; Mohd Razak, S.; Cho, Y.; Liu, H.H.; Vaidya, R.; Jafarpour, B. Neural Network-Assisted Clustering for Improved Production Predictions in Unconventional Reservoirs. In Proceedings of the SPE Western Regional Meeting, Anchorage, AK, USA, 22–25 May 2023.
18. Razak, S.M.; Cornelio, J.; Cho, Y.; Liu, H.H.; Vaidya, R.; Jafarpour, B. Embedding physical flow functions into deep learning predictive models for improved production forecasting. In Proceedings of the Unconventional Resources Technology Conference, Houston, TX, USA, 20–22 June 2022.
19. Baca, R.G.; Arnett, R.C.; Langford, D.W. Modelling fluid flow in fractured-porous rock masses by finite-element techniques. *Int. J. Numer. Methods Fluids* **1984**, *4*, 337–348. [CrossRef]
20. Xia, Y.; Jin, Y.; Oswald, J.; Chen, M.; Chen, K. Extended finite element modeling of production from a reservoir embedded with an arbitrary fracture network. *Int. J. Numer. Methods Fluids* **2018**, *86*, 329–345. [CrossRef]
21. Jin, Y.; Chen, K.; Chen, M. An asymptotic solution for fluid production from an elliptical hydraulic fracture at early-times. *Mech. Res. Commun.* **2015**, *63*, 48–53. [CrossRef]
22. Oswald, J.; Gracie, R.; Khare, R.; Belytschko, T. An extended finite element method for dislocations in complex geometries: Thin films and nanotubes. *Comput. Methods Appl. Mech. Eng.* **2009**, *198*, 1872–1886. [CrossRef]
23. Karimi-Fard, M.; Durlofsky, L.J.; Aziz, K. An efficient discrete fracture model applicable for general purpose reservoir simulators. *SPE J.* **2004**, *9*, 227–236. [CrossRef]
24. Reichenberger, V.; Jakobs, H.; Bastian, P.; Helmig, R. A mixed-dimensional finite volume method for two-phase flow in fractured porous media. *Adv. Water Resour.* **2006**, *29*, 1020–1036. [CrossRef]
25. Noorishad, J.; Mehran, M. An upstream finite element method for solution of transient transport equation in fractured porous media. *Water Resour. Res.* **1982**, *18*, 588–596. [CrossRef]
26. Xia, Y.; Jin, Y.; Chen, M.; Chen, K. An enriched approach for modeling multiscale discrete-fracture/matrix interaction for unconventional-reservoir simulations. *SPE J.* **2019**, *24*, 349–374. [CrossRef]
27. Karimi-Fard, M.; Firoozabadi, A. Numerical simulation of water injection in fractured media using the discrete-fractured model and the galerkin method. *SPE Res. Eval. Eng.* **2003**, *6*, 117–126. [CrossRef]
28. Melenk, J.M.; Babuška, I. The partition of unity finite element method: Basic theory and applications. *Comput. Methods Appl. Mech. Eng.* **1996**, *139*, 289–314. [CrossRef]
29. Berrone, S.; Pieraccini, S.; Scialò, S. On simulations of discrete fracture network flows with an optimization-based extended finite element method. *SIAM J. Sci. Comput.* **2013**, *35*, A908–A935. [CrossRef]
30. Wang, S.; Chen, Z.; Chen, S. Applicability of deep neural networks on production forecasting in Bakken shale reservoirs. *J. Petrol. Sci. Eng.* **2019**, *179*, 112–125. [CrossRef]
31. Luo, G.; Tian, Y.; Bychina, M. Production optimization using machine learning in Bakken shale. In Proceedings of the Unconventional Resources Technology Conference, Houston, TX, USA, 23–25 July 2018; Society of Exploration Geophysicists: Tulsa, OK, USA; American Association of Petroleum Geologists: Tulsa, OK, USA; Society of Petroleum Engineers: Richardson, TX, USA, 2018.
32. Panja, P.; Velasco, R.; Pathak, M.; Deo, M. Application of artificial intelligence to forecast hydrocarbon production from shales. *Petroleum* **2018**, *4*, 75–89. [CrossRef]
33. Hochreiter, S.; Schmidhuber, J. Long short-term memory. *Neural Comput.* **1997**, *9*, 1735–1780. [CrossRef]
34. Sutskever, I.; Vinyals, O.; Le, Q.V. Sequence to sequence learning with neural networks. *Adv. Neural Inf. Process. Syst.* **2014**, *2*, 3104–3112.
35. Glorot, X.; Bengio, Y. Understanding the difficulty of training deep feedforward neural networks. In Proceedings of the Thirteenth International Conference on Artificial Intelligence and Statistics, Chia Laguna Resort, Sardinia, Italy, 13–15 March 2010.
36. Kingma, D.P.; Ba, J. Adam: A method for stochastic optimization. In Proceedings of the International Conference for Learning Representations, San Diego, CA, USA, 7–9 May 2015.

energies

MDPI

Article

Rock Typing Approaches for Effective Complex Carbonate Reservoir Characterization

Sergey Krivoshchekov *, Alexander Kochnev, Nikita Kozyrev, Andrey Botalov, Olga Kochneva and Evgeny Ozhgibesov

Petroleum Geology Department, Perm National Research Polytechnic University, Komsomolsky Prospect, 29, 614990 Perm, Russia; sashakoch93@gmail.com (A.K.); botalov97@yandex.ru (A.B.)
* Correspondence: krivoshchekov@pstu.ru

Abstract: For highly heterogeneous complex carbonate reef reservoirs, rock typing with respect to depositional conditions, secondary processes, and permeability and porosity relationships is a useful tool to improve reservoir characterization, modeling, prediction of reservoir volume properties, and estimation of reserves. A review of various rock typing methods has been carried out. The basic methods of rock typing were applied to a carbonate reservoir as an example. The advantages and disadvantages of the presented methods are described. A rock typing method based on a combination of hydraulic flow units and the R35 method is proposed. Clustering methods for rock typing are used. The optimum clustering method is identified, and for each rock type, the permeability–porosity relationships are built and proposed for use in the geomodelling stage.

Keywords: permeability; porosity; complex carbonate reservoirs; rock typing; petrophysical correlations

Citation: Krivoshchekov, S.; Kochnev, A.; Kozyrev, N.; Botalov, A.; Kochneva, O.; Ozhgibesov, E. Rock Typing Approaches for Effective Complex Carbonate Reservoir Characterization. *Energies* **2023**, *16*, 6559. https://doi.org/10.3390/en16186559

Academic Editor: Dameng Liu

Received: 6 July 2023
Revised: 30 August 2023
Accepted: 7 September 2023
Published: 12 September 2023

1. Introduction

Carbonate reservoirs are characterized by a high heterogeneity of void space. The heterogeneity is caused by both abrupt changes in facies depositional conditions and widespread secondary processes that partially or completely rearrange the internal structure. The secondary changes are often selective. Reservoir properties can vastly vary both vertically and horizontally. Predicting properties in the reservoir interwell space is very important. Many studies address this issue, including facies analysis, determination of depositional conditions, rock typing, and sequence stratigraphy [1].

Identifying zones with different secondary processes and, hence, with different reservoir properties is essential for understanding reservoir structure and its static modeling. Detailed distribution of properties plays a significant role in calculating reserves, static and dynamic reservoir model preparation, planning geological and engineering activities, etc. The relationship between geological heterogeneity, reservoir quality, and reservoir performance [2] is paramount. Understanding these key relationships enables effective reservoir management [3].

The relationship between reservoir performance and quality is mostly fully reflected by permeability. Permeability is the most important property that determines fluid filtration in a reservoir. Many studies focused on permeability predictions in the interwell space [4,5]. For terrigenous reservoirs, permeability predictions in the interwell zone are simpler tasks than for carbonate reservoirs [6–8]. For terrigenous reservoirs, the standard approach with core-well-log correlation identification and property interpolation using geostatistic methods usually shows satisfied results of permeability distribution across reservoir volume [9]. For a complex carbonate reservoir, the problem of permeability distribution across the volume is extremely difficult [10,11]. Different approaches or their combinations are used for complex carbonate reservoir permeability predictions: rock typing, multi-scale data combining, seismotypes identification, scale effect estimation, anisotropy study, uncertainties analysis, etc. [12–15]. The permeability parameter is determined from various sources: core samples, well-test studies, nuclear magnetic resonance, acoustic logging, and others. The results of core studies are usually combined with logging curves to predict permeability in intervals without coring. However, for carbonate reservoirs, this prediction is generally ineffective. For more reliable predictions of permeability and reservoir quality, rock typing is used [16–18].

G. Archie [19] was one of the first to propose the definition of a rock type: a rock strata whose parts were deposited under the same conditions and were subjected to the same secondary transformation processes (fracture, cementation, or dissolution). A particular rock type must have a specific pore size distribution and, therefore, individual capillary pressure curves. The pore size distribution controls porosity and is related to permeability and water saturation [20].

There are many different approaches for typing carbonate reservoirs. The division into rock types is based on different physical characteristics of the rock: its hydraulic flow units (HFU), pore space geometry and structure, pore channel size, type and size of grains, their relation to the binder mass, etc. [21]. The hydraulic flow unit method is widely used for typing the carbonate reservoirs [22–24]. A hydraulic flow unit is an interval with certain properties prevailing in a reservoir. These intervals are controlled by both geological and petrophysical attributes that enable the prediction of reservoir characteristics. Accordingly, the diagenetic transformations prevailing in a particular part of the reservoir are an essential parameter controlling the hydraulic flow units [2]. The method is also applicable for rock typing in geological and dynamic modeling [25,26]. The method is based on the calculation of an integrated parameter—a flow zone indicator (FZI). However, there are uncertainties associated with combining hydraulic flow units into classes. Various techniques, such as GHE, machine learning, and the discrete rock typing (DRT) method, are used for this purpose [21,22].

A rather common method is to identify rock types based on pore channel size (Winland) [27,28]. Typing by pore channel radius has been proposed [29]:

- Macropores (2 μm < R35 < 10 μm);
- Mesopores (0.5 μm < R35 < 2 μm);
- Micropores (0.1 μm < R35 < 0.5 μm).

In [30], it is proposed to distinguish rock types by calculating pore geometry (PG) and pore structure (PS). The calculated parameters are plotted on one graph in a bi-logarithmic scale, and the points lying on one straight line will correspond to one class. As the class increases, the quality of the reservoir will increase. The classification of P.W. Choquette and L.C. Pray is a classical lithological classification. It includes the following characteristics: basic porosity types, genetic factors, grain size, and grain number factors [31]. Another common classification is the approach developed by R.J. Dunham [32]. It is a structural classification that considers the following factors: the presence and type of form elements or grains, their relationship with the binder mass, and the structure of the binder mass. The Archie classification is based on the study of the structural features of the rock matrix and the nature of the visible void space. G.E. Archie studied a significant amount of actual data to show that the rock types are characterized by certain petrophysical properties (porosity, permeability, capillary pressure, and electrical resistivity), well-logging results, and petrophysical relationships. In other words, a comprehensive approach to rock-type identification is necessary [19]. The authors [33–35] have proposed an approach based on the analysis of accumulated correlations between porosity and permeability, enabling the estimation of the porosity–permeability relationship as the void space increases. Lorenz curves have been successfully used to assess reservoir heterogeneity [36,37]. Lorenz curves are also widely used in classifying reservoirs by void type. By highlighting sharp changes in the plot, it is possible to account for the inclusion of highly permeable intervals or fractures.

Thus, a review of the main rock typing methods shows that for the purposes of geological and dynamic models and permeability modeling in the reservoir volume, the methods based on the permeability and porosity relationships are the most applicable. These methods are based on hydraulic flow units, pore channel radius, and Lorenz curves, but the methods are rather generalized and need to be adapted to the peculiarities of a particular reservoir.

This study will typify the reservoir of Alpha oilfield using different methods, perform a comparison of these methods, propose an approach to rock typing using machine learning methods (clustering), identify the best method, and build dependencies to predict the permeability of different rock types.

2. Geological Settings

The Alpha field is located in the Timan-Pechora oil and gas province, on the territory of the Denisovsky Depression (Figure 1). Reef structures were widespread in the area during the Yelets and Zadonian times.

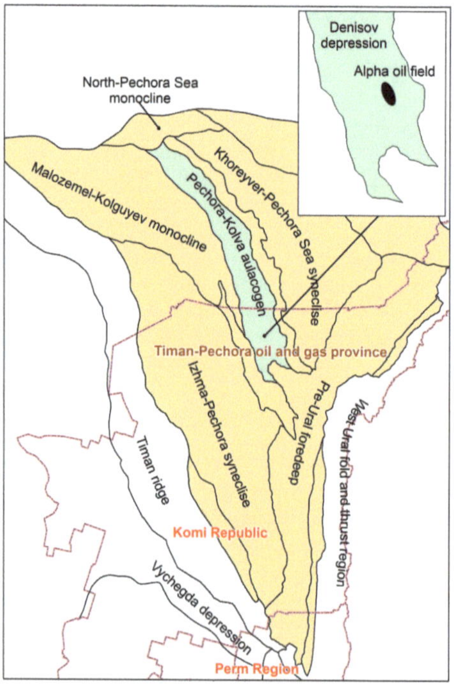

Figure 1. Region tectonic map (Alpha oilfield location).

Rocks of carbonate strata of the Yelets and Zadonian age (Devonian) are characterized by core predominantly in their upper part (D3el) (Figure 2). According to the core, the formation is mostly limestone with areas of abundant secondary dolomitization. Leaching is widespread, with caverns of various sizes ranging from leaching pores to large voids up to several centimeters in diameter. Leaching has also been observed in limestones, but cavernosity is particularly strong in dolomites.

Tilting and sub-vertical fracturing are presented, but often, the tilting fractures are resistive and impermeable, which are characterized by the upper part of the sequence. Sub-horizontal fractures are abundant in the upper part of the sequence and mostly remain open, even when naturally occurring.

The reservoir rocks are classified as fracture-cavernous and pore-cavernous. Given the widespread development of leaching pores, all reservoirs in the section can be classified as a pore-fracture-cavernous type (Figure 3). Accordingly, caverns are the main part of pore space in the reservoir.

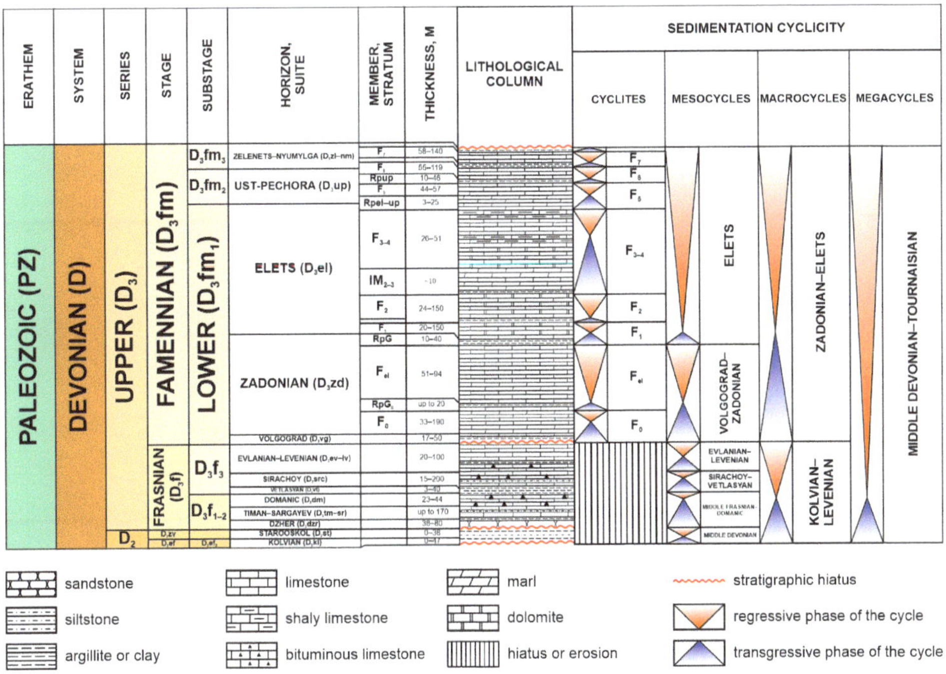

Figure 2. Regional stratigraphic column.

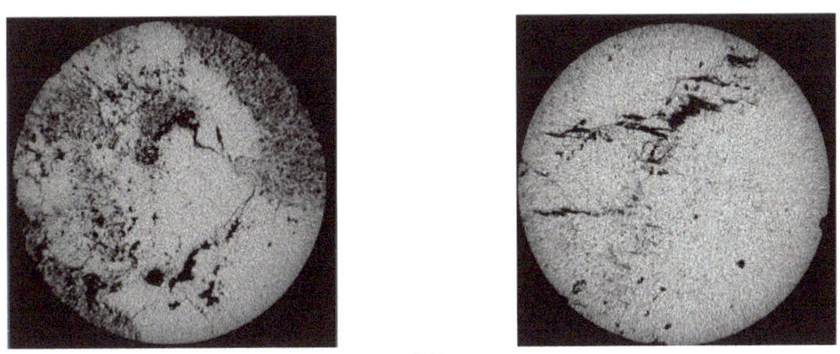

(a)

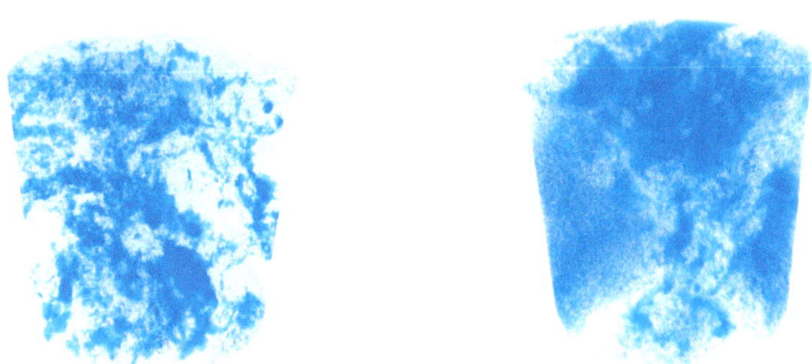

(b)

Figure 3. Example of core samples (Yelets strata) with fracture-cavernous type (tomography results): (a) XY view; (b) 3D pore space view.

The range of porosity varies from 1 to 18.3%. Permeability varies from 0.001 to 722.8 (mD), with an average value of 20.3 (mD).

According to the lithological description of the core samples, three lithotypes are distinguished: microbial limestones, peloid limestones, and secondary dolomites (Figure 4). Microbial limestones are characterized by a clotted granular microstructure and areas with fenestrae and are partially dolomitized, stylolitized, porous, cavernous-porous, and fractured. Secondary processes are noted—calcitization, recrystallization, sulfatization, dolomitization, and rarely, pyritization. Peloid limestones consist of peloids, interclasts, and ooids. Stylolitization and fracturing are observed. Secondary dolomites are medium-coarse-grained, porous, and cavernous-porous.

(a) (b) (c)

Figure 4. Examples of lithotypes: (**a**) microbial limestones; (**b**) peloid limestones; and (**c**) secondary dolomites.

3. Materials and Methods

This study used various methods for rock typing of core samples (461 core samples) based on the assessment of permeability and porosity relationships.

The hydraulic flow unit method is based on calculating a complex parameter—the flow zone indicator (*FZI*) (1):

$$FZI = \frac{RQI}{\varphi_z} \quad (1)$$

where *RQI* is the reservoir quality index, μm, and φ_z is the normalized porosity index, u.f.

RQI is defined by the equation:

$$RQI = 0.0314\sqrt{\frac{K_{pr}}{K_p}} \quad (2)$$

where K_{pr} is the permeability coefficient, mD, and K_p is the porosity coefficient, u.f.

Φ_z describes the ratio of the void volume to the solid rock volume and is defined by the equation:

$$\varphi_z = \frac{K_p}{(1 - K_p)} \quad (3)$$

It is assumed that when *RQI* and φz values are plotted on a bi-logarithmic scale, sample points with close *FZI* values will be located near the same straight line and, thus, be characterized by similar pore channel features and, thus, form a hydraulic flow unit. Different approaches exist for combining and grouping points into one class: cluster analysis, neural networks, cumulative frequency analysis, discrete type method, and global hydraulic unit class method. In the first stage, the global hydraulic unit classes (GHE) method is proposed, which suggests establishing class boundaries based on the generalization of a large number of field studies [37].

The following *DRT* method is based on converting a continuous *FZI* value into a discrete one, allowing the geological model's grid cell value to be set and then the petrophysical dependence for each rock type to be defined [38]:

$$DRT = 2\ \ln(FZI) + 10.6 \tag{4}$$

A rather common method is to isolate rock types based on pore channel size (Winland) [27,28]:

$$lgR35 = 0.732 + 0.588lgK_{pr} - 0.864lgK_p \tag{5}$$

where *R35* is the pore channel radius corresponding to 35% pore volume saturation with nonwetting phase, µm, K_{pr} is the permeability coefficient, mD, and K_p is the porosity coefficient, %.

Next, the typing was performed using the Lorenz plots method. Lorenz curves are an alternative graphical representation of the distribution function to estimate the degree of heterogeneity of a reservoir [36,39,40]. The degree of heterogeneity is assessed by comparing the areas under the curves with the area of a triangle cut off by a line of equal values. The accumulated porosity values (Storativity) are plotted on the abscissa axis, and the corresponding accumulated permeability values (Transmissivity) are plotted on the ordinate axis.

The next applied approach is also based on the analysis of accumulated porosity and permeability values. The approach allows evaluation of the relationship between porosity and permeability as the void space increases [34,35]. The methodology consists of sorting samples in ascending order of porosity values, calculating the correlation coefficient between permeability and porosity parameters at value number $n = 3$, then at $n = 4$, and so on, that is, the accumulated correlation coefficient between the parameters is determined, and then this coefficient is plotted on ordinate axis, and porosity values on abscissa axis.

In the next stage of the research, clustering techniques were used to self-select the classes that would be more adapted to the current properties of the field. Two basic clustering algorithms are used in this work: *k*-means and EM.

The *k*-means method is a cluster analysis method that aims to divide m observations (from space R^n) into k clusters, with each observation belonging to the cluster to whose center (centroid) it is closest [41].

The Euclidean distance is used as a measure of proximity:

$$\rho(x, y) = \|x - y\| = \sqrt{\sum_{p=1}^{n} (x_p - y_p)^2}, \text{ where } x, y \in R^n \tag{6}$$

Let us consider a number of observations $\left(x^{(1)}, x^{(2)}, \ldots, x^{(m)}\right)$, $x^{(i)} \in R^n$.

The *k*-means method divides m observations into k groups (or clusters) ($k \leq m$) $S = (S_1, S_2, \ldots, S_k)$, in order to minimize the total squared deviation of cluster points from the centroids of these clusters:

$$min\left[\sum_{i=1}^{k} \sum_{x^{(j)} \in S_i} \|x^{(j)} - \mu_i\|^2\right], \text{ where } x^{(j)} \in R^n, \mu_i \in R^n \tag{7}$$

μ_i—centroid for cluster S_i.

The work of the *k*-means algorithm can be roughly divided into four main stages [42]: identifying the *k* centers of the clusters, determining whether objects belong to clusters, identifying the centroids of k clusters, and comparing the cluster centers and centroids.

The algorithm is guaranteed to converge in a finite number of iterations. The clustering error and the number of iterations depend on the initial choice of centroids, so it is common practice to run *k*-means several times with different initial centroid candidates [43].

The EM (expectation–maximization) clustering method is an algorithm that allows efficient handling of large amounts of data, unlike the previous method. The idea of the EM algorithm is based on the assumption that any observation belongs to all clusters but with different probabilities. Therefore, two additional columns are generated in the output: cluster number and probability of belonging. The object must be assigned to the cluster for which this probability is higher. Some of the advantages of the EM algorithm are as follows [44]: efficient processing of big data, resistance to noise and data omissions, possibility to build the desired number of clusters, and fast convergence with successful initialization.

4. Results

Firstly, after the *FZI* calculation, rock typing was carried out using the Global Hydraulic Unit (*GHE*) and Discrete Rock Typing (*DRT*) methods. Figure 5 compares the GHE and DRT classifications using the Alpha field as an example.

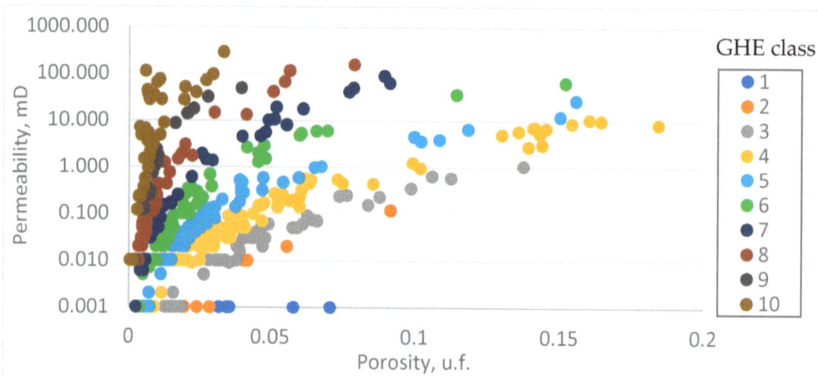

(a)

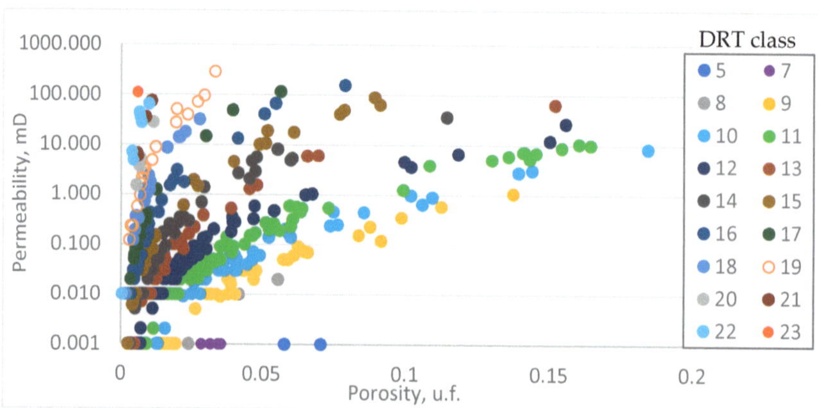

(b)

Figure 5. Comparison of (**a**) GHE and (**b**) DRT classifications.

In the *DRT* discrete rock type classification, a significantly higher number of rock types is distinguished (19) compared to the *GHE* typing (10). This distinction results in narrower rock types, while it succeeds in increasing the coefficient of determination for certain classes. Table 1 presents a comparison of equations and coefficients of determination by type. The *DRT* values from 5 to 23 correspond to classes 1–19. For classes 1–3 and 19, it was impossible to build dependencies due to the small number of samples in the class.

Table 1. Comparison of GHE and DRT classifications.

***GHE* Classification**			***DRT* Classification**		
Class	**Regression Dependence**	**Determination Coefficient**	**Class**	**Regression Dependence**	**Determination Coefficient**
2	$y = 0.0003 \times e^{71.888x}$	$R^2 = 0.98$	4	$y = 0.0001 \times e^{96.725x}$	$R^2 = 0.93$
3	$y = 0.0008 \times e^{66.062x}$	$R^2 = 0.72$	5	$y = 0.0007 \times e^{64.187x}$	$R^2 = 0.85$
4	$y = 0.0065 \times e^{49.341x}$	$R^2 = 0.55$	6	$y = 0.0037 \times e^{50.779x}$	$R^2 = 0.90$
5	$y = 0.0047 \times e^{70.066x}$	$R^2 = 0.91$	7	$y = 0.005 \times e^{54.849x}$	$R^2 = 0.85$
6	$y = 0.0118 \times e^{83.039x}$	$R^2 = 0.78$	8	$y = 0.0049 \times e^{68.011x}$	$R^2 = 0.92$
7	$y = 0.0151 \times e^{112.92x}$	$R^2 = 0.81$	9	$y = 0.0103 \times e^{76.634x}$	$R^2 = 0.98$
8	$y = 0.0417 \times e^{135.73x}$	$R^2 = 0.61$	10	$y = 0.0107 \times e^{105.42x}$	$R^2 = 0.89$
9	$y = 0.0748 \times e^{220.27x}$	$R^2 = 0.65$	11	$y = 0.015 \times e^{110.85x}$	$R^2 = 0.82$
10	$y = 0.506 \times e^{224.47x}$	$R^2 = 0.78$	12	$y = 0.0328 \times e^{135.02x}$	$R^2 = 0.83$
-	-	-	13	$y = 0.0444 \times e^{162.76x}$	$R^2 = 0.88$
-	-	-	14	$y = 0.0386 \times e^{293.85x}$	$R^2 = 0.80$
-	-	-	15	$y = 0.2359 \times e^{224.17x}$	$R^2 = 0.97$
-	-	-	16	$y = 0.0865 \times e^{523.56x}$	$R^2 = 0.90$
-	-	-	17	$y = 0.3076 \times e^{506.17x}$	$R^2 = 0.98$
-	-	-	18	$y = 0.0451 \times e^{890.15x}$	$R^2 = 0.58$

y—permeability coefficient, mD, *x*—porosity coefficient, u.f.

Table 1 shows an increase in the determination coefficients when using the *DRT* approach. Determination coefficients for all classes, except for class 18, are higher than 0.8. For class 18, the determination coefficient is 0.58; however, this type belongs to the fracture type and is characterized by a large scatter of properties, and the number of values is only six. This methodology allows us to obtain high coefficients of determination but distinguishes many classes, which will complicate the distribution of rock types in the geomodelling process.

In the next stage of the work, the pore channel radius corresponding to 35% saturation of the pore volume with a nonwetting phase was calculated using the Winland equation. The reservoir was typified according to the method classification [16], i.e., four classes were identified;

Class 1—micropores (0.1 μm < *R35* < 0.5 μm);
Class 2—mesopores (0.5 μm < *R35* < 2 μm);
Class 3—macropores (2 μm < *R35* < 10 μm);
Class 0—caverns (>10 μm).

Figure 6 shows the results of rock typing using this method.

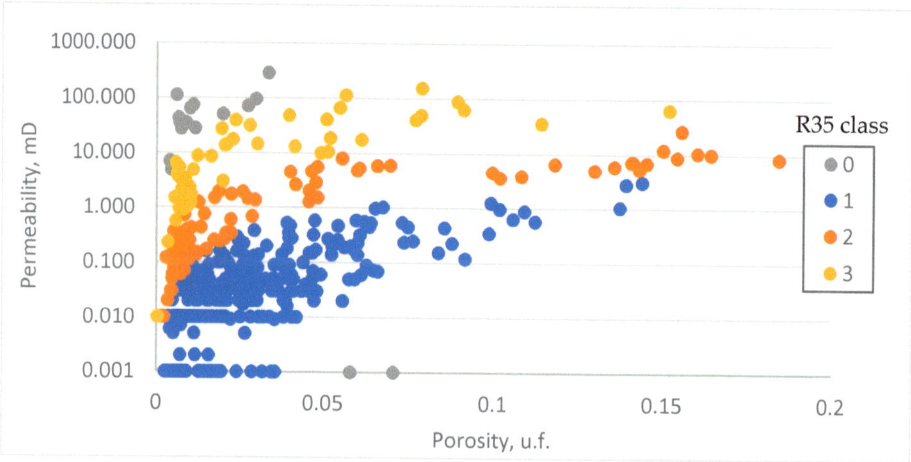

Figure 6. Classification by pore channel radius. The Alpha field.

Apparently, distinguishing only four classes does not provide reliable reservoir typing with high coefficients of determination between parameters in each class. However, the pore channel radius parameter can be used as one of the rock typing criteria.

A typing exercise based on the results of the Lorenz curve is presented in Figure 7.

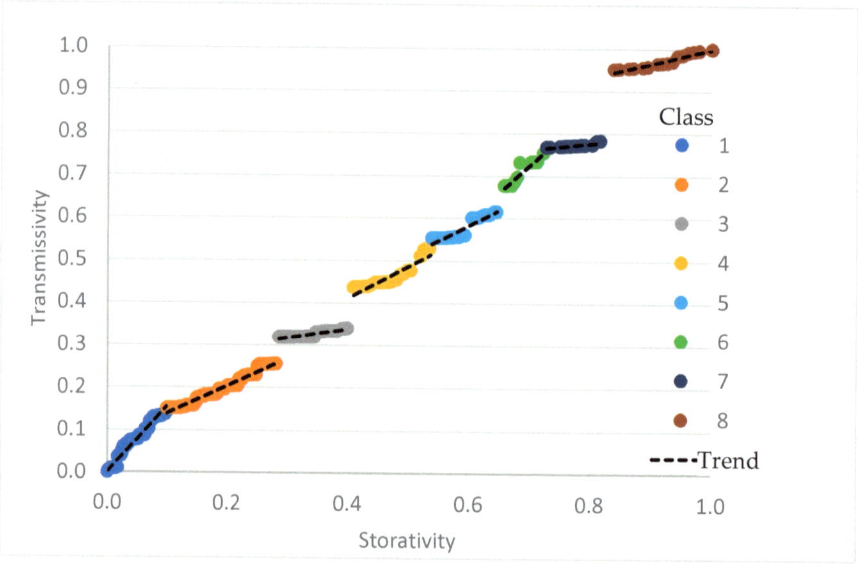

Figure 7. Building a Lorenz curve for sample typing. The Alpha field.

Describing this plot (Figure 7), one can conventionally identify zones of sharp changes in the trends of the curve of accumulated values. These changes are characterized by a change in rock type.

Thus, the Lorenz curve method of rock typing distinguishes eight classes in petrophysical dependency (Figure 8).

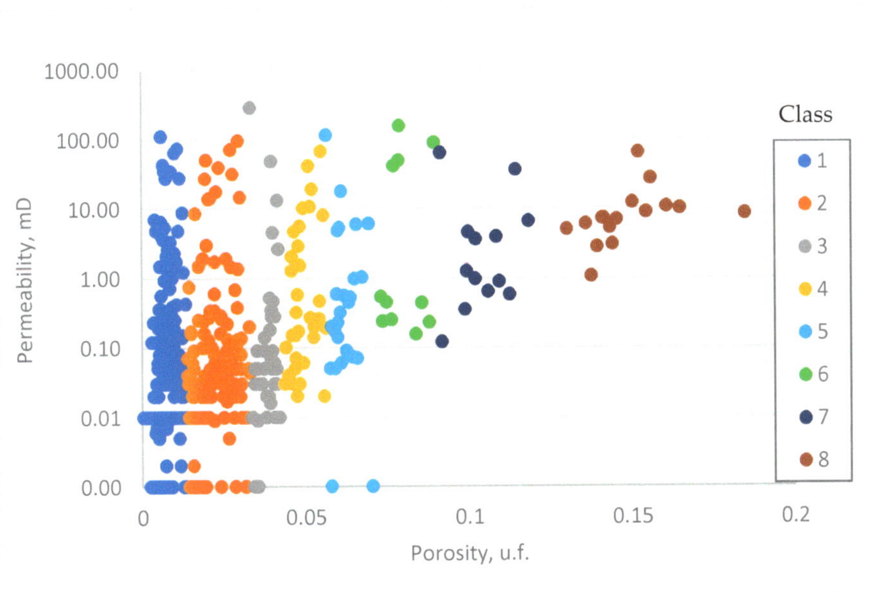

Figure 8. Lorenz curve typing of reservoirs. The Alpha field.

Figure 8 identifies eight rock types, but the prevailing parameter in this classification is rock porosity. It can be seen that within rock types, it is not possible to obtain high values of the coefficient of determination between the parameters. Such an approach is appropriate when justifying the boundary values of porosity in modeling. Often, in order to recreate a cloud of petrophysical dependency values, the model is divided by porosity samples (bins), and the dependencies are plotted within the resulting classes. Determination coefficients in this approach remain low, but the model cubes allow to reproduction of the initial cloud of petrophysical dependence points. The extended practice is to distinguish porosity intervals based on arbitrary criteria, and this approach allows this to be conducted in a justified manner.

The next applied approach is also based on the analysis of accumulated values of the correlation coefficient between porosity and permeability as porosity values increase. The typing using this methodology is shown in Figure 9.

In Figure 9, six types of rocks can be distinguished according to changes in trends. Determination coefficients were calculated for the classes, ranging from 0.07 to 0.972. The first class is characterized by a low coefficient of determination, as the relationship between permeability and porosity values is practically absent and unstable. This class can be attributed to the matrix component with the inclusion of samples with fractures and samples with practically no reservoir properties. For the second class, there is an upward trend in the correlation value between the parameters, but some samples, on the contrary, are out of the trend, probably because of the inclusion of fractures in the matrix type of the reservoir. For the third class, with increasing porosity values, there is a decreasing trend in the correlation coefficient between porosity and permeability, i.e., there are samples with fractured reservoir type—with generally low porosity, a number of samples have high permeability. The next class, four, is characterized by a general growth trend in the porosity–permeability relationship. This class may be classified as a pore reservoir. The fifth class, on the contrary, manifests a general downward trend of porosity-permeability relationship, and the cavernous component can be observed. The last class also belongs to the pore-cavernous or pore-cavernous-fracture type.

The results of rock typing are shown in Figure 10.

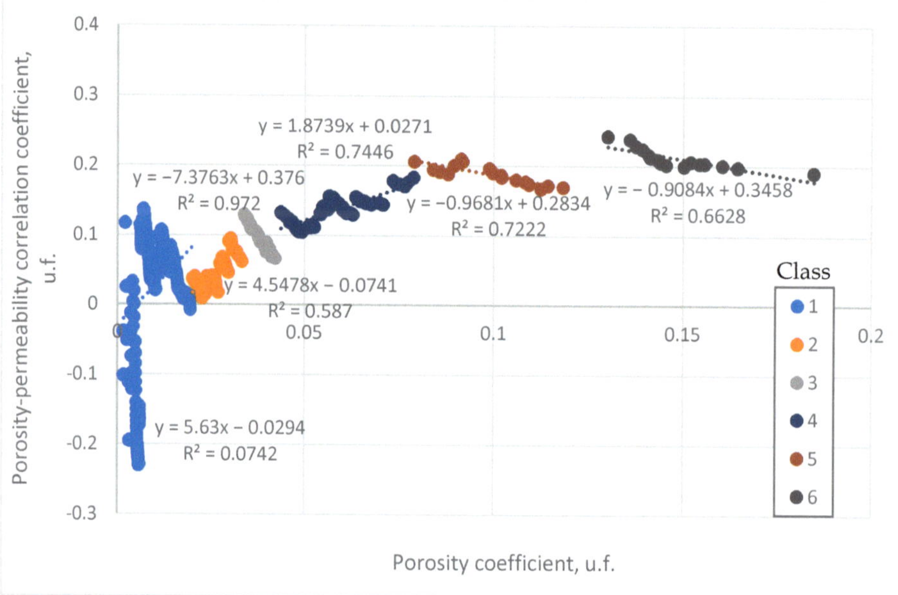

Figure 9. Correlation coefficient between permeability and porosity versus porosity. The Alpha field.

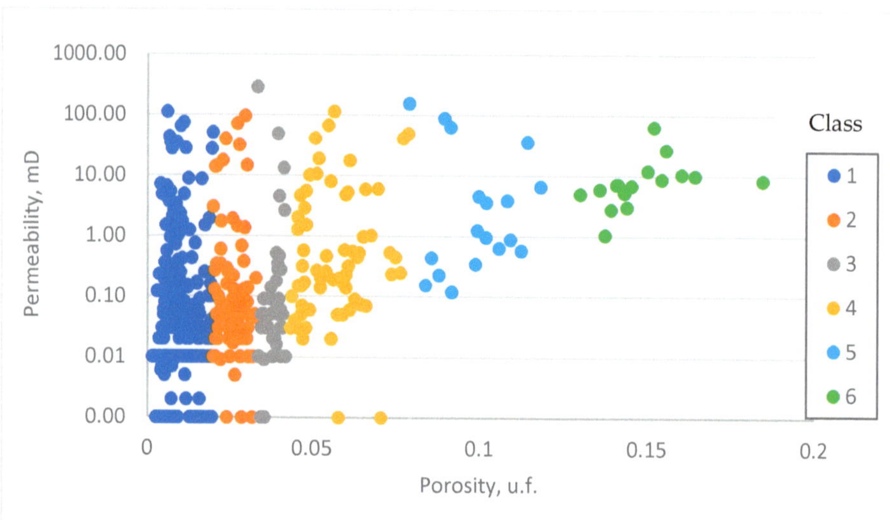

Figure 10. Rock typing by the accumulated correlation plot. The Alpha field.

The previous approach is based on accumulated parameter values; this approach plots the division of samples into rock types predominantly by porosity. This approach can be applied to reconstruct the petrophysical dependency cloud in a geological model, but for correct rock typing, it is necessary to use the *FZI* parameter, which allows the correct segregation of rock types based on both porosity and permeability values. This is especially relevant when separating low-quality and fractured reservoirs. Therefore, the *FZI* parameter is the basis for rock typing in further study.

Previously, boundaries for rock types were calculated using the *GHE* and *DRT* methods, which provided fairly good results, but these boundaries are based on the experience of researchers and are calculated for fields of different territories. It is proposed to select boundaries for rock typing of reservoirs of the fields of the Denisovsky Depression. Machine learning methods have been used for this purpose.

The main indicators for clustering are the calculated *FZI* parameter and *R35*. Clustering was carried out for the calculated *FZI* values, as well as jointly for *FZI* and *R35*. The *k*-means and EM methods were used for clustering.

Figures 11 and 12 below compare the two clustering methods, *k*-means and EM, by *FZI* parameter.

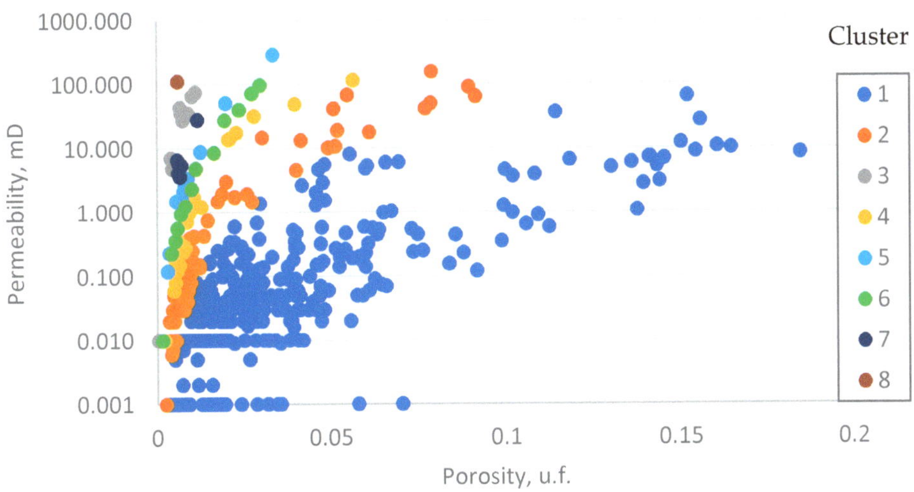

(**a**)

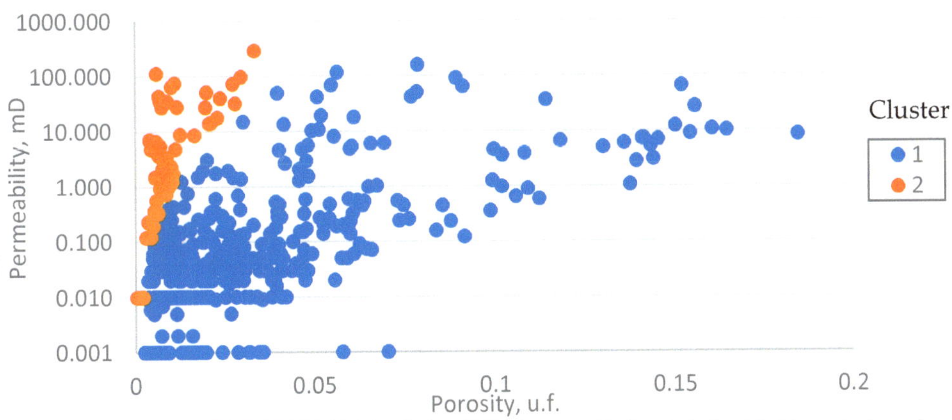

(**b**)

Figure 11. Comparison of: (**a**) k-means; (**b**) EM rock typing by FZI parameter.

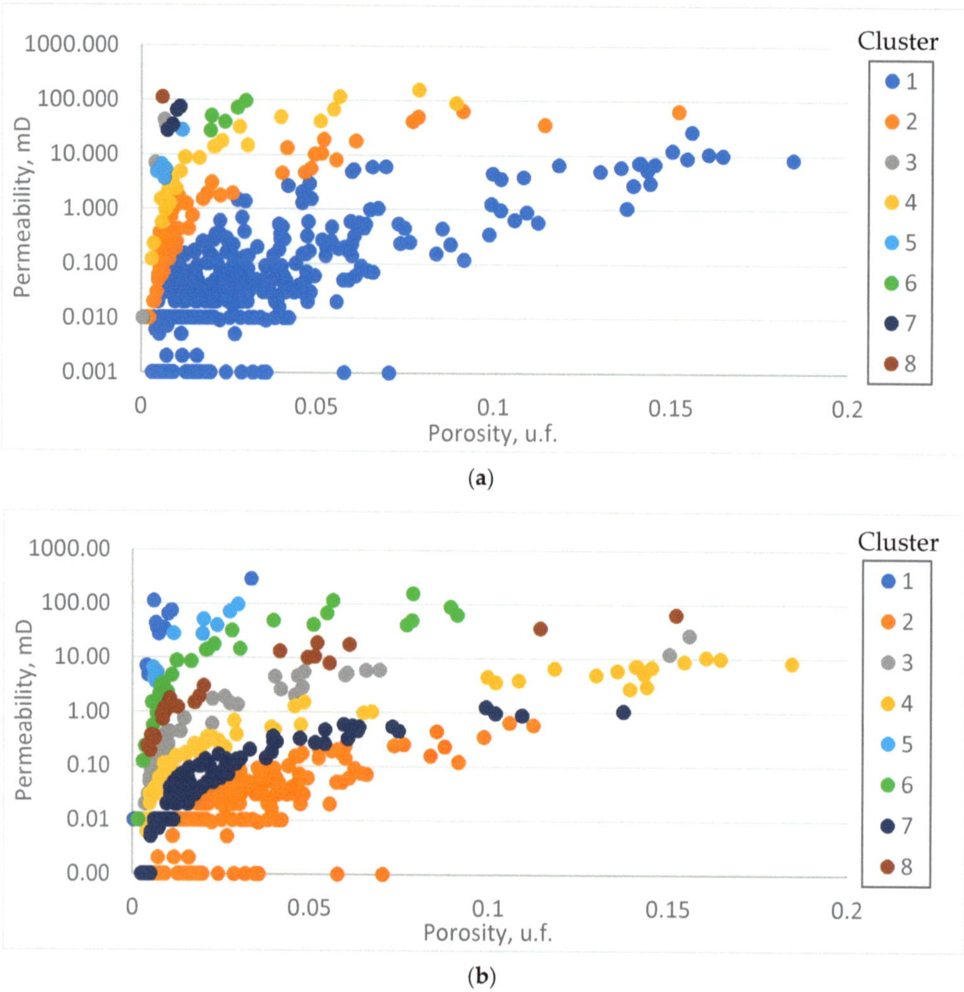

Figure 12. Comparison of: (**a**) k-means; (**b**) EM rock typing by FZI and R35 parameters.

Figure 11 shows that clustering by the *FZI* parameter does not reliably separate samples into rock types. The first class has a dominant influence on clustering. Therefore, it was decided to add the pore channel radius parameter calculated earlier to improve clustering quality (Figure 12).

With the *R35* parameter added for the *k*-means method, there is an improvement in the classification of rock types, but the first class remains predominant. The EM algorithm allowed the samples to be divided into almost equal parts, of course, with a preference for the first class, but this was not critical. Therefore, it is proposed that the EM algorithm and the distinguished criteria for FZI and R35 typing be used in further analysis. In the next step, calculations with $n = 10$ and $n = 12$ were performed to select the optimal number of clusters. The results of the classifications comparison are shown in Figure 13.

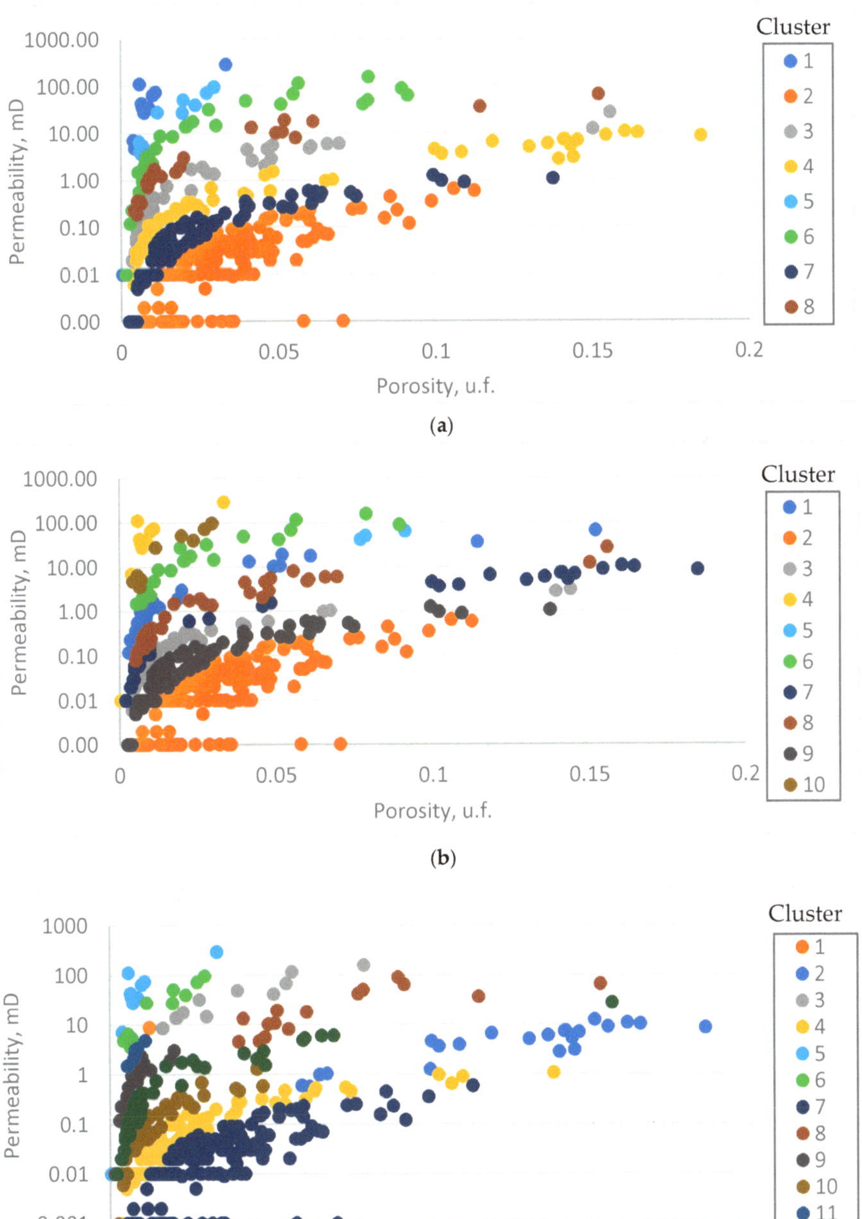

Figure 13. Comparison of rock typing by EM method to select the number of clusters (rock types). Number of clusters: (**a**) 8; (**b**) 10; (**c**) 12.

As a result, clustering models have been derived that allow rock types to be distinguished with a high degree of confidence, as confirmed by the coefficients of determination between permeability and porosity parameters within classes.

The comparison of the coefficients of determination for different numbers of clusters is presented in Table 2.

Table 2. Comparison of determination coefficients for different numbers of clusters. EM algorithm.

Cluster	8 Clusters		10 Clusters		12 Clusters	
	Equation	R^2	Equation	R^2	Equation	R^2
1	$y = 8902.5x - 17.533$	$R^2 = 0.88$	$y = 0.3547 \times e^{47.106x}$	$R^2 = 0.77$	(1 value)	(1 value)
2	$y = 3.3793x - 0.0587$	$R^2 = 0.61$	$y = 0.002 \times e^{60.971x}$	$R^2 = 0.68$	$y = 0.2162 \times e^{23.336x}$	$R^2 = 0.57$
3	$y = 122.08x - 0.9967$	$R^2 = 0.85$	$y = 0.0335 \times e^{42.529x}$	$R^2 = 0.87$	$y = 5.6505 \times e^{45.206x}$	$R^2 = 0.86$
4	$y = 48.027x - 0.5159$	$R^2 = 0.89$	$y = 3.7495 \times e^{181.16x}$	$R^2 = 0.85$	$y = 8.0369x - 0.0742$	$R^2 = 0.95$
5	$y = 3120.9x - 15.714$	$R^2 = 0.88$	$y = 5.9112 \times e^{26.089x}$	$R^2 = 0.91$	$y = 3.7495 \times e^{181.16x}$	$R^2 = 0.85$
6	$y = 1099.4x - 4.7463$	$R^2 = 0.70$	$y = 1557.2x - 12.372$	$R^2 = 0.82$	$y = 2.6719 \times e^{124.16x}$	$R^2 = 0.93$
7	$y = 9.0492x - 0.083$	$R^2 = 0.92$	$y = 0.0571 \times e^{34.385x}$	$R^2 = 0.65$	$y = 0.0018 \times e^{63.192x}$	$R^2 = 0.58$
8	$y = 383.7x - 3.6482$	$R^2 = 0.94$	$y = 0.3547 \times e^{47.106x}$	$R^2 = 0.77$	$y = 649.79x - 16.278$	$R^2 = 0.58$
9	-	-	$y = 9.0622x - 0.0838$	$R^2 = 0.92$	$y = 153.37x - 0.3226$	$R^2 = 0.82$
10	-	-	$y = 2.6729 \times e^{124.89x}$	$R^2 = 0.93$	$y = 18.335x - 0.0983$	$R^2 = 0.80$
11	-	-	-	-	$y = 0.4716 \times e^{214.35x}$	$R^2 = 0.94$
12	-	-	-	-	$y = 0.1061 \times e^{53.025x}$	$R^2 = 0.86$

From Table 2, it can be concluded that, in general, it was possible to improve rock typing compared to standard approaches (*GHE, DRT, R35*). When using the EM clustering algorithm, determination coefficients for all cases are higher than 0.57, even for classes characterizing the fractured type of reservoir. Based on the analysis of Table 2, differences in approximation formulas are observed. For calculation with the number of clusters equal to eight, all dependencies by classes are approximated linearly with high coefficients of determination. Essentially, this approach simplifies the dependencies between permeability and porosity, but in the modeling phase, it can have the effect of overestimating the permeability. Generally, exponential relationships are adopted to approximate porosity and permeability relationships.

For a calculation with the number of clusters equal to 10, only two classes stand out that need to be approximated linearly; the remaining dependencies are exponential. Importantly, the linear dependencies are obtained only for classes that are characterized by a high correlation between porosity and permeability. That is, when porosity increases, permeability values also increase. The types of reservoirs are pore and pore-cavernous. For more complex void space types, exponential relationships are used.

For calculations with the number of clusters equal to 12, the obtained coefficients of determination are lower, but nevertheless not lower than 0.57. Three classes characterizing pore, pore-cavernous, and cavernous-pore reservoir types are also distinguished, which are approximated linearly.

The rock types corresponding to the main secondary rock transformations that have influenced the change in void space structure have been identified. Rocks of rock types 4–5 are fractured, partly dolomitized; 6, 1, 5 fractured and leached, partly dolomitized; 7–8 mostly leached, partly dolomitized; rock types 3–9 are mostly porous, partly dolomitized; rock type 2 is porous, occasionally cavernous, dolomitization and partial leaching processes are present.

According to clustering algorithms implementation, two criteria for rock typing were indicated: *FZI* and *R35*. The use of these criteria combination significantly enhances the rock classification through clustering methods. Table 3 shows the comparison between all feasible rock typing methods for reservoir modeling.

Table 3. Comparison of different rock typing methods by statistical analysis of determination coefficients.

Method	Rock Types Numbers	Mean R^2	Min R^2	Max R^2	Variance	St. Deviation	Variation Coefficient
DRT	18	0.73	0.58 *	0.98	0.12	0.35	47.64
GHE	10	0.68	0.61 *	0.99	0.07	0.27	40.31
EM (8 clusters)	8	0.83	0.61	0.94	0.01	0.11	14.01
EM (10 clusters)	10	0.82	0.65	0.93	0.009	0.09	12.02
EM (12 clusters)	12	0.73	0.57 *	0.95	0.07	0.27	36.81

* Exclude zero values of R^2 that characterize classes with 1–4 number of core samples.

In terms of comparison, the results indicate that the EM clustering algorithm with 10 clusters exhibits one of the highest mean and minimal values of R^2, as well as the lowest variance and coefficient of variation. This suggests the least difference in R^2 between rock types and enhances the reliability of predicting the correlation between porosity and permeability. It is important to mention that, for the other methods, classes with a limited number of values stand out, preventing a dependable identification of the parameter correlation. Despite the high maximum R^2 values for the *DRT* and *GHE* methods, these are observed only for classes with a number of core samples less than 10 or due to the division into a large number of classes. While the *DRT* and *GHE* methods are consistent, the EM clustering algorithm is better suited to the geological settings of the Alpha oilfield.

Based on the study results, it is found that the *FZI* and *R35* criteria, 10 clusters (classes), and the EM clustering algorithm should be used for optimal and automatic rock typing.

5. Conclusions

A review of the main rock typing methods was carried out in this research. It was found that a significantly higher number of rock types (19) is identified on the basis of classification using the *DRT* discrete rock types method compared to *GHE* typing (10). This separation makes it possible to identify narrower rock types while increasing the coefficient of determination for certain classes. The *DRT* method produces high determination coefficients, but many classes are distinguished, which would make it difficult to distribute rock types when averaging properties onto a model grid and upscaling it.

The pore channel radius corresponding to 35% saturation of the pore volume with the nonwetting phase (*R35*) was calculated. A classification based on the R35 parameter has been carried out, but the method will not produce a reliable reservoir typology with high coefficients of determination between the parameters in each class.

Typing was carried out using the method of Lorenz plots. This method does not produce high permeability-porosity correlations between the parameters within the rock types.

Rock typing has been carried out based on the calculation of the accumulated correlation coefficient. This approach can be applied to reconstruct the petrophysical dependence cloud in the geological model, but in the authors' opinion, for the most correct rock typing, it is necessary to use the *FZI* parameter, which allows to distinguish rock types more correctly, especially when separating low-quality and fractured reservoirs.

By using different rock typing methods and evaluating the relationship between porosity and permeability, it was found that *FZI* and *R35* are the necessary criteria for selecting rock types, and their combination allows for improved quality of type predictions. On the basis of these criteria, clustering was carried out using machine learning methods. The clustering based on the EM algorithm allowed us to identify 10 rock types with high coefficients of determination. The identified rock types corresponded to the main secondary rock transformations that affected the structure of the void space. Rocks of rock types 4–5 are prone to fracturing, partly dolomitized; 1, 5, 6 are fractured and leached, partly dolomitized; 7–8 are mainly leached, partly dolomitized; rock types 3, 9 are mainly pore type, partly dolomitized; rock type 2 is porous, occasionally cavernous, dolomitization and partial leaching processes are present. These rock types have been chosen as the basis for further geological and dynamic modeling.

In future research, the authors plan to use machine learning algorithms for combining core and well-logging studies to determine rock types in intervals not characterized by core sampling. This will allow rock types to be distributed in the reservoir volume in the most reliable manner for both core-sampled intervals and the rest of the section. In the intervals of rock types, it is proposed to propagate the obtained permeability–porosity relationships for permeability array modeling, taking into account the high heterogeneity of the reservoir.

Author Contributions: Conceptualization, S.K. and A.K.; methodology, S.K.; software, A.K.; validation, N.K., A.B. and O.K.; formal analysis, A.K.; investigation, A.B. and E.O.; resources, S.K.; data curation, N.K.; writing—original draft preparation, A.K.; writing—review and editing, A.B.; visualization, N.K. and E.O.; supervision, O.K.; project administration, S.K.; funding acquisition, O.K. All authors have read and agreed to the published version of the manuscript.

Funding: This study was conducted under the Russian Science Foundation grant №. 22-17-00111.

Data Availability Statement: Not applicable.

Conflicts of Interest: The authors declare no conflict of interest.

References

1. Tavakoli, V. *Carbonate Reservoir Heterogeneity: Overcoming the Challenges*; Springer International Publishing: Tehran, Iran, 2019; pp. 1–108.
2. Sharifi-Yazdi, M.; Rahimpour-Bonab, H.; Nazemi, M.; Tavakoli, V.; Gharechelou, S. Diagenetic impacts on hydraulic flow unit properties: Insight from the Jurassic carbonate Upper Arab Formation in the Persian Gulf. *J. Pet. Explor. Prod. Technol.* **2020**, *10*, 1783–1802. [CrossRef]
3. Tabrizinejadas, S.; Fahs, M.; Hoteit, H.; Younes, A.; Ataie-Ashtiani, B.; Simmons, C.; Carrayrou, J. Effect of temperature on convective-reactive transport of CO_2 in geological formations. *Int. J. Greenh. Gas Control* **2023**, *128*, 103944. [CrossRef]
4. Al-Otaibi, M.; Abdullah, E.; Hanafy, S. Prediction of Permeability from Logging Data Using Artificial Intelligence Neural Networks. In Proceedings of the SPE Western Regional Meeting, Anchorage, AK, USA, 22–25 May 2023. [CrossRef]
5. Mustafa, R.; Sameera, H. Studying the Effect of Permeability Prediction on Reservoir History Matching by Using Artificial Intelligence and Flow Zone Indicator Methods. *Iraqi Geol. J.* **2023**, *56*, 9–21. [CrossRef]
6. Zolotukhin, A.B.; Gayubov, A.T. Machine learning in reservoir permeability prediction and modelling of fluid flow in porous media. *IOP Conf. Ser. Mater. Sci. Eng.* **2019**, *700*, 012023. [CrossRef]
7. Efremova, E.I.; Putilov, I.S. On the question of hydraulic flow units use in terrigenous deposits taking into account facies (on the example of the Sof'inskoye field in the Perm Krai). *J. Pet. Min. Eng.* **2022**, *22*, 52–57.
8. Antoniuk, V.; Bezrodna, I.; Petrokushyn, O. Multiple regressions and ann techniques to predict permeability from pore structure for terrigenous reservoirs, west-shebelynska area. *Eur. Assoc. Geosci. Eng.* **2019**, *2019*, 1–5. [CrossRef]
9. Srinivasan, S.; Leung, J.Y. *Petroleum Reservoir Modeling and Simulation: Geology, Geostatistics, and Performance Prediction*, 1st ed.; McGraw Hill: New York, NY, USA, 2022. Available online: https://www.accessengineeringlibrary.com/content/book/9781259834295 (accessed on 1 February 2023).
10. Haddad, A.; Lathion, R.; Courgeon, S.; Fabre, G.; Martinuzzi, V.; Pedraza, S.; Hauvette, L.; Games, F. Multi-Scale Karstic Reservoir Characterization: An Innovative Approach to Improve Reservoir Model Predictions and Decision Making. In Proceedings of the International Petroleum Technology Conference, Riyadh, Saudi Arabia, 21–23 February 2022. [CrossRef]
11. Mulyanto, B.; Dewanto, O.; Yuliani, A.; Yogi, A.; Wibowo, R. Porosity and permeability prediction using pore geometry structure method on tight carbonate reservoir Porosity and permeability prediction using pore geometry structure method on tight carbonate reservoir. In Proceedings of the 9th International Conference on Theoretical and Applied Physics (ICTAP), Bandar Lampung, Indonesia, 15 July 2020. [CrossRef]
12. Krivoshchekov, S.; Kochnev, A.; Kozyrev, N.; Ozhgibesov, E. Factoring Permeability Anisotropy in Complex Carbonate Reservoirs in Selecting an Optimum Field Development Strategy. *Energies* **2022**, *15*, 8866. [CrossRef]
13. Lathion, R.; Courgeon, S.; Martinuzzi, V.; Fabre, G.; Haddad, A.; Games, F. Identification and integrated characterization of large-scale karstic network: Critical impact on reservoir understanding and modeling. In Proceedings of the 8ème Congrès Français de Sédimentologie—ASF 2022, Brest, Belarus, 28–30 September 2022.

14. Shishaev, G.; Demyanov, V.; Arnold, D.; Vygon, R. History Matching and Uncertainty Quantification of Reservoir Performance with Generative Deep Learning and Graph Convolutions. In Proceedings of the International Petroleum Technology Conference, Riyadh, Saudi Arabia, 21–23 February 2022. [CrossRef]
15. Robail, F.; Sanyal, S.; Azudin, A.; Koh, K.; Nizar, F.; Rosli, U. Machine Learning for Facies Distribution of Large Carbonate Reservoir Models—A Case Study. In Proceedings of the International Petroleum Technology Conference, Bangkok, Thailand, 1–3 March 2023. [CrossRef]
16. Elarouci, F.; Maalouf, C.; Espinoza, I.; Al-jaberi, S.; Amer, M.; Smith, S.; Jalalh, A. Integrated Reservoir Characterization to Approach the Carbonate Permeability Distribution Challenges: Case Study from Offshore Abu Dhabi. In Proceedings of the Abu Dhabi International Petroleum Exhibition & Conference, Abu Dhabi, United Arab Emirates, 7–10 November 2016. [CrossRef]
17. Albreiki, M.; Geiger, S.; Corbett, P. Impact of modelling decisions and rock typing schemes on oil in place estimates in a giant carbonate reservoir in the Middle East. *Pet. Geosci.* **2021**, *28*, 1–21. [CrossRef]
18. Palabiran, M.; Sesilia, N.; Akbar, M.N.A. An Analysis of Rock Typing Methods in Carbonate Rocks For Better Carbonate Reservoir Characterization: A Case Study of Minahaki Carbonate Formation, Banggai Sula Basin, Central Sulawesi. In Proceedings of the 41th Scientific Annual Meeting of Indonesian Association of Geophysicists, Lampung, Indonesian, 26–29 September 2016.
19. Archie, G.E. Classification of Carbonate Reservoir Rocks and Petrophysical Considerations. *AAPG Bull.* **1952**, *36*, 278–298. [CrossRef]
20. Raznicyn, A.; Putilov, I. Development of a Methodological Approach to Identifying Petrophysical Types of Complicated Carbonate Rocks According to Laboratory Core Studies. *Perm J. Pet. Min. Eng.* **2021**, *21*, 109–116. [CrossRef]
21. Ghanbarian, B.; Lake, L.; Sahimi, M. Insights Into Rock Typing: A Critical Study. *SPE J.* **2018**, *24*, 230–242. [CrossRef]
22. Ellabad, Y.; Corbett, P.W.M.; Straub, R. Hydraulic Units approach conditioned by well testing for better permeability modeling in a North African Oil Field. In Proceedings of the 2001 International Symposium of the Society of Core Analysts, Edinburgh, UK, 17–19 September 2001.
23. Lazim, S.A.; Hamd-Allah, S.M.; Jawad, A. Permeability Estimation for Carbonate Reservoir (Case Study/South Iraqi Field). *Iraqi J. Chem. Pet. Eng.* **2018**, *19*, 41–45. [CrossRef]
24. El-Sawy, M.Z.; Abuhagaza, A.A.; Nabawy, B.; Lashin, A. Rock typing and hydraulic flow units as a successful tool for reservoir characterization of Bentiu-Abu Gabra sequence, Muglad basin, southwest Sudan. *J. Afr. Earth Sci.* **2020**, *171*, 103961. [CrossRef]
25. Nabawy, B.; Rashed, M.; Mansour, A.; Afifi, W. Petrophysical and microfacies analysis as a tool for reservoir rock typing and modeling: Rudeis Formation, off-shore October Oil Field, Sinai. *Mar. Pet. Geol.* **2018**, *97*, 260–276. [CrossRef]
26. Irawan, D.; Wihdany, F.I.; Idea, K. Permeability Anisotropy Effect in Reservoir Characterization: New Rock Typing Approach. In Proceedings of the Forty-Third Annual Convention & Exhibition, Jakarta, Indonesia, 4–6 September 2019.
27. Kolodzie, S. Analysis of pore throat size and use of the Waxman–Smits equation to determine OOIP in Spindle Field, Colorado. In Proceedings of the SPE Annual Technical Conference and Exhibition, Dallas, TX, USA, 21 September 1980. [CrossRef]
28. Pittman, E.D. Relationship of porosity and permeability to various parameters derived from mercury injection-capillary pressure curves for sandstone. *Am. Assoc. Pet. Geol. Bull.* **1992**, *76*, 191–198. [CrossRef]
29. Sousa, A.M.; Ines, N.; Bizarro, P.; Ribeiro, M.T. Improving Carbonate Reservoir Characterization and Modelling through the Definition of Reservoir Rock Types by Integrating Depositional and Diagenetic Trends. In Proceedings of the Abu Dhabi International Petroleum Exhibition and Conference, Abu Dhabi, United Arab Emirates, 10–13 November 2014. [CrossRef]
30. Permadi, P.; Susilo, A. Permeability Prediction and Characteristics of Pore Structure and Geometry as Inferred from Core Data. In Proceedings of the 2009 SPE/EAGE Reservoir Characterization and Simulation Conference, Abu Dhabi, United Arab Emirates, 19–21 October 2009. [CrossRef]
31. Choquette, P.W.; Pray, L.C. Geologic Nomenclature and Classification of Porosity in Sedimentary Carbonates. *Am. Assoc. Pet. Geol. Bull.* **1970**, *54*, 207–250. [CrossRef]
32. Dunham, R.J. Classification of Carbonate Rocks According to Depositional Texture. In *Classification of Carbonate Rocks—A Symposium*; The American Association of Petroleum Geologists: Tulsa, OK, USA, 1962; pp. 108–121.
33. Repina, V.A.; Galkin, V.I.; Galkin, S.V. Complex petrophysical correction in the adaptation of geological hydrodynamic models (on the example of Visean pool of Gondyrev oil field). *J. Min. Inst.* **2018**, *231*, 268–274. [CrossRef]
34. Galkin, V.I.; Ponomareva, I.N.; Repina, V.A. Study of the process of oil recovery in reservoirs of various types of voids using multivariate statistical analysis. *Bull. Perm Natl. Res. Polytech. Univ. Geol. Oil Gas Eng. Min.* **2016**, *15*, 145–154.
35. Putilov, I.; Kozyrev, N.; Demyanov, V.; Krivoshchekov, S.; Kochnev, A. Factoring in Scale Effect of Core Permeability at Reservoir Simulation Modeling. *SPE J.* **2022**, *27*, 1930–1942. [CrossRef]
36. Galvao, P.; Halihan, T.; Hirata, R. The karst permeability scale effect of Sete Lagoas, MG, Brazil. *J. Hydrol.* **2016**, *532*, 149–162. [CrossRef]
37. Corbett, P.W.M.; Potter, D.K. Petrotyping: A basemap and atlas for navigating through permeability and porosity data for reservoir comparison and permeability prediction. In Proceedings of the International Symposium of the Society of Core Analysts, Abu Dhabi, United Arab Emirates, 5–9 September 2004.
38. Garrouch, A.A.; Al-Sultan, A.A. Exploring the link between the flow zone indicator and key open-hole log measurements: An application of dimensional analysis. *Pet. Geosci.* **2019**, *25*, 219–234. [CrossRef]
39. Lake, L.; Jensen, J. A review of heterogeneity measures used in Reservoir Characterisation. *In Situ* **1991**, *15*, 409–440.

40. Gunter, G.W.; Finneran, J.M.; Hartmann, D.J.; Miller, J.D. Early Determination of Reservoir Flow Units Using an Integrated Petrophysical Method. In Proceedings of the SPE Annual Technical Conference and Exhibition, San Antonio, TX, USA, 5 October 1997. [CrossRef]
41. Korablev, N.M.; Fomichev, A.A. K-means data clustering using artificial immune systems. *Bionics Intell. Sci. Mag.* **2011**, *3*, 102–106.
42. Mirkin, B.G. *Clustering for Data Mining. A Data recovery Approach*; Chapman and Hall/CRC: New York, NY, USA, 2005; pp. 269–296. [CrossRef]
43. Bradley, P.S.; Fayyad, U.M. Refining Initial Points for K-Means Clustering. In Proceedings of the International Conference on Machine Learning, San Francisco, CA, USA, 24–27 July 1998.
44. Cherezov, D.S.; Tukachev, N.A. Classification and clasterization base methods review. *Bull. Voronezh State Univ. Ser. Syst. Anal. Inf. Technol.* **2009**, *2*, 25–29.

energies

Article

Well-Test Interpretation Model of Water-Injection Well in a Low-Permeability Reservoir and Its Application

Jianxiong Li [1], Kai Wang [1], Sheng Su [1], Kejun Huang [1], Haoran Feng [2,*], Ziyi Jin [3], Cong Peng [2], Jinbao Liu [2] and Pin Jia [2]

1 The 4th Oil Production Plant of Changqing Oilfield Branch of CNPC, Yinchuan 750005, China; sus01_cq@petrochina.com.cn (S.S.)
2 College of Petroleum Engineering, China University of Petroleum (Beijing), Beijing 102249, China; 2022210272@student.cup.edu.cn (C.P.); jiapin1990@163.com (P.J.)
3 PetroChina Southwest Oil and Gas Field Company, Sichuan Shale Gas Exploration and Development Co., Ltd., Chengdu 610057, China; jzy_ynwa2020@163.com
* Correspondence: 2022210271@student.cup.edu.cn

Abstract: For low-permeability reservoirs, water-flooding development is usually adopted, which leads to induced fractures near the wellbore, increasing reservoir heterogeneity, and making water-flooding development more complex. This paper focuses on low-permeability reservoirs, considering the characteristics of induced fractures and elliptic-flow composite, and the well-test model for injection wells is established. The mathematical model in Laplace space is obtained through dimensionless transformation and Laplace transformation. Subsequently, the Mathieu function is introduced to obtain the bottom hole pressure, and the pressure response curve is drawn. The six flow stages of the curve are defined, and the sensitivity of parameters such as half-length of induced fractures, range of lateral-swept area, permeability in unswept area, and outer boundary distance at constant pressure are analyzed. The results show that the half-length of the fracture mainly affects the linear flow of the fracture, the range of the lateral wave-affected area mainly affects the radial flow of the swept area, the permeability of the unswept area mainly affects the radial flow of the unswept area, and the outer boundary distance at constant pressure mainly affects the boundary flow. Based on the production performance of a certain injection well in J Oilfield, a series of key parameters are obtained through analytical solution model inversion, including the induced-fracture half-length of 10.32 m, the lateral-swept range of elliptic partition flow of 128.95 m, the permeability of the swept area of 6.87 mD, and the mobility ratio of 119.92, which show the superiority of the analytical solution model.

Keywords: low-permeability reservoir; induced fracture; elliptic-flow composite; well-test model

Citation: Li, J.; Wang, K.; Su, S.; Huang, K.; Feng, H.; Jin, Z.; Peng, C.; Liu, J.; Jia, P. Well-Test Interpretation Model of Water-Injection Well in a Low-Permeability Reservoir and Its Application. *Energies* **2024**, *17*, 5663. https://doi.org/10.3390/en17225663

Academic Editor: Nikolaos Koukouzas

Received: 1 November 2023
Revised: 12 March 2024
Accepted: 28 May 2024
Published: 13 November 2024

1. Introduction

For low-permeability reservoirs, although water-injection-induced fractures help to improve the injection capacity, due to the contradiction between the low permeability of the formation and the high conductivity of the induced fractures, the injected water flows along the fractures at high speed, only affecting the narrow and long areas, resulting in a serious imbalance of waterflooding, aggravating the dynamic heterogeneity of low-permeability reservoirs, making the water-flooding process more complicated [1–4]. The study of water-injection-induced fractures started in the 1980s. Hagoort [5] clearly put forward the concept of water-injection-induced fractures in his doctoral thesis and proposed that induced fractures will be formed when the injection pressure exceeds the formation fracture pressure. Furthermore, he used the numerical simulation method to simulate the fracture extension process caused by water injection. Liu et al. [6] believed that high water-injection volume and injection pressure would lead to the opening, extension, and connection of natural micro-fractures. They produced high-permeability strips in the direction of the connection line between the water-injection well and the oil-production well, resulting in a one-way penetration of injected water, ultimately affecting recovery. Guo et al. [7] took a water-injection well in the WY area as an example to analyze the characteristics of bottom hole pressure response under different test times and found that the cracks induced around the water-injection well were dynamic, reflecting the process of cracks gradually extending under the action of continuous water injection. In the same year, Wang et al. [8] and Xie et al. [9] considered that micro-fractured reservoirs are generally developed in low-permeability conditions, and they are closed in the original formation, which is not easy to identify in the early stage. These closed fractures will continue to open and extend with the increase in water-injection pressure and formation pressure, forming dynamic fractures.

In view of the well-test interpretation model for the formation of induced fractures in water-injection wells, it is believed that due to the absence of proppant in water-injection-induced fractures, when the well is shut down for a pressure test, the fracture will gradually close with the decrease in bottom hole pressure. The water previously stored in the fracture is squeezed into the formation, and the early pressure response after the well shuts down presents a straight line with a pressure derivative unit slope [5]. Spivey and Lee [10] proposed a dual-volume wellbore storage model considering fractures, wellbore storage, and skin effects. The bottom hole pressure solution was given, two reservoir sections were obtained, the pressure derivatives coincided, and the slope was one unit. BinAkresh and Rahman [11,12] analyzed the test data of water-injection wells, verified the law shown by the theoretical model, and clarified the reservoir effect of fractures after shut-in. Koning [13,14] considered fracture closure because of the influence of well storage and considered different fracture shapes under an elliptic coordinate system. The pressure solution before fracture closure and the pressure solution after fracture closure were obtained by partial integration. The mathematical model of fracture closure after the injection-well shut-in was given in the form of a piecewise function.

For water-flooding reservoirs, the fluid in the formation changes from single-phase flow to oil–water two-phase flow due to the injection of water. At this time, well-test data of the injection wells need to be based on the mathematical model of oil–water two-phase flow. The flow theory of composite reservoirs is usually used for well-test analysis. A composite reservoir has the special characteristic that the discontinuity of the reservoir or the fluid's properties in the radial direction is caused by the injection of fluid so that the reservoir is divided into two or more flow areas in the radial direction. In the 1950s, the water-flooding well-test interpretation of water-injection wells was mainly based on the radial-composite reservoir model. The reservoir was divided into two to three regions in the radial direction. The reservoir and fluid properties were the same in a single region, but the properties between regions were different [15]. Considering the wellbore storage and skin effect, Li et al. [16] established a mathematical model of an oil–water two-phase unsteady flow well test considering the influence of water cut and drew a new typical curve chart of oil–water two-phase flow well test. Liu et al. [17] established a well-test mathematical model of a two-zone composite reservoir with water-injection wells considering the influence of water cut and solved it analytically. The effects of different water cuts, water saturations, permeability ratios, water drive front radii, and other parameters on the well-test curve were analyzed. Jia et al. [18] established a mathematical model of an oil–water two-phase flow well test in dual-media reservoirs. Based on the B-L water-flooding theory, the water-saturation distribution was obtained at any time, and the pressure solution model was divided into multiple ring regions in the radial direction. Then, the model was solved analytically, and the characteristics of the well-test curve of oil–water two-phase flow in water-injection wells were analyzed. According to the one-dimensional B-L water-flooding equation, the position of the oil–water two-phase interface could be obtained at any time, and the water-saturation distribution during the water-injection process could be further solved. In addition, Soliman et al. [19,20] and Craig and Blasingame [21,22] have also performed much research on the well-test interpretation model of water-injection wells in low-permeability reservoirs.

At present, there are two main defects in the research of the well-test interpretation model for water-injection wells in low-permeability reservoirs. Firstly, there is no effective method for characterization and inversion of water-injection-induced fractures. The current research is mainly to analyze the influence of water-injection-induced fractures on the development of low-permeability reservoirs. Although some scholars have proposed mathematical models to characterize this process, it is still impossible to diagnose and interpret the pressure data of most water-injection wells in the field. Secondly, a circular partition is usually considered for the composite partition problem, while the water-injection-induced fractures often form elliptic partitions, and there are no effective diagnostic and discriminant criteria for fracture extension and propagation.

In order to make up for the shortcomings of the existing research, this paper will establish a well-test interpretation model for water-injection wells considering induced fractures and elliptic water-flooding zones on the basis of predecessors; analyze the influence of well-test pressure template curve characteristics and sensitivity parameters; and form a well-test interpretation method for induced fractures and elliptic water-flooding flow characteristics, so as to realize quantitative characterization of fractures and quantitative inversion of dynamic parameters. The key parameters of the reservoir are inverted by the real dynamic data of a water-injection well in a low-permeability reservoir to verify the model results.

2. Physical Model

There is a water-injection well in the center of the heterogeneous reservoir. After long-term water injection, an oil–water two-phase zone is formed in the near-well waterflood-swept area, and a single-phase oil zone is formed in the far-well waterflood-unswept area. Due to the distribution of water saturation, the mobility of the swept area is different from that of the unswept area, and the reservoir shows the characteristics of elliptic zone during the shut-in test. At the same time, considering the dynamic change of fracture induced by water injection, the pressure is conducted from the swept area to the unswept area, and the outer boundary of the unswept area is disturbed by the adjacent injection well.

As shown in Figure 1, the basic assumptions of the model are:

(1) The formation in the two regions is horizontally homogeneous and isotropic;
(2) The outer boundary of the reservoir is a finite elliptic constant-pressure outer boundary;
(3) Vertical fractures with infinite conductivity are developed around the injection well, which connect the reservoir and the wellbore, and the closure of fractures are not considered;
(4) The formation fluid and rock are slightly compressible, and the compression coefficient is constant. The inner area is oil–water two-phase, and the outer area is single-phase;
(5) The flow of formation fluid satisfies the linear Darcy's law;
(6) The influence of wellbore storage and skin are considered, while the influence of gravity and capillary force are ignored.

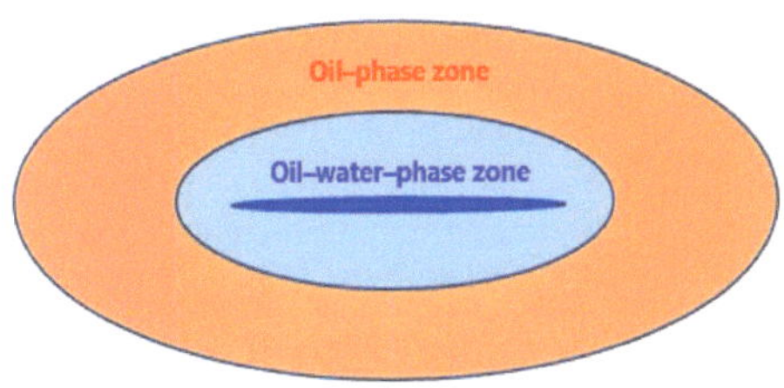

Figure 1. Schematic diagram of physical model of induced fracture by injection-well and elliptic-flow composite.

3. Mathematical Model

For the elliptic partition characteristics, the mobility ratio is introduced to represent the ratio of the flow coefficient between the near-well waterflood-swept zone and the far-well waterflood-unswept zone, which characterizes the difference in physical properties between the two zones. The larger the value, the greater the mobility difference between the inner and outer zones [23,24].

$$M = \frac{k_1/\mu_1}{k_2/\mu_2} \tag{1}$$

where k_1 and k_2 (mD) are the permeabilities of the swept zone and the unswept zone, and μ_1 and μ_2 (mPa·s) are the viscosities of the fluid in the swept area and the unswept area, respectively.

For the unsteady flow mathematical model of reservoir fluid with water-injection-induced fracture and elliptic partition characteristics, according to the principle of conservation of mass, the establishment of the basic differential equation for the unsteady flow mathematical model needs to be integrated into the following equation [25].

For slightly compressible fluid, the two-dimensional flow equation in the rectangular coordinate system is:

$$\frac{\partial^2 p}{\partial x^2} + \frac{\partial^2 p}{\partial y^2} = \frac{\phi \mu C_t}{k}\frac{\partial p}{\partial t} \tag{2}$$

By introducing two variables and using the Mathieu function, Equation (2) is transformed into the elliptic coordinate system, which is defined as [26]:

$$z = x + iy = L\cosh(\xi + i\eta) \tag{3}$$

$$\overline{z} = x - iy = L\cosh(\xi - i\eta) \tag{4}$$

The Mathieu function transformation generates the diffusion equation of the slightly compressible fluid in the elliptic coordinate system:

$$\frac{\partial^2 p}{\partial \xi^2} + \frac{\partial^2 p}{\partial \eta^2} = \frac{L^2}{2}(\cosh 2\xi - \cos 2\eta)\frac{\phi \mu C_t}{k}\frac{\partial p}{\partial t} \tag{5}$$

where ξ and η are the spatial coordinates in the elliptic-flow geometry as shown in the schematic diagram of the elliptic coordinate system.

In order to obtain the transient pressure characteristics of water-injection wells and reservoirs in a large time range under boundary conditions, the dimensionless solution of the diffusion equation in the swept area and the unswept area is needed. Therefore, the following dimensionless parameters are defined:

$$t_D = \frac{2.634 \times 10^{-4} k_1 t}{\phi_1 \mu_1 C_{t1} L^2} \tag{6}$$

$$D = \frac{k_1/(\phi_1 \mu_1 C_{t1})}{k_2/(\phi_2 \mu_2 C_{t2})} \tag{7}$$

where t_D is the dimensionless time; D is the pressure coefficient ratio; Φ_1 and Φ_2 are the porosities of the swept area and the unswept area, respectively; C_{t1} and C_{t2} (MPa^{-1}) are the comprehensive compression coefficients of the swept area and the unswept area, respectively; and L is the half-length of the water-injection-induced fracture.

According to the inner boundary conditions, the dimensionless pressure is defined as:

$$p_{Dj} = \frac{k_1 h}{q B_1 \mu_1}(p_i - p_j) \tag{8}$$

where j = 1 and 2 represent the swept area and the unswept area, respectively.

Then, the dimensional Equation (5) is transformed into the dimensionless form. The mathematical model of the swept area is given by:

$$\frac{\partial^2 p_{D1}}{\partial \xi^2} + \frac{\partial^2 p_{D1}}{\partial \eta^2} = \frac{1}{2}(\cosh 2\xi - \cos 2\eta)\frac{\partial p_{D1}}{\partial t_D} \tag{9}$$

The mathematical model of the unswept area is given by:

$$\frac{\partial^2 p_{D2}}{\partial \xi^2} + \frac{\partial^2 p_{D2}}{\partial \eta^2} = \frac{D}{2}(\cosh 2\xi - \cos 2\eta)\frac{\partial p_{D2}}{\partial t_D} \tag{10}$$

The initial condition is given by:

$$p_{D1}(\xi, \eta, 0) = p_{D2}(\xi, \eta, 0) = 0 \tag{11}$$

The inner boundary condition is given by:

$$\frac{\partial p_{D1}}{\partial \xi}(\xi, \eta, t_D)|_{\xi=\xi_w} = 1 \tag{12}$$

The continuity conditions are given by:

$$p_{D1}(\xi_0, \eta, t_D) = p_{D2}(\xi_0, \eta, t_D) \tag{13}$$

$$\frac{\partial p_{D2}}{\partial \xi}(\xi_0, \eta, t_D) = M\frac{\partial p_{D1}}{\partial \xi}(\xi_0, \eta, t_D) \tag{14}$$

The constant-pressure outer boundary condition is given by:

$$p_{D2}\Big(\xi, \eta, t_D)|_{\xi=\xi_e} = 0 \tag{15}$$

4. Model Solving

Based on the Laplace transform of the above equations, we obtain:

$$\overline{p}(s) = \int_0^\infty p_D(t_D)e^{-st}dt_D \tag{16}$$

Since the pressure is only a function of the elliptic space variable u, the transformed equation is solved by separating the variables:

$$\overline{p}_D(\xi, \eta) = X(\xi)Y(\eta) \tag{17}$$

The following equations are obtained by this method [27]:

$$\frac{\partial^2 Y}{\partial \eta^2} + (a + 2\lambda \cos 2\eta)Y = 0 \tag{18}$$

$$\frac{\partial^2 X}{\partial \xi^2} - (a + 2\lambda \cos 2\xi)X = 0 \tag{19}$$

Equations (18) and (19) are called Mathieu differential equations [24]. Their solutions are given by the Mathieu function. It should be noted that the solution under actual production conditions should satisfy the following conditions:

(1) The solution must be periodic, and the period is π;
(2) The solution should be continuous along the orthogonal interval line:

$$\overline{p}_D(0, \eta) = \overline{p}_D(0, -\eta)$$

At the same time, the pressure gradient should meet the following conditions:

$$\frac{\partial}{\partial \xi}[\overline{p}_D(\xi,\eta)]\bigg|_{\xi\to 0} = -\frac{\partial}{\partial \xi}[\overline{p}_D(\xi,-\eta)]\bigg|_{\xi\to 0}$$

(3) The solution should be symmetric with respect to the long axis and the short axis of the ellipse;
(4) The solution should be bounded.

The Mathieu function pairs satisfying these four conditions are:

$$Ce_{2n}(\xi,-\lambda)ce_{2n}(\eta,-\lambda),\ Fek_{2n}(\xi,-\lambda)ce_{2n}(\eta,-\lambda)$$

Then, the solutions of Equations (18) and (19) are given by this set of linearly independent functions:

$$Y(\eta) = ce_{2n}(\eta,-\lambda) \tag{20}$$

$$X(\xi) = Ce_{2n}(\xi,-\lambda) + Fek_{2n}(\xi,-\lambda) \tag{21}$$

Next, we substitute (20) and (21) into (17) to obtain:

$$\begin{gathered}\overline{p}_{D1}(\xi,\eta) = C_{2n}Ce_{2n}(\xi,-\lambda)ce_{2n}(\eta,-\lambda)\\ +F_{2n}Fek_{2n}(\xi,-\lambda)ce_{2n}(\eta,-\lambda)\end{gathered} \tag{22}$$

For the swept area, C_{2n} and F_{2n} in Equation (22) are two separate Fourier coefficients used to satisfy the internal and external boundary conditions. Due to being replaced, the pressure solutions of the unswept area are slightly different, and the expression is:

$$\begin{gathered}\overline{p}_{D2}(\xi,\eta) = B_{2n}Ce_{2n}(\xi,-q)ce_{2n}(\eta,-q)\\ +E_{2n}Fek_{2n}(\xi,-q)ce_{2n}(\eta,-q)\end{gathered} \tag{23}$$

Equations (22) and (23) represent the general solutions of the pressure in the swept area and the unswept area, respectively, and then the special solutions are obtained by the boundary condition.

For the internal boundary conditions, according to Darcy's law, the flow rate from the reservoir into the induced fracture is obtained by the following formula:

$$q = \frac{kh}{\mu}\int_0^{2\pi}\left(\frac{\partial p}{\partial \xi}\right)_{\xi=\xi_w} d\eta \tag{24}$$

After dimensionless processing and Laplace transform, we obtain:

$$\overline{q}_D = \frac{2}{\pi}\int_0^{2\pi}\left(\frac{\partial p_D}{\partial \xi}\right)_{\xi=\xi_w} d\eta \tag{25}$$

Then the derivation of (22) is substituted into (25), and the integration is carried out according to the periodicity of the Mathieu function, and the following results are obtained:

$$\int_0^{2\pi} ce_{2n}(\eta,-\lambda)d\eta = \frac{\pi}{2}A_0^{2n} \tag{26}$$

The following expression of dimensionless yield can be obtained:

$$\overline{q}_D = \sum_{n=0}^{\infty} A_0^{2n}\left[C_{2n}Ce'_{2n}(\xi_w,-\lambda) + F_{2n}Fek'_{2n}(\xi_w,-\lambda)\right] \tag{27}$$

According to Duhamel's principle:

$$1/s^2 = \overline{p}_{wD}\overline{q}_D \tag{28}$$

In Laplace space:

$$1/(4\lambda)^2 = \overline{p}_{wD}\overline{q}_D \tag{29}$$

According to the Laplace transform, the solution of the model is obtained:

$$\overline{p}_{wD} = \frac{s^2}{\sum_{n-0}^{\infty} A_0^{2n}\left[C_{2n}Ce'_{2n}(\xi_w, -s/4) + F_{2n}Fek'_{2n}(\xi_w, -s/4)\right]} \tag{30}$$

Considering wellbore storage and skin effect:

$$\overline{p}_{wD} = \frac{s\overline{p}_{wD} + S}{s + C_{wD}s^2(s\overline{p}_{wD} + S)} \tag{31}$$

5. Model Verification

In order to verify the accuracy of the established model, the numerical well-test module of the commercial software Saphir 5.10® was used to establish a constant-injection production model of a water-injection well in the center of a circular formation with a constant-pressure boundary. The formation was homogeneous and of equal thickness, with induced fractures near the well; a radial-composite zonal flow was formed, and the fluid was single-phase oil and slightly compressible. The relevant parameters of the reservoir and fluid are shown in Table 1.

Table 1. Model input parameters.

Parameter	Value
Initial pressure/MPa	25
Wellbore radius/m	0.15
Porosity	0.1
Oil viscosity/(mPa·s)	0.8
Half-length of induced fracture/m	20
Composite radius/m	120
Mobility	80
Distance of constant-pressure boundary/m	1000
Injection duration before shut-in/h	800
Injection rate before shut-in/($m^3 \cdot d^{-1}$)	25

By changing the influence parameters, the log–log theoretical curves of pressure and derivative under the analytical solution model are drawn and then compared with those of the commercial software.

As can be seen from Figure 2, the analytical solution model fits well with the theoretical curve drawn by the commercial software, and the log–log curve shows typical induced-fracture and radial-composite characteristics, which verifies the reliability of the analytical solution model. The analytical solution model can also explain the range of lateral sweep of elliptic partition flow which cannot be obtained by commercial software, which shows the superiority of the analytical solution model.

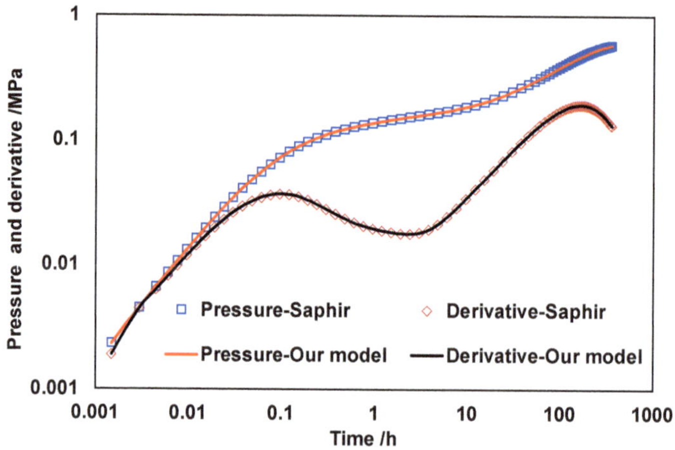

Figure 2. Analytical solution model and commercial software log–log curve fitting results.

6. Curve Analysis

6.1. Flow Regimes

For the analytical solution in Laplace space, the Stehfest numerical algorithm is used for numerical inversion, and the solution in real space can be obtained. Thus, the dimensionless well-test pressure pattern curve is drawn by programming. Finally, the dimensionless time and dimensionless bottom hole pressure P_{wD} of the water-injection-well induced-fracture elliptic composite partition model and the double logarithmic curve of the corresponding pressure derivative under dimensionless time are obtained, as shown in Figure 2.

It can be seen from Figure 3 that there are six typical stages in the well-test pressure pattern curve of the water-injection-well-induced fracture with elliptic composite system model:

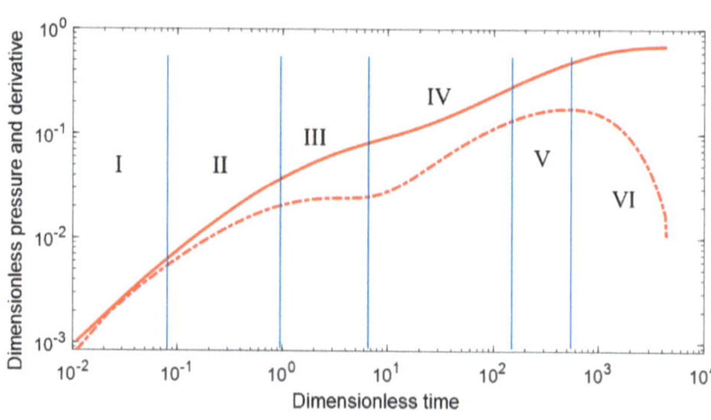

Figure 3. Well-test pressure curve of induced fracture by injection well and elliptic-flow composite.

Section I is the wellbore storage stage; the curve is two straight lines with a slope of 1. The well-storage effect is expressed by the well-storage coefficient C, with the unit m^3/MPa, which reflects the volume change of the wellbore fluid affected by the elastic compressibility with the increase or decrease in the pressure of the wellbore fluid;

Section II is the stage of induced-fracture linear flow. At this stage, the pressure derivative curve presents a straight line with a one-half slope;

Section III is the radial flow stage of the swept area. The pressure wave spreads outward in a circle, and the pressure derivative curve on the well-test pressure pattern curve is a horizontal line, which is the characteristic response of the radial flow in the swept area;

Section IV is the transition flow stage. The pressure wave spreads from the swept area to the unswept area, and the degree of change of the pressure curve depends on the mobility ratio between the swept area and unswept area;

Section V is the radial flow stage of the unswept area. As the pressure wave in the unswept area gradually spreads outward, the fluid of the matrix system in the unswept area flows radially into the swept area. The pressure derivative curve on the well-test pressure pattern curve is a horizontal line, which is the characteristic response of the radial flow in the unswept area. The height of the two horizontal lines depends on the mobility ratio between the swept area and unswept area.

Section VI is the constant-pressure boundary response stage. After the pressure wave is transmitted to the adjacent water-injection well, the flow will eventually reach a stable flow state due to the constant pressure. The pressure derivative curve on the well-test pressure pattern curve drops sharply, which is the characteristic response of the constant-pressure outer boundary.

The schematic diagram of each flow stage of the model is shown in Figure 4. The arrows represent the direction of flow.

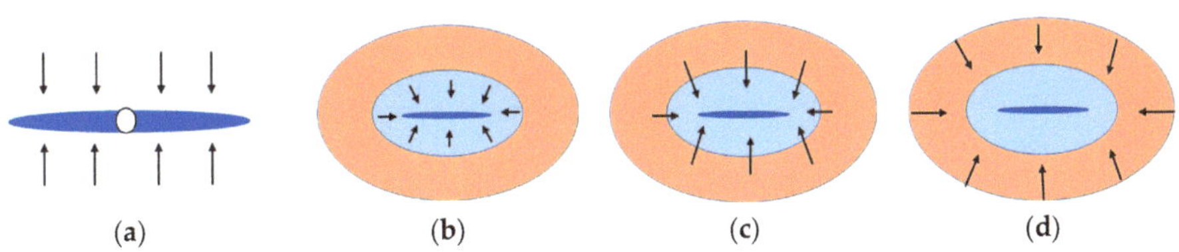

Figure 4. Schematic diagram of flow stage of induced fracture by injection well and elliptic-flow composite model. (**a**) Induced-fracture linear flow; (**b**) Radial flow; (**c**) Transition flow; (**d**) Boundary-dominated flow.

6.2. Analysis of Influencing Factors

The well-test pressure response characteristics of the induced fractures and elliptic-flow model mainly consider four influencing factors: the half-length of the induced fracture, the range of the lateral-swept area, the permeability of the unswept area, and the outer boundary distance of constant pressure. Figures 5–8 show the well-test pressure pattern curves of each influencing factor. The lines refer to the dimensionless pressure curves and the dashed lines refer to the pressure derivative curves.

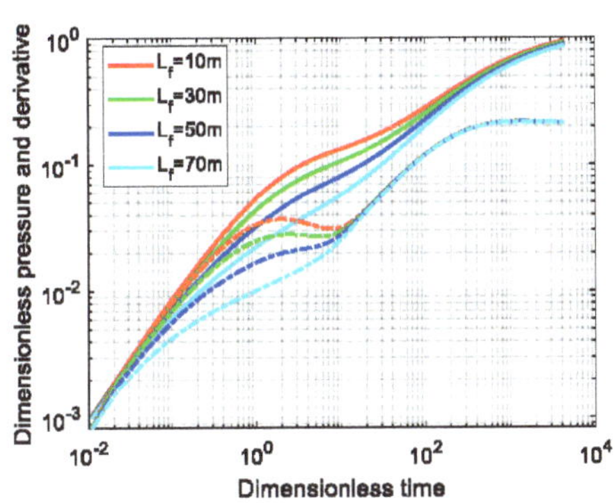

Figure 5. The influence of half-length of induced fracture on well-test pressure curve.

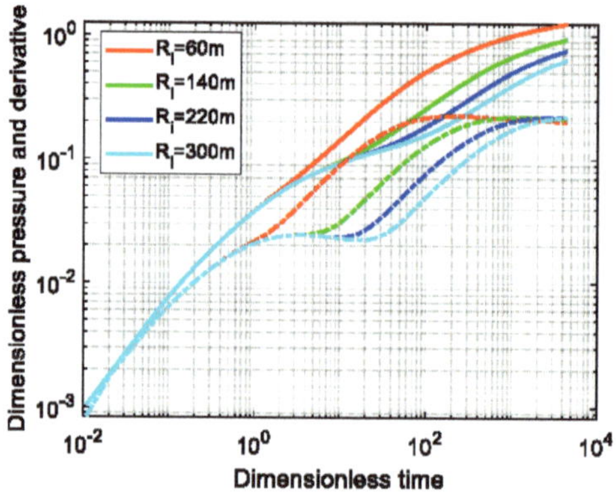

Figure 6. The influence of the range of lateral-swept area on well-test pressure curve.

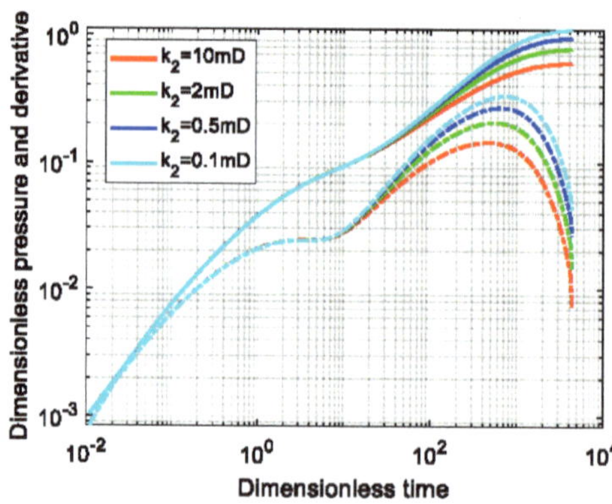

Figure 7. The influence of the permeability of the unswept area on well-test pressure curve.

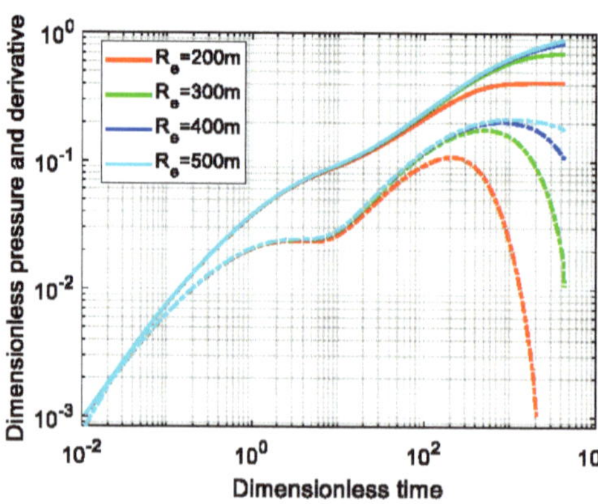

Figure 8. The influence of the outer boundary distance of constant pressure on well-test pressure curve.

(1) The half-length of induced fracture L_f

The half-length of the induced fracture mainly affects phase II of the fracture linear flow, indicating that a fractured flow channel has been formed around the injection well, injected water has been following unidirectional displacement along the fracture, and the pressure wave has gradually propagated outward. As can be seen from Figure 5, the larger the half-length of the induced fracture, the more obvious the fracture linear flow characteristics, and the shorter the radial-flow phase III duration in the corresponding swept area.

(2) The range of lateral-swept area R_i

The range of a lateral-swept area mainly affects the radial-flow phase III in the sweep area. The pressure wave propagates outward in a circle, and radial flow occurs in the near-well area. As can be seen from Figure 6, the larger the range of lateral sweep, the longer the duration of the radial-flow phase in the sweep region, and the earlier the occurrence of the pressure derivative warping up.

(3) The permeability of the unswept area k_2

The permeability of the unswept area mainly affects the radial-flow phase V in the unswept region. As the pressure wave in the unswept region gradually diffuses outward, the matrix system fluid in the unswept region flows radially into the swept region. As can be seen from Figure 7, the greater the permeability in the unswept region, the lower the pressure derivative of the radial flow in phase V in the unswept region.

(4) The outer boundary distance of constant pressure R_e

The outer boundary distance of constant pressure mainly affects phase VI of boundary flow. With the increase in time, after the pressure wave is transmitted to the nearby water-injection well, the seepage will eventually reach a stable flow state. As can be seen from Figure 8, the smaller the outer boundary distance is, the earlier the boundary-flow phase VI appears, and the less obvious the characteristics of the corresponding radial-flow phase V in the unswept region.

7. Application Analysis

A low-permeability reservoir in J Oilfield in northwest China was selected as the research object. The buried depth of the reservoir in J Oilfield is 1500–1800 m, the average effective thickness of the main reservoir formation is 15.6 m, the average porosity is 11.5%, and the average permeability is 0.46 mD. It is a typical low-permeability reservoir with poor reservoir physical properties, strong heterogeneity, and fracture. After the pressure data points obtained from the production field are preprocessed, the pressure derivative is solved by the Bourdet algorithm, and the actual curve is drawn. Finally, the theoretical curve and the actual curve are fitted to perform parameter inversion by the least square method and other algorithms.

After 351.3 h of performing the pressure drop test, the formation pressure of well L28-38 decreased by 0.63 MPa. The basic data from the model are shown in Table 2, the production performance of the test section of well L28-38 is shown in Figures 9 and 10, and the fitting results of the analytical solution model are shown in Figure 11.

Table 2. Parameters input in model.

Parameter	Value
Effective thickness/m	23.4
Initial pressure/MPa	21.86
Wellbore radius/m	0.108
Porosity	0.12
Viscosity/(mPa·s)	1.00
Crude oil density/(g·cm^{-3})	0.74
Volumetric coefficient/(m^3·m^{-3})	1.00
Composite compressibility/MPa^{-1}	1.05×10^{-3}
Injection rate before shut-in/(m^3·d^{-1})	27.8

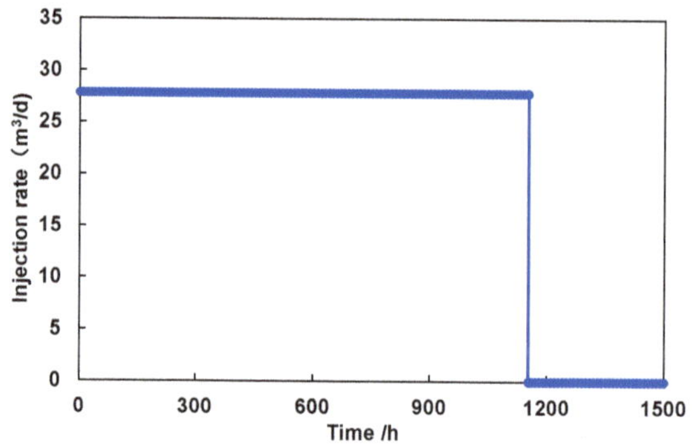

Figure 9. Production curve of test interval on case well.

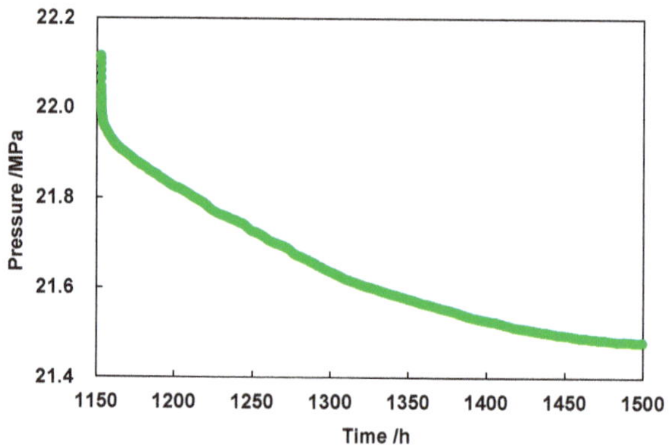

Figure 10. Pressure curve of test interval on case well.

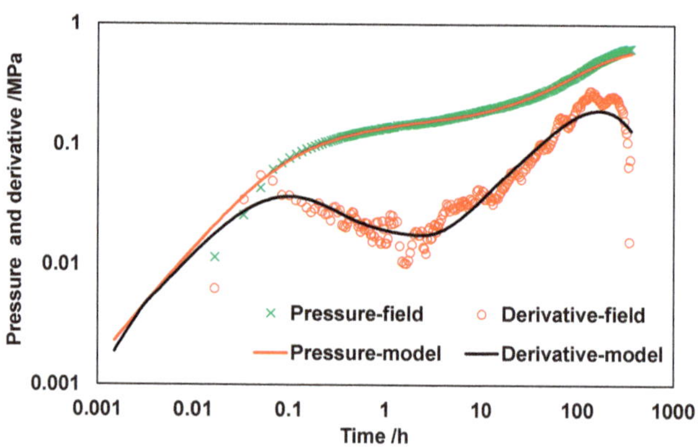

Figure 11. Pressure drop log–log fitting curve of analytical solution.

From Figure 11 and Table 3, it can be seen that the L28-38 well shows slight fracture characteristics, the half-length of the induced fracture is 10.32 m, and the distance of the lateral-swept area is 128.95 m. The reservoir permeability of the swept area is 34.65 mD, and the mobility ratio of the inner and outer zones reaches 119.92. The physical properties of the inner and outer zones are very different, indicating that waterflooding significantly improves the reservoir's physical properties near the wellbore. At the end, the pressure derivative curve began to fall, indicating that the constant-pressure boundary was reached. This boundary effect may be due to the pressure-wave effect of the surrounding water-injection wells, resulting in water overflow between the water-injection wells, and the effective displacement between the injection and production wells could not be formed, which had a great impact on the later development of the oilfield.

Table 3. Inversion results of analytical solution.

Parameter	Inversion Results
Wellbore storage constant/($m^3 \cdot MPa^{-1}$)	0.72
Permeability of the swept area/mD	34.65
Half-length of induced fracture/m	10.32
Distance of lateral-swept area/m	128.95
Area of lateral-swept area/m^2	5.22×10^4
Mobility ratio	119.92
Distance of constant-pressure boundary/m	1180.22
Current formation pressure/MPa	23.51

8. Conclusions

In this paper, aimed at the well-test model of water-injection wells in low-permeability reservoirs and considering the characteristics of the induced fractures and elliptic composite partitions, respectively, we obtained the mathematical model in Laplace space by dimensionless and Laplace transform. Then, the Mathieu function was introduced to obtain the analytical solution of bottom hole pressure, and the reservoir parameters inversion were realized by least square fitting.

There are six typical stages in the well-test pressure pattern curve of the model: wellbore storage stage, induced-fracture linear flow stage, radial-flow stage of the swept area, transition flow stage, radial-flow stage of the unswept area, and constant-pressure outer boundary response stage.

The larger the fracture half-length is, the more obvious the linear flow is, and the shorter the radial flow duration in the swept area is. The farther the range of the lateral sweep is, the longer the duration of the radial flow is, and the earlier the pressure derivative is upturned. The higher the permeability of the unswept area is, the lower the pressure derivative is. The smaller the distance of constant-pressure outer boundary is, the earlier the boundary flow appears, and the less obvious the radial flow in the unswept area is.

Based on a case study from J Oilfield, the dynamic production of a water-injection well was inversely calculated by analytical model. The obtained key parameters include a half-length of the induced fracture of 10.32 m, a range of the lateral-swept area of the elliptic partition flow of 128.95 m, permeability in the water-flooding swept area of 34.65 mD, and a mobility ratio of 119.92. Further validation of the application analysis demonstrates the accuracy and superiority of the analytical model.

Author Contributions: Conceptualization, J.L. (Jianxiong Li) and K.W.; Software, K.W., S.S., H.F. and C.P.; Validation, K.H. and Z.J.; Formal analysis, S.S. and H.F.; Investigation, K.W., K.H. and C.P.; Resources, J.L. (Jianxiong Li); Data curation, Z.J. and J.L. (Jinbao Liu); Writing—original draft, J.L. (Jianxiong Li) and K.W.; Writing—review and editing, S.S. and H.F.; Visualization, J.L. (Jinbao Liu); Supervision, J.L. (Jianxiong Li), S.S. and P.J.; Project administration, K.H. and P.J.; Funding acquisition, K.H. All authors have read and agreed to the published version of the manuscript.

Funding: The authors declare that this study received funding from CNPC Major Technology Project "Theory and Key Technologies of Exploration and Development of Daqing Gulong Shale Oil", Topic 2 "Theory and Key Technologies of Beneficial Development of Gulong Shale Oil". The funder was not involved in the study design, collection, analysis, interpretation of data, the writing of this article or the decision to submit it for publication.

Data Availability Statement: The original contributions presented in the study are included in the article, further inquiries can be directed to the corresponding author.

Conflicts of Interest: Authors Jianxiong Li, Kai Wang, Sheng Su, Kejun Huang were employed by the company the 4th Oil Production Plant of Changqing Oilfield Branch of CNPC. Author Ziyi Jin was employed by the company PetroChina Southwest Oil and Gas Field Company, Sichuan Shale Gas Exploration and Development Co., Ltd. The remaining authors declare that the research was conducted in the absence of any commercial or financial relationships that could be construed as a potential conflict of interest.

References

1. Jansen, J.D.; Van den Hoek, P.J.; Zwarts, P.J.; Al-Masfry, R.; Hustedt, B.; van Schijndel, L. Behavior and Impact of Dynamic Induced Fractures in Waterflooding and EOR. Presented at the 42nd U.S. Rock Mechanics Symposium (USRMS), San Francisco, CA, USA, 29 June–2 July 2008; ARMA-08-135.
2. Van den Hoek, P.J.; Mahani, H.; Husedt, B.; Zwarts, D.; Sobera, M. Dynamic Induced Fractures in Waterflooding and EOR. Presented at the SPE Russian Oil and Gas Technical Conference and Exhibition, Moscow, Russia, 28–30 October 2008; SPE-115204-MS.
3. Aniskin, A.; Chmuzh, I.; Sakhibgareev, R.R. Waterflood Induced Fracture Modeling in West Salym Field. Presented at the SPE Russian Oil and Gas Conference and Exhibition, Moscow, Russia, 26–28 October 2010; SPE-136560-MS.
4. Bhardwaj, P.; Hwang, J.; Manchanda, R.; Sharma, M.M. Injection Induced Fracture Propagation and Stress Reorientation in Waterflooded Reservoirs. Presented at the SPE Annual Technical Conference and Exhibition, Dubai, United Arab Emirates, 26–28 September 2016; SPE-181883-MS.
5. Hagoort, J. Waterflood-Induced Hydraulic Fracturing. Ph.D. Thesis, Delft Technical University, Delft, The Netherlands, 1981.
6. Liu, H.; Zhang, S.; Zhong, S.; Gu, J.; Lu, J.; Deng, W. Research on hydraulic mechanism of water flooding in fractured reservoir. *Drill. Prod. Technol.* **2006**, *29*, 57–60.
7. Guo, F.; Wang, H.; An, M.; Zhang, H.; Hu, X. Dynamic fracture variation characteristics and its influence in Ansai ultra-low permeability oilfield. *J. Yangtze Univ. (Nat. Sci. Ed.)* **2015**, *12*, 77–79.
8. Wang, Y.; Song, X.; Tian, C.; Shi, C.; Li, J.; Hui, G.; Hou, J.; Gao, C.; Wang, X.; Liu, P. Dynamic fracture is a new development geological attribute in water flooding of ultra-low permeability reservoirs. *Pet. Explor. Dev.* **2015**, *42*, 222–228. [CrossRef]
9. Xie, J.; Long, G.; Tian, C.; Hou, J.; Li, J.; Wang, Y. Genesis of dynamic fracture in ultra-low permeability sandstone reservoir and its influence on waterflood development: A case study of Chang6 Oil Formation in Wangyao Area, Ansai Oilfield. *Pet. Geol. Recovery Effic.* **2015**, *22*, 106–110.
10. Spivey, J.P.; Lee, W.J. Variable Wellbore Storage Models for a Dual-Volume Wellbore. Presented at the SPE Annual Technical Conference and Exhibition, Houston, TX, USA, 3–6 October 1999; SPE-56615-MS.
11. BinAkresh, S.A.; Rahman, N.M.A. Challenges in Interpreting Well Testing Data from Fractured Water Injection Wells with a Dual Storage Phenomenon. Presented at the SPE Middle East Oil and Gas Show and Conference, Manama, Bahrain, 25–28 September 2011; SPE-139587-MSS.
12. BinAkresh, S.A.; Rahman, N.M.A. Modeling Pressure-Transient Data for Characterizing the Formation Damage in Water Injection Wells Operating above the Fracturing Pressure. Presented at the SPE European Formation Damage Conference and Exhibition, Budapest, Hungary, 3–5 June 2015; SPE-174278-MS.
13. Koning, E.J.L.; Niko, H. Fractured water-injection wells: A pressure falloff test for determining fracture dimensions. Presented at the SPE Annual Technical Conference and Exhibition, Las Vegas, NV, USA, 22–26 September 1985; SPE-14458-MS.
14. Koning, E.J.L. Waterflooding under Fracturing Conditions. Ph.D. Thesis, Technical University of Delft, Delft, The Netherlands, 1988.
15. Williams, J.L. The Influence of Induced Vertical Fractures on Overall Reservoir Performance. Presented at the SPE Permian Basin Oil and Gas Recovery Conference, Midland, TX, USA, 10–11 March 1977; SPE-6381-MS.
16. Li, X.; Liu, Q. Typical analysis curve and application of new unstable well test in oil-water two-phase Wells. *Oil Gas Well Test* **2003**, *12*, 5–9.
17. Liu, J.; Meng, Y.; Li, G.; Wei, N.; Bi, B.; Li, W. Well test model of formation fluid complex injection well based on oil-water two-phase seepage. *J. Oil Gas Technol.* **2014**, *36*, 128–132.
18. Jia, P.; Cheng, L.; Huang, S.; Cao, R.; Xue, Y. Analysis model of unsteady pressure in finite conductivity fractured directional well. *Acta Pet. Sin.* **2015**, *36*, 496–503.
19. Soliman, M.Y.; Azari, M.; Ansah, J.; Kabir, C.S. Design, Interpretation, and Assessment of Short-Term Pressure-Transient Tests. Presented at the SPE Annual Technical Conference and Exhibition, Houston, TX, USA, 26–29 September 2004; SPE-90837-MS.
20. Soliman, M.Y.; Kabir, C.S. Testing Unconventional Formations. *J. Pet. Sci. Eng.* **2012**, *92–93*, 102–109. [CrossRef]
21. Craig, D.P.; Blasingame, T.A. A New Refracture-Candidate Diagnostic Test Determines Reservoir Properties and Identifies Existing Conductive or Damaged Fractures. Presented at the SPE Annual Technical Conference and Exhibition, Dallas, TX, USA, 9–12 October 2005; SPE-96785-MS.
22. Craig, D.P.; Blasingame, T.A. Application of a New Fracture-Injection/Falloff Model Accounting for Propagating, Dilated, and Closing Hydraulic Fractures. Presented at the SPE Gas Technology Symposium, Calgary, AB, Canada, 15–17 May 2006; SPE-100578-MS.
23. Abbaszadeh, M.; Kamal, M. Pressure-Transient Testing of Water-Injection Wells. *SPE Res. Eng.* **1989**, *4*, 115–124. [CrossRef]
24. Barua, J.; Horne, R.N. Computerized Analysis of Thermal Recovery Well Test Data. *SPE Form Eval.* **1987**, *2*, 560–566. [CrossRef]

25. Liu, N. *Practical Modern Well Test Interpretation Method*; Petroleum Industry Press: Beijing, China, 2008.
26. McLachlan, N.W. *Theory and Application of Mathieu Functions*; Oxford University Press: London, UK, 1947.
27. Obut, T. Determination of the Front Position with the Pressure Transient Analysis in the Presence of an Infinite Conductivity Vertical Fracture. Master's Thesis, Pennsylvania State U., University Park, PA, USA, 1983.

energies

Article

Failure Mechanism of Integrity of Cement Sheath under the Coupling Effect of Formation Creep and Temperature during the Operation of Salt Rock Gas Storage

Heng Yang [1,2], Yuhuan Bu [1,2,*], Shaorui Jing [1,2], Shenglai Guo [1,2] and Huajie Liu [1,2]

1 Key Laboratory of Unconventional Oil & Gas Development, China University of Petroleum (East China), Qingdao 266580, China; yanghenghd@foxmail.com (H.Y.); jingsrhd@foxmail.com (S.J.); guoshenglai@upc.edu.cn (S.G.); liuhuajieupc@163.com (H.L.)
2 School of Petroleum Engineering, China University of Petroleum (East China), Qingdao 266580, China
* Correspondence: buyuhuan@163.com

Abstract: Maintaining the integrity of the cement sheath is essential for the sealing of underground gas storage. The formation creep, temperature changes, and operating pressure changes during the operation of underground gas storage can cause changes in the stress of the cement sheath, which probably induces the failure of the cement sheath's integrity. A creep model taking the effects of stress and temperature into account is developed to study the cement sheath's integrity in creep formation, and the feasibility of creep simulation via finite element method is verified. The finite element method is used to analyze the effects of formation creep, temperature, operating pressure, and the cement sheath's elastic modulus on the cement sheath's integrity. The result shows that the coupling effect of formation creep and temperature increases the cement sheath's failure risk; both the formation creep and the decrease in cement sheath temperature increase the Von Mises stress on the cement sheath, increasing the risk of the cement sheath's shear failure. The decrease in cement sheath temperature decreases the circumferential compressive stress on the cement sheath and raises the risk of the cement sheath's tensile failure. Shear failure of the cement sheath occurs at high operating pressure upper limits. The operating pressure is less than 70 MPa, or the cement sheath's elastic modulus is less than 3 GPa, which can prevent the failure of the cement sheath's integrity during the operation of underground gas storage.

Keywords: formation creep; downhole temperature change; finite element method; integrity of cement sheath; operating pressure; elastic modulus of cement sheath

Citation: Yang, H.; Bu, Y.; Jing, S.; Guo, S.; Liu, H. Failure Mechanism of Integrity of Cement Sheath under the Coupling Effect of Formation Creep and Temperature during the Operation of Salt Rock Gas Storage. *Energies* **2023**, *16*, 7089. https://doi.org/10.3390/en16207089

Academic Editors: Mofazzal Hossain, Fabio Montagnaro, Haifeng Zhao and Yang Xia

Received: 11 August 2023
Revised: 11 September 2023
Accepted: 13 October 2023
Published: 13 October 2023

1. Introduction

Salt rock exhibits extremely low permeability and porosity [1], making it a commonly employed choice as a sealing caprock for depleted oil and gas reservoirs [2]. Under deviatoric stress, the salt rock formation will creep and compress the cement sheath, potentially resulting in the failure of the cement sheath's integrity (CSI). The operation of underground gas storage (UGS) includes gas injection and gas production phases. The gas injected into the wellbore undergoes heat exchange with the assembly comprising the downhole casing, cement sheath, and formation. Alterations in the assembly's temperature can subsequently impact the stress state of the cement sheath [3], potentially leading to compromised CSI. Variations in the operating pressure of UGS can also cause changes in the stress state of the cement sheath, potentially leading to compromised integrity. Weakening of the CSI will diminish its capacity to seal effectively. This could potentially result in the escape of gas from the UGS, carrying negative consequences for safety, the environment, and the economy [4]. The failure types of CSI include shear failure (i.e., compressive failure), debonding failure, and tensile failure (i.e., radial crack) [5–7]. The Von Mises criterion [8–13], Tresca criterion [14,15], Mohr–Coulomb criterion [5,16,17], and

Drucker–Prager criterion [18,19] are often used to judge the cement sheath's shear failure. The cement sheath's shear failure can produce plastic strain. The cement sheath's tensile failure is evaluated by the maximum tensile stress criterion [20–22]. When the cement sheath's circumferential stress surpasses its ability to withstand tension, tensile failure occurs. The cement sheath's debonding failure is evaluated by the interfacial radial stress. When the interfacial radial tensile stress surpasses its bonding strength, debonding failure occurs [23,24].

Liu [25] and Orlic [26,27] studied the impact of formation creep on the closure of the wellbore. Their research results indicated that wellbore closure can be triggered by salt formation creep, and the greater the wellbore's Von Mises stress, the faster the wellbore closure rate. Melo [28], Yu [29], and Velilla [30] have investigated the effect of salt formation's creep on casing. The result indicated that salt rock creep can cause increased casing stress and may cause casing yield failure. Jandhyala [9] investigated the effect of formation creep on the cement sheath. The result indicated that the cement with a higher elasticity has a stronger bearing capacity. Jesus [5], Raoof [31], and Yin [3] investigated the effect of downhole temperature changes on CSI. The finding showed that the decrease in temperature may cause the debonding failure. Zhang [16] analyzed the CSI during the operation of UGS using elastoplastic theory. The findings indicated that shear failure may occur when the casing is pressurized; during casing pressure relief, debonding failure may occur. Yang [2] used the finite element method (FEM) to investigate the effect of the creep of salt formation on CSI within UGS, considering non-uniform in situ stress conditions. The finding showed that salt rock creep under non-uniform in situ stress can cause the cement sheath's shear failure. The greater the non-uniform in situ stress, the earlier the shear failure occurs. In summary, scholars' research mainly focuses on the influences of the salt formation's creep on wellbore closure and casing failure. However, few studies have investigated the failure of CSI in the salt formation of UGS. In investigations concerning the influence of salt formation creep on CSI, the influence of downhole temperature fluctuations on both formation creep and CSI has been overlooked. In investigations concerning the failure of CSI during the operation of UGS, the impact of downhole temperature fluctuations on the cement sheath's integrity has been overlooked. Therefore, studying the effects of formation creep, downhole temperature changes, and the operating pressure on CSI throughout UGS operations holds immense importance in ensuring UGS safety.

In this study, a creep model for salt rock was developed by conducting creep experiments that involved subjecting the salt rock to varying temperature and pressure loads. The feasibility of simulating salt rock creep using the FEM is verified by comparing the FEM-simulated steady-state creep rate of salt rock with the steady-state creep rate obtained in actual salt rock creep experiments. By establishing an FEM model of casing–cement sheath–creep formation assembly, the effects of salt rock formation creep, temperature changes, and the upper limit of operating pressure on the CSI throughout the operation of UGS were studied. The influences of these factors on the CSI were determined. By examining how variations in the cement sheath's elastic modulus impact its integrity, measures to avert the failure of the CSI during the operation of UGS were obtained. The findings from this study offer valuable insights into the potential integrity issues concerning cement sheaths in UGS wells. These results hold significant importance in terms of averting cement sheath integrity failures and ensuring the secure and effective functioning of the operation of UGS.

2. Establishment of FEM Model for Casing–Cement Sheath–Creep Formation Combination

In this section, a salt rock creep model is obtained from the salt rock creep experiments at firs; then, the feasibility of salt rock creep simulations in FEM is verified by comparing the steady-state creep rates between FEM simulations and salt rock creep experiments. Finally, an FEM model of the combined casing–cement sheath–creep formation is developed based on the salt rock creep model obtained from experiments.

2.1. Salt Rock Creep Model

In order to develop the FEM model for the casing–cement sheath–creep formation assembly, it is necessary to acquire the creep model of the creep formation. Consequently, we conducted creep experiments on salt rock under various temperatures and pressures. The salt rock creep model can be derived by fitting the results of salt rock creep experiments. The experimental equipment used in the salt rock creep experiment is GCTS RTR-2000. The length-to-diameter ratio for the experimental salt rock creep samples is 2:1, which is consistent with the experimental criteria of rock mechanics. The temperature and pressure loads applied to the samples are shown in Table 1. The deviator stress is determined as the disparity between the axial pressure and the confining pressure. Sample numbers 1–5 refer to creep experiments conducted on salt rock under various deviatoric stress conditions, and sample numbers 3, 6, 7, and 8 refer to creep experiments conducted on salt rock under various temperatures. The figures illustrate the creep behavior of salt rock at varying deviatoric stresses (Figure 1) and varying temperatures (Figure 2). As depicted in Figure 1, higher deviator stress corresponds to increased axial strain. Moreover, elevated deviator stress leads to a more rapid increase in axial strain. As depicted in Figure 2, higher temperature corresponds to increased axial strain. Furthermore, elevated temperature leads to a more rapid increase in axial strain. Table 2 displays the steady-state creep rates for diverse samples. Higher deviator stress corresponds to an increased steady-state creep rate, indicating that the deviator stress promotes creep. Higher temperature corresponds to an increased steady-state creep rate, indicating that temperature promotes creep.

Table 1. Temperature and pressure loads in salt rock creep experiments.

Sample Number	Confine Pressure/MPa	Axial Pressure/MPa	Deviatoric Stress/MPa	Temperature/°C
1	5	30	25	24
2	10	30	20	24
3	15	30	15	24
4	20	30	10	24
5	25	30	5	24
6	15	30	15	50
7	15	30	15	70
8	15	30	15	90

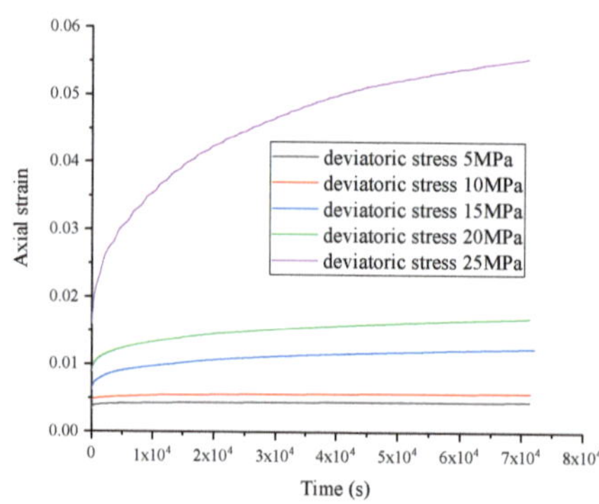

Figure 1. Creep curves of salt rock at varying deviatoric stresses.

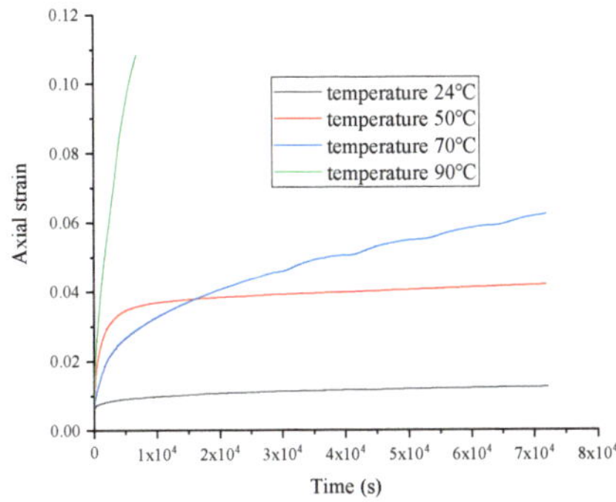

Figure 2. Creep curves of salt rock at varying temperatures.

Table 2. Steady-state creep rates for diverse samples.

Sample Number	Steady-State Creep Rate/s^{-1}
1	1.782×10^{-7}
2	1.192×10^{-7}
3	4.199×10^{-8}
4	3.687×10^{-8}
5	2.543×10^{-9}
6	6.655×10^{-8}
7	3.753×10^{-7}
8	8.534×10^{-6}

During the long-term creep process, compared to transition and accelerated creep stages, the duration of steady-state creep is the longest, and the creep strain of long-term creep is mainly contributed by steady-state creep. Therefore, the salt rock creep model mainly focuses on steady-state creep. Currently, the commonly used creep model that represents the influences of temperature and stress on the steady-state creep rate is the hyperbolic sine law model [30,32–38], as shown in Equation (1):

$$\varepsilon^{\bullet cr} = A(\sinh Bq)^{n} \exp(-\frac{\Delta H}{R(\theta - \theta^{Z})}), \quad (1)$$

where $\varepsilon^{\bullet cr}$ represents the steady-state creep rate ($^{-1}$); q represents the Von Mises stress (MPa); θ represents the temperature (°C); θ^{Z} represents the absolute zero in temperature (−273.15 °C); R represents the universal gas constant (8314.3 mJ/(mol·°C)); ΔH represents the activation energy (mJ/mol); and A, B, and n are material parameters.

The Von Mises stress in Equation (1) equals to the deviator stress in the salt rock creep experiment. We use MATLAB curve fitting tool to fit the data in Table 2, and the fitting result is shown in Equation (2), where R^2 = 0.9442 indicates that the goodness of fit is close to 1, showing a favorable fitting outcome. Equation (2) effectively portrays the interplay between temperature, stress, and the steady-state creep rate:

$$\varepsilon^{\bullet cr} = 32.31(\sinh 0.2186q)^{1.197} \exp(-\frac{6.097 \times 10^{7}}{8314.3(\theta + 273.15)}), R^2 = 0.9442 \quad (2)$$

2.2. Feasibility Verification of Salt Rock Creep Simulation in FEM

Using Equation (2), we establish an FEM model in ABAQUS that replicates the dimensions of the salt rock creep experimental specimen, as depicted in Figure 3. The entire model is loaded with temperature load, the top surface is loaded with axial pressure, the side surface is loaded with confining pressure, and the bottom of the model is immovable. The steady-state creep rates of salt rock creep experiments and finite element simulations under different temperature and pressure loads are illustrated in Figure 4. The steady-state creep rates of salt rock creep experiments and the steady-state creep rates of finite element simulation have a good consistency, indicating that finite element simulation can well reflect the steady-state creep rate of salt rock at varying temperatures and pressure loads, verifying the feasibility of salt rock creep simulation in FEM.

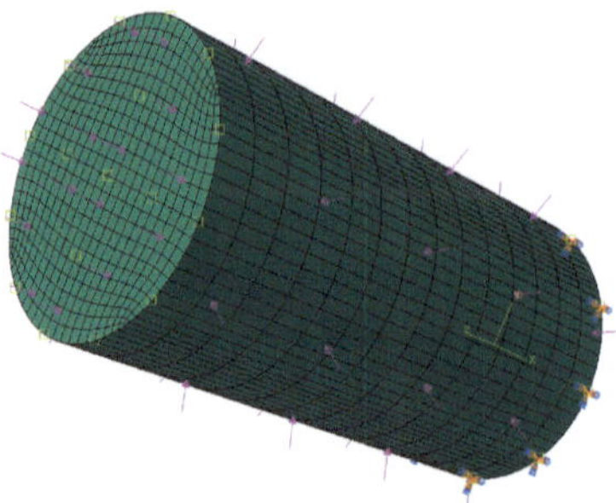

Figure 3. FEM model for creep simulation.

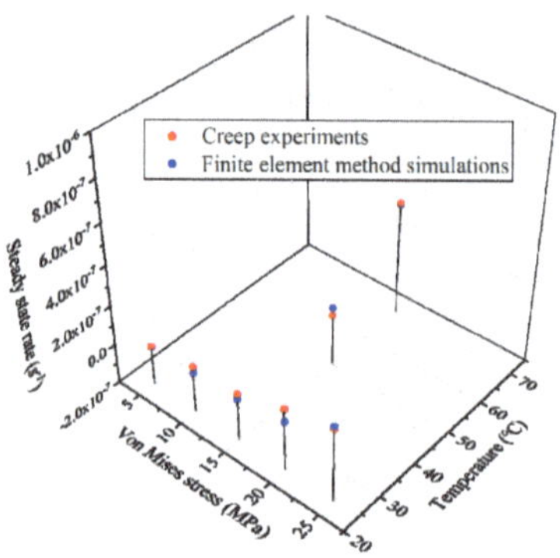

Figure 4. Steady-state creep rate of salt rock at varying temperatures and pressure loads.

2.3. FEM Model of Casing–Cement Sheath–Creep Formation Combination

The FEM model of the casing–cement sheath–formation combination in a salt rock formation at a well depth of 2500 m in an UGS is established. The axial strain of the casing–cement sheath–formation combination is very small and can be ignored. According to the theory of elasticity, the casing–cement sheath–formation combination can be simplified as a plane strain model [2], as illustrated in Figure 5. The model's geometric parameters are detailed in Table 3, and the model's physical attributes are detailed in Table 4. The mesh type is CPE4T. The contact methodology adopted for interactions between the casing and cement sheath, as well as for those between the cement sheath and the formation, are as follows: normal behavior is hard contact; tangential behavior is frictionless. The casing's yield strength is 758 MPa. The cement sheath's compressive strength is 42.13 MPa, along with a tensile strength of 3.99 MPa. The bonding strength between cement sheath and casing is 1.23 MPa. The formation creep model is shown in Equation (2). The formation temperature and uniform in situ stress are exerted onto the upper and right boundaries of the model. The casing internal pressure (i.e., operating pressure) and gas injection temperature are exerted onto the casing's inner wall. Symmetric constraints are exerted onto the model's lower and left boundaries. The uniform in situ stress and temperature of the model come from a real case. The uniform in situ stress is 50 MPa. The formation temperature is 90 °C. The gas injection temperature is 20 °C. The model's initial temperature is 90 °C.

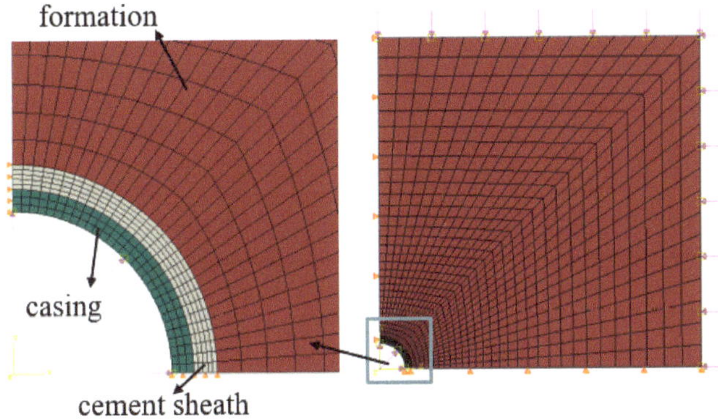

Figure 5. FEM model of casing–cement sheath–creep formation combination.

Table 3. Model geometric parameters [2].

Material	Inner Diameter/mm	Outer Diameter/mm
Casing	247.94	282.58
Cement sheath	282.58	320
Formation	320	3200 × 3200

Table 4. Physical property parameters of the model [2,39].

Material	Elastic Modulus/GPa	Poisson's Ratio	Density/ (kg·m^{-3})	Coefficient of Expansion /10^{-5} °C^{-1}	Specific Heat/ (J kg^{-1}·°C^{-1})	Thermal Conductivity/ (W·m^{-1}·°C^{-1})
Casing	210	0.3	7800	1.22	460	45
Cement sheath	10.61	0.22	1800	1.05	865	0.9
Formation	1.80	0.38	2300	1.03	896	2.2

The operating pressure of UGS for one cycle is shown in Figure 6. The operating cycle of UGS includes four stages: constant low pressure, pressurization gas injection, constant high pressure, and depressurization gas production. The upper and lower limit operating pressure are crucial parameters. The duration of one cycle of UGS operation is one year, and the duration of each operation phase is three months.

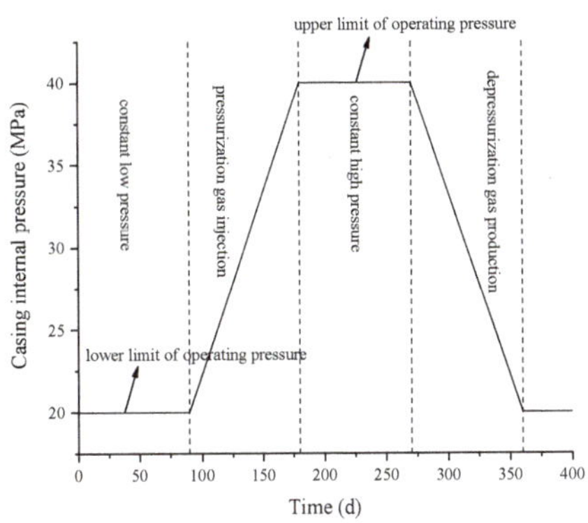

Figure 6. Change of casing internal pressure during one cycle of UGS operation.

3. Results and Discussion

3.1. Variation of Formation Temperature and Stress during Operation of UGS

Creep refers to the increase in the strain of an object over time under a fixed load. The formation in the combination model in this paper is a salt rock formation with creep characteristics. When subjected to a constant load, the salt rock formation will undergo creep, exerting pressure on the cement sheath. This continuous pressure escalation on the cement sheath could potentially result in an eventual integrity failure. Therefore, it is imperative to investigate the variation of temperature and stress over time in the formation near the borehole during the operation of UGS to determine when the salt rock formation will creep during the UGS's operation.

The remaining parameters remain unchanged, regardless of formation creep. The UGS has an upper operating pressure limit of 40 MPa and a lower operating pressure limit of 20 MPa. To simulate a full UGS operation cycle, we study the temperature and stress changes in the formation near the borehole and determine when the salt rock formation will occur creep. Figure 7 illustrates the alteration in temperature within the inner wall of the formation over a duration of UGS operation, while Figure 8 presents the corresponding changes in Von Mises stress. The observations from Figures 7 and 8 reveal that during the constant low pressure stage, the temperature and Von Mises stress acting on the formation's inner wall always remain unchanged. During the pressurized gas injection stage, the temperature of the formation's inner wall rapidly decreases, resulting in a large thermal stress and leading to a rapid escalation of Von Mises stress on the formation's inner wall. Subsequently, with gradual stabilization of the formation temperature and a concurrent increase in casing internal pressure, the Von Mises stress on the formation's inner wall gradually decreases. During the constant high pressure stage, the temperature of the formation's inner wall gradually recovers and eventually reaches a stable state, leading to a gradual decline and subsequent stabilization of the Von Mises stress of the formation's inner wall. During the depressurization gas production stage, the temperature of the formation's inner wall remains unchanged, while the casing internal pressure gradually decreases, resulting in a gradual increase in Von Mises stress on the formation's inner wall. Based on the preceding analysis, it is evident that during the constant low pressure stage, the Von Mises stress of the formation always remain unchanged, so the formation will occur creep at this stage. During the pressurized gas injection stage, the formation's Von Mises stress is not constant, so the formation will not occur creep at this stage. During the later period of the constant high pressure stage, the formation's Von Mises stress remains unchanged, and the formation will occur creep during this time period. During the depressurization gas production stage, the formation's Von Mises stress is not constant, so the formation will not occur creep at this stage.

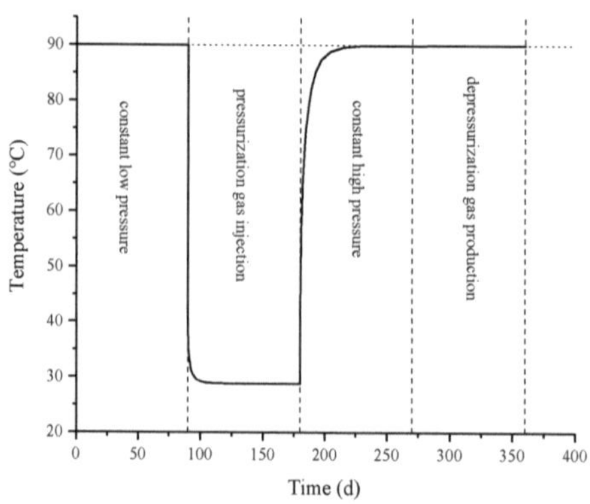

Figure 7. Temperature variation of formation inner wall of UGS during one cycle of operation without formation creep.

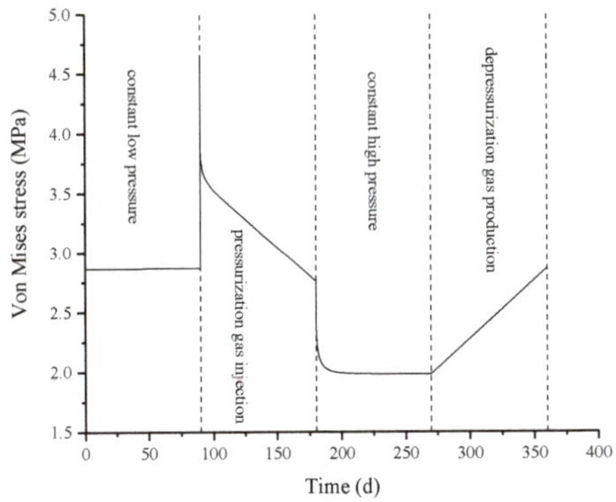

Figure 8. Von Mises stress variation of formation inner wall of UGS during one cycle of operation without formation creep.

The other parameters of the model remain constant, taking into account the formation creep that occurs during the stages of constant low pressure and constant high pressure. The UGS is operated for one cycle; the comparison of the Von Mises stress of the formation's inner wall with and without formation creep is illustrated in Figure 9. As depicted in Figure 9, in the case of formation creep, the formation's Von Mises stress experiences a decline during both the constant low-pressure and constant high-pressure stages, gradually converging toward zero. The formation creep compresses the cement sheath, which plays a resistance role to the formation creep, causing the radial stress of the formation's inner wall to increase, the Von Mises stress of the formation's inner wall to decrease, and the ability of formation creep to decrease.

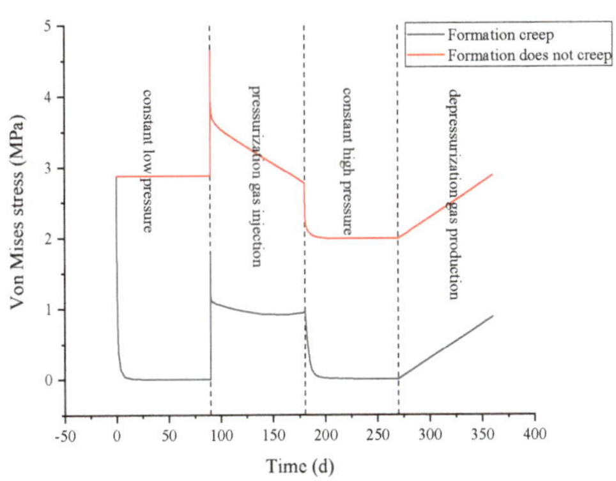

Figure 9. Comparison of Von Mises stress of formation inner wall with and without formation creep.

3.2. *Effects of Downhole Temperature Changes and Formation Creep on Integrity Failure of Cement Sheath*

To investigate the effect of downhole temperature fluctuations, formation creep, and the combined influence of these two factors on the CSI during the operation of UGS, four different combination models were established, as shown in Table 5. Model A: the impact of downhole temperature changes and formation creep is not considered; Model B: not considering downhole temperature changes but considering the impact of formation creep; Model C: not considering formation creep but considering the influence of downhole temperature changes; Model D: considering the influence of downhole temperature changes and formation creep. The comparison between Model A and Model B can determine the impact of formation creep on the CSI. The comparison between Model A and Model C can determine the influence of downhole temperature changes on the CSI. The comparison between Model A and Model D can determine the impact of the combined effects of downhole temperature changes and formation creep on the CSI. In this study, the cement sheath's shear failure is assessed by employing equivalent plastic strain (PEEQ) and Von Mises stress as evaluation criteria. The evaluation of the cement sheath's tensile failure is conducted using circumferential stress. The evaluation of the cement sheath's debonding failure is conducted using radial stress at the cement sheath interface. The UGS has an upper operating pressure limit of 40 MPa and a lower operating pressure limit of 20 MPa. We simulate the operation of the UGS for one cycle and assess the CSI failure of four different models.

Table 5. Comparison of different models.

Model	Formation Creep	Downhole Temperature Change
Model A	×	×
Model B	✓	×
Model C	×	✓
Model D	✓	✓

During one complete UGS operational cycle, the temperature of the cement sheath's inner wall (CSIW) of Model C and Model D varies over time, as depicted in Figure 10. The temperature curves of the CSIW of Model C and Model D overlap approximately, showing that the formation creep has almost no influence on the cement sheath's temperature. During the constant low pressure stage, the cement sheath's temperature remains constant. During the pressurized gas injection stage, the temperature of the cement sheath rapidly decreases and then stabilizes. During the constant high pressure stage, the cement sheath's temperature gradually recovers and subsequently reaches a stable state. During the depressurization gas production stage, the cement sheath's temperature remains unchanged.

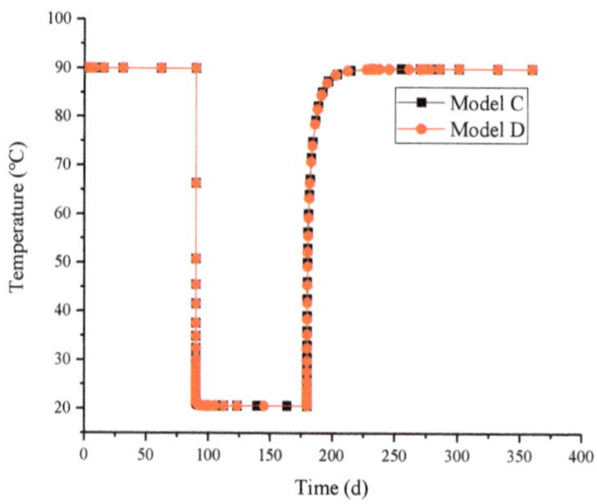

Figure 10. Variation of temperature on the CSIW with time for diverse models.

The Von Mises stress on the CSIW of different models fluctuates over time during the operation of the UGS for one cycle, as illustrated in Figure 11. As depicted in Figure 11a, during the operation of UGS, the Von Mises stress on the CSIW for the four models remains below the cement sheath's compressive strength, indicating that the cement sheath of the four models did not occur shear failure. The order of maximum Von Mises stress on the CSIW for the four models is Model D > Model C > Model B > Model A, which indicates that Model D carries the highest risk of the cement sheath's shear failure. Combining Figures 9 and 11b, it is evident that formation creep leads to an elevation in Von Mises stress on the CSIW during the constant low pressure stage. Subsequently, as the Von Mises stress on the formation's inner wall decreases and reaches a stable state, the ability of formation creep weakens, leading to a gradually stabilization of the Von Mises stress on the CSIW gradually stabilizes. Combining Figures 10 and 11c, it is evident that during the initial stage of pressurized gas injection, there is a swift decline in the temperature of the CSIW, leading to a pronounced surge in Von Mises stress on the CSIW. Subsequently, the temperature of the CSIW stabilizes, while the casing internal pressure gradually increases, consequently causing a sustained elevation in the Von Mises stress on the CSIW. During the constant high pressure stage, the temperature of the CSIW gradually increases and then stabilizes. Consequently, the Von Mises stress on the CSIW gradually decreases and then stabilizes. Combining Figure 11d with Figures 9 and 10, it can be seen that the formation creep that occurs during the constant low pressure stage causes the Von Mises stress on the CSIW to first increase and then stabilize. Moreover, the cement sheath's temperature during the pressurized gas injection stage decreases, causing an increase in Von Mises on the CSIW. Therefore, among the four models, Model D exhibits the highest Von Mises stress on the CSIW.

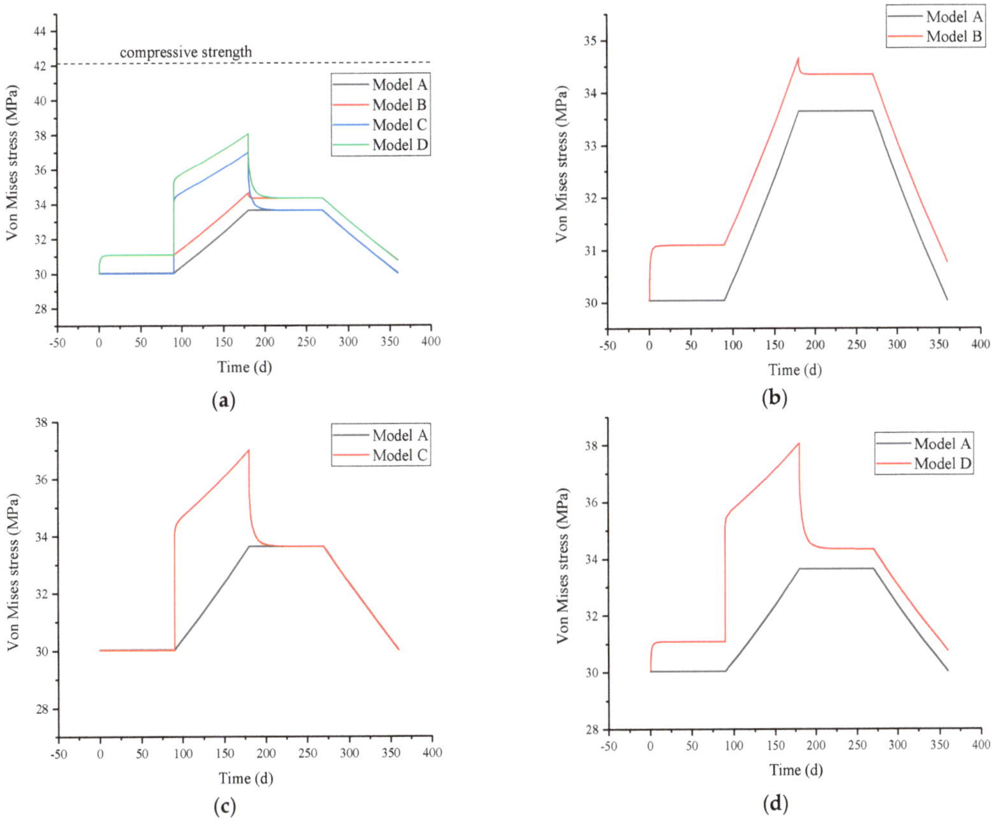

Figure 11. Variation of Von Mises stress on the CSIW with time for different models. (**a**) comparison of all models; (**b**) comparison between Model A and Model B; (**c**) comparison between Model A and Model C; (**d**) comparison between Model A and Model D.

The circumferential stress on the CSIW of different models fluctuates over time during a complete cycle of UGS operation, as illustrated in Figure 12. As depicted in Figure 12a, the circumferential stresses in all four models remain negative (indicating compressive stress) and are below the cement sheath's tensile strength. This observation implies that the cement sheath of the four models does not occur circumferential tensile failure. Among the four models, the maximum circumferential stress on the CSIW follows this order: Model C > Model D > Model A > Model B. This ranking indicates that Model C presents the greatest risk of the cement sheath's tensile failure. As depicted in Figure 12b, during the constant low pressure stage, formation creep induces a 0.7 MPa increase in the circumferential compressive stress on the CSIW. Conversely, in Figure 12c, during the pressurized gas injection stage, the temperature of the cement sheath decreases, leading to a reduction of 7.8 MPa in circumferential compressive stress on the CSIW. Consequently, Figure 12d shows a decrease in the maximum circumferential compressive stress on the CSIW.

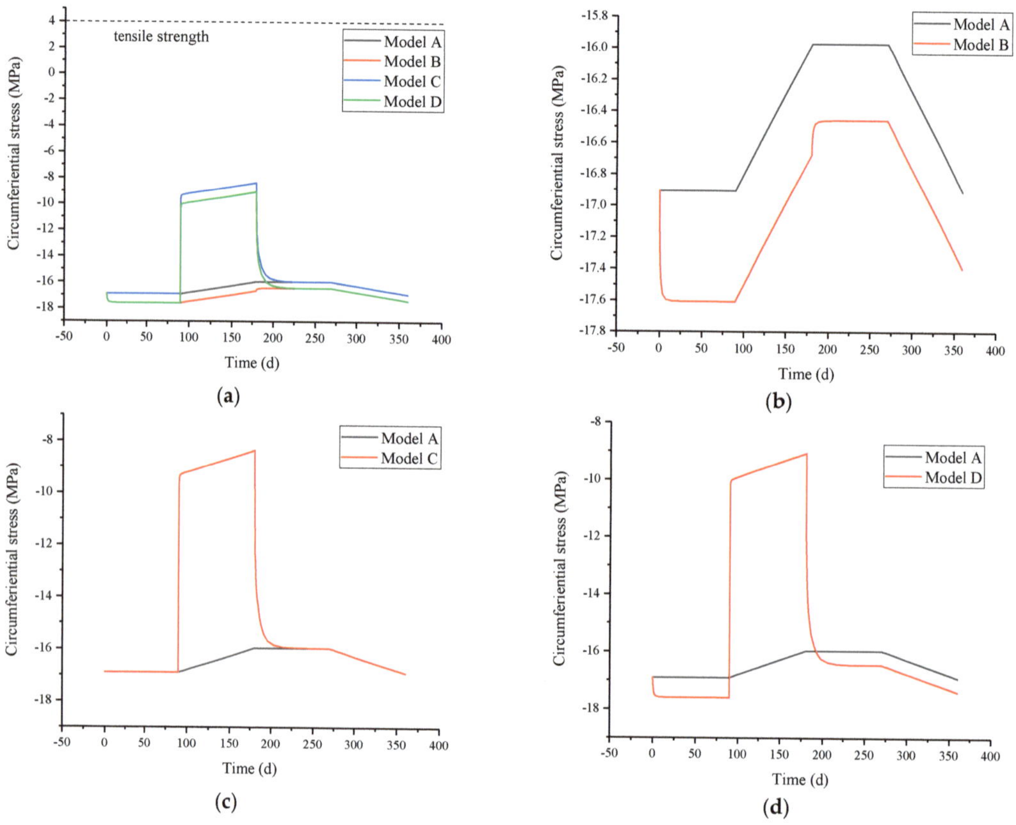

Figure 12. Variation of circumferential stress on the CSIW with time for four models: (**a**) comparison of all models; (**b**) comparison between Model A and Model B; (**c**) comparison between Model A and Model C; (**d**) comparison between Model A and Model D.

The radial stress on the CSIW of different models varies over time during a complete cycle of UGS operation, as illustrated in Figure 13. As depicted in Figure 13a, the radial stresses in all four models are all negative (indicating compressive stress), showing the absence of debonding failure in the cement sheaths of these models. The maximum radial stress on the CSIW follows this order for the four models: Model C > Model A > Model D > Model B. Figure 13b shows that the formation creep increases the radial compressive stress on the CSIW. In Figure 13c, during the pressurized gas injection stage, the temperature of the cement sheath decreases, resulting in decreased radial compressive stress on the CSIW. Conversely, during the constant high pressure stage, the temperature increase in the cement sheath elevates the radial compressive stress on the CSIW. Figure 13d shows that the combined influence of formation creep and temperature changes in the cement sheath leads to an increase in the radial compressive stress on the CSIW.

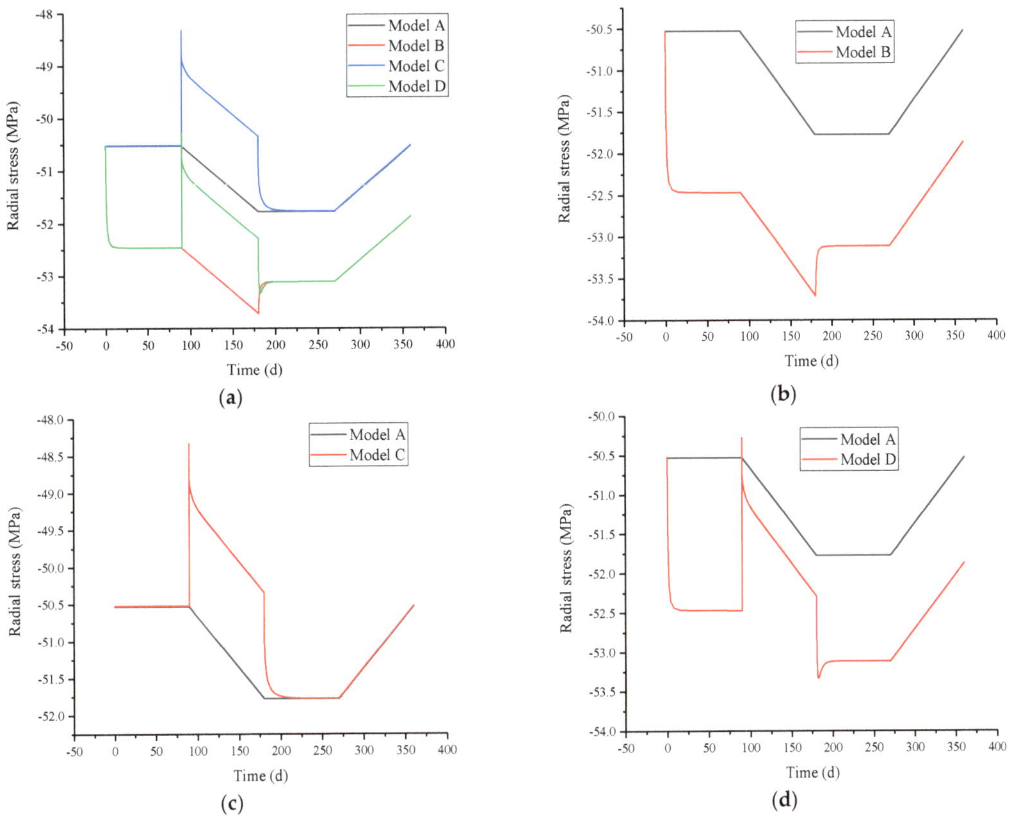

Figure 13. Variation of radial stress on the CSIW with time for four models: (**a**) comparison of all models; (**b**) comparison between Model A and Model B; (**c**) comparison between Model A and Model C; (**d**) comparison between Model A and Model D.

To summarize, the Von Mises stress on the CSIW reaches its peak under the combined effect of formation creep and downhole temperature changes, consequently posing the highest risk of shear failure for the cement sheath. The circumferential compressive stress on the CSIW is the lowest under the influence of underground temperature changes, thus elevating the risk of tensile failure for the cement sheath.

3.3. The Influence of the Upper Limit of Operating Pressure of UGS on the Integrity of Cement Sheath

Considering the comprehensive effect of formation creep and downhole temperature changes, while keeping other parameters constant, the impact of the upper limit of operating pressure on the CSI is investigated. Figure 14 illustrates the Von Mises stress and PEEQ on the CSIW with the different upper limits of operating pressure during a complete cycle of UGS operation. As depicted in Figure 14, the higher the upper limit of operating pressure, the greater the maximum Von Mises stress and PEEQ on the CSIW. Notably, the Von Mises stress on the CSIW reached the cement sheath's compressive strength at upper operating pressure limits of 70 MPa and 80 MPa, leading to shear failure for the cement sheath. The variation of circumferential stress on the CSIW with the different upper limits of operating pressure over time is illustrated in Figure 15, and the variation of radial stress over time is illustrated in Figure 16. The circumferential stress on the CSIW with different upper operating pressure limits are all circumferential compressive stresses, and the tensile failure for the cement sheath is not observed. Likewise, the radial stresses on the CSIW with different upper operating pressure limits are all radial compressive stresses, and the debonding failure for the cement sheath is not observed. In summary, under the high upper limit of operating pressure conditions, the CSI will fail in the form of shear failure. Therefore, to guarantee the safe and efficient operation of UGS, the upper limit of operating pressure should be controlled to not exceed 70 MPa.

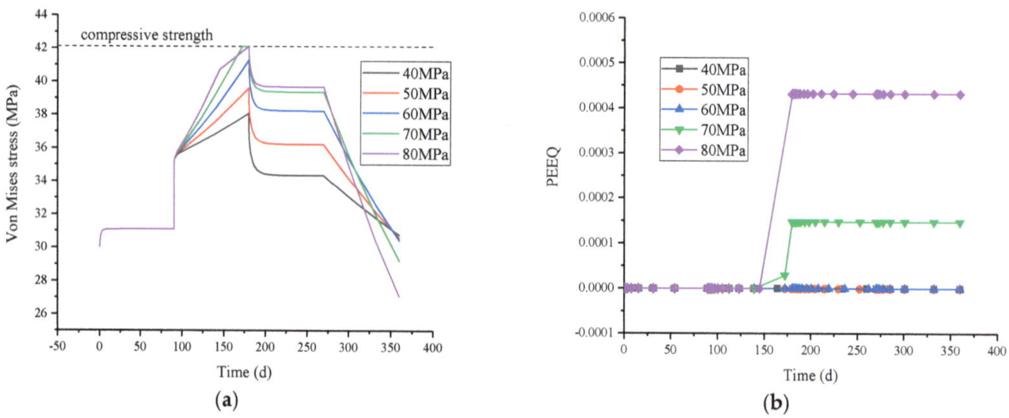

Figure 14. Variation of Von Mises stress and PEEQ on the CSIW with time for different upper limit of operating pressure: (**a**) Von Mises stress; (**b**) PEEQ.

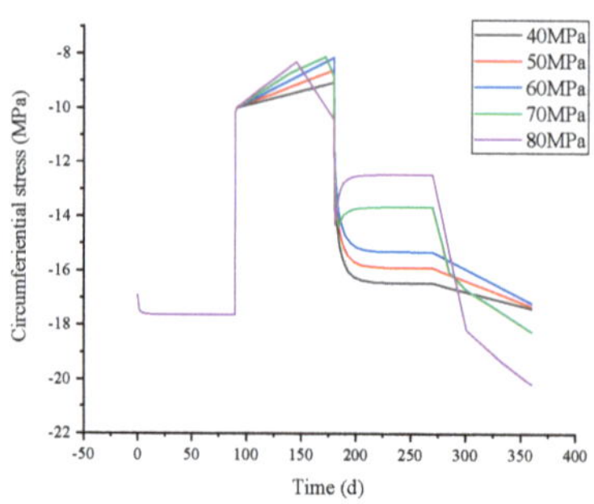

Figure 15. Variation of circumferential stress on the CSIW with time for different upper limit of operating pressure.

3.4. Measure to Prevent the Integrity Failure of Cement Sheath

From the perspective of regulating the cement sheath's elastic modulus to prevent the cement sheath's shear failure during the operation of UGS under the combined effects of formation creep and downhole temperature changes, considering the formation creep and downhole temperature changes, the upper limit of the operating pressure is 80 MPa, keeping all other parameters unchanged. The UGS is operated for one cycle, and the cement sheath's elastic modulus varies at 3 GPa, 6 GPa, 9 GPa, 12 GPa, and 15 GPa, respectively. We investigate how the cement sheath's elastic modulus influences its susceptibility to shear failure. After one cycle of operation of the UGS, Figure 17 shows the PEEQ cloud diagrams of the cement sheath with different elastic moduli. As depicted in Figure 17, a higher elastic modulus corresponds to a larger PEEQ in the cement sheath and a greater extent of failure area. When the elastic modulus is 3 GPa, the PEEQ of the cement sheath registers at 0, indicating the absence of shear failure.

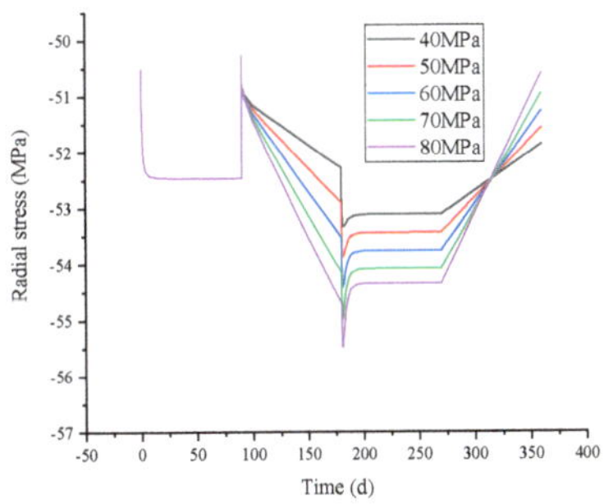

Figure 16. Variation of radial stress on the CSIW with time for different upper limit of operating pressure.

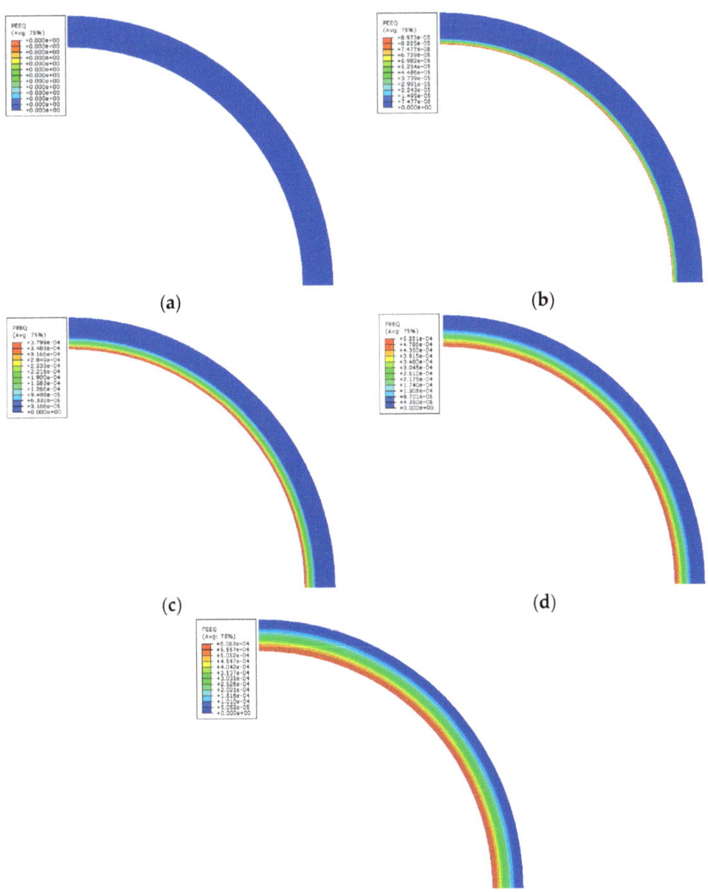

(**a**) (**b**) (**c**) (**d**) (**e**)

Figure 17. PEEQ cloud diagrams of cement sheath with different cement sheath elastic moduli: (**a**) 3 GPa; (**b**) 6 GPa; (**c**) 9 GPa; (**d**) 12 GPa; (**e**) 15 GPa.

The Von Mises stress and PEEQ on the CSIW with different elastic moduli vary over time during the operation of the UGS for one cycle, as illustrated in Figure 18. In Figure 18, increasing the cement sheath's elastic modulus results in higher Von Mises stress and PEEQ on its inner wall. A higher cement sheath's elastic modulus leads to an earlier attainment of Von Mises stress, approaching the cement sheath elastic modulus inner wall, thus resulting in quicker PEEQ generation and earlier onset of cement sheath shear failure. In summary, reducing the cement sheath's elastic modulus can maintain the CSI during the operation of the UGS.

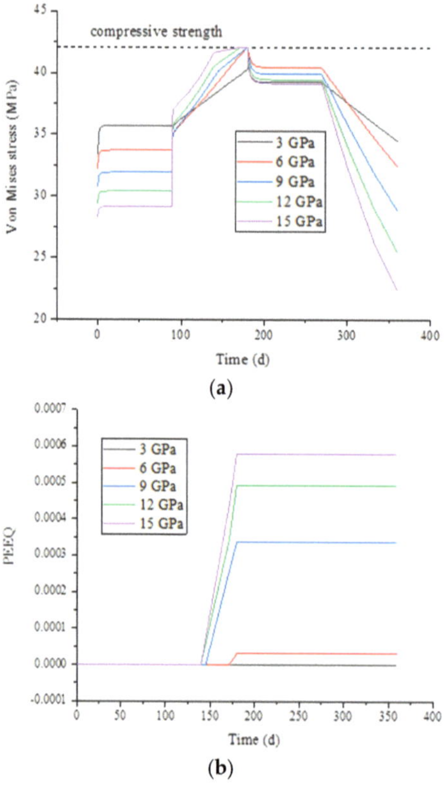

Figure 18. Variation of Von Mises stress and PEEQ on the CSIW with time for different cement sheath elastic moduli: (**a**) Von Mises stress; (**b**) PEEQ.

4. Conclusions

This study performs creep experiments on salt rock to acquire the steady-state creep rate of salt rock across different temperatures and deviatoric stress levels. Experimental data are fitted to obtain the hyperbolic sine law mode for salt rock creep, reflecting the influence of temperature and deviatoric stress on the steady-state creep rate. Using the salt rock creep model as a foundation, an FEM model of the casing–cement sheath–creep formation assembly was developed in ABAQUS to study the effects of formation creep, temperature changes, and the upper limit of operating pressure on the CSI. The results demonstrate that both formation creep and a decrease in the cement sheath's temperature increase the cement sheath's Von Mises stress, consequently elevating the risk of shear failure. Reduced cement sheath temperature, conversely, diminishes circumferential compressive stress in the cement sheath, thereby elevating the risk of tensile failure. Under a high upper limit of operating pressure, shear failure manifests within the cement sheath. A higher cement sheath's elastic modulus leads to increased maximum Von Mises stress, larger shear failure area, and higher PEEQ values. Lowering the cement sheath's elastic modulus and limiting operating pressure can maintain the CSI under the combined effect of formation creep and downhole temperature changes during the operation of the UGS.

Author Contributions: Conceptualization, methodology and writing—original draft preparation, H.Y.; resources and supervision, Y.B.; investigation and validation, S.J.; visualization and data curation, S.G.; writing—reviewing and editing and formal analysis, H.L. All authors have read and agreed to the published version of the manuscript.

Funding: This work was funded by the National Natural Science Foundation of China [51974355].

Data Availability Statement: Not applicable.

Conflicts of Interest: The authors declare no conflict of interest.

Abbreviations

Finite element method (FEM); equivalent plastic strain (PEEQ); underground gas storage (UGS); cement sheath integrity (CSI); cement sheath inner wall (CSIW).

References

1. Yang, C.; Daemen, J.J.K.; Yin, J.H. Experimental investigation of creep behavior of salt rock. *Int. J. Rock Mech. Min. Sci.* **1999**, *36*, 233–242. [CrossRef]
2. Yang, H.; Bu, Y.; Guo, S.; Liu, H.; Du, J.; Cao, X. Effects of in-situ stress and elastic parameters of cement sheath in salt rock formation of underground gas storage on seal integrity of cement sheath. *Eng. Fail. Anal.* **2021**, *123*, 105258. [CrossRef]
3. Yin, F.; Hou, D.; Liu, W.; Deng, Y. Novel assessment and countermeasure for micro-annulus initiation of cement sheath during injection/fracturing. *Fuel* **2019**, *252*, 157–163. [CrossRef]
4. Bu, Y.; Du, J.; Guo, S.; Liu, H.; Liang, Y.; Su, Z.; Li, Y. Effects of formation water influx on the bonding strength between oil well cement and the formation. *Constr. Build. Mater.* **2020**, *251*, 118989. [CrossRef]
5. De Andrade, J.; Sangesland, S. Cement Sheath Failure Mechanisms: Numerical Estimates to Design for Long-Term Well Integrity. *J. Pet. Sci. Eng.* **2016**, *147*, 682–698. [CrossRef]
6. Bu, Y.; Yang, H.; Zhao, L.; Guo, S.; Liu, H.; Ma, X. Stress concentration of perforated cement sheath and the effect of cement sheath elastic parameters on its integrity failure during shale gas fracturing. *Front. Mater.* **2022**, *9*, 980920. [CrossRef]
7. He, T.; Wang, T.; Wang, D.; Xie, D.; Daemen, J. Failure mechanism of gas storage salt cavern cement sheath under the coupling effect of plasticity and fatigue damage. *Constr. Build. Mater.* **2022**, *354*, 129179. [CrossRef]
8. Zhao, L.; Yang, H.; Wei, Y.; Bu, Y.; Jing, S.; Zhou, P. Integrity and Failure Analysis of Cement Sheath Subjected to Coalbed Methane Fracturing. *Fluid Dyn. Mater. Process.* **2023**, *19*, 329–344. [CrossRef]
9. Jandhyala, S.R.K.; Barhate, Y.R.; Anjos, J.; Fonseca, C.E.; Ravi, K. Cement Sheath Integrity in Fast Creeping Salts: Effect of Well Operations. In Proceedings of the SPE Offshore Europe Oil and Gas Conference and Exhibition, Aberdeen, UK, 3–6 September 2013; Society of Petroleum Engineers: Alexandria, VA, USA. [CrossRef]
10. Tian, L.; Bu, Y.; Liu, H.; Lu, C.; Guo, S.; Xu, H.; Ren, Y. Effects of the mechanical properties of a cement sheath and formation on the sealing integrity of the cement-formation interface in shallow water flow in deep water. *Constr. Build. Mater.* **2023**, *369*, 130496. [CrossRef]
11. Zhou, S.; Liu, R.; Zeng, H.; Zeng, Y.; Zhang, L.; Zhang, J.; Li, X. Mechanical characteristics of well cement under cyclic loading and its influence on the integrity of shale gas wellbores. *Fuel* **2019**, *250*, 132–143. [CrossRef]
12. Li, C.; Guan, Z.; Zhao, X.; Yan, Y.; Zhang, B.; Wang, Q.; Sheng, Y. A new method to protect the cementing sealing integrity of carbon dioxide geological storage well: An experiment and mechanism study. *Eng. Fract. Mech.* **2020**, *236*, 107213. [CrossRef]
13. Yang, H.; Bu, Y.; Jing, S.; Guo, S.; Liu, H. Adaptability of Different Shear Failure Criteria in Shear Failure Evaluation of Cement Sheaths. *Arab. J. Sci. Eng.* **2023**, 1–10. [CrossRef]
14. Patel, H.; Salehi, S. Development of an Advanced Finite Element Model and Parametric Study to Evaluate Cement Sheath Barrier. *J. Energy Resour. Technol.-Trans. ASME* **2019**, *141*, 092902. [CrossRef]
15. Patel, H.; Salehi, S. Structural integrity of liner cement in oil & gas wells: Parametric study, sensitivity analysis, and risk assessment. *Eng. Fail. Anal.* **2021**, *122*, 105203. [CrossRef]
16. Zhang, H.; Shen, R.; Yuan, G.; Ba, Z.; Hu, Y. Cement sheath integrity analysis of underground gas storage well based on elastoplastic theory. *J. Pet. Sci. Eng.* **2017**, *159*, 818–829. [CrossRef]
17. Yan, X.; Jun, L.; Gonghui, L.; Qian, T.; Wei, L. A new numerical investigation of cement sheath integrity during multistage hydraulic fracturing shale gas wells. *J. Nat. Gas Sci. Eng.* **2018**, *49*, 331–341. [CrossRef]
18. Liu, K.; Gao, D.; Zeng, J.; Wang, Z. Study on Cement Sheath Integrity in Horizontal Wells during Hydraulic Fracturing Process. In Proceedings of the 52nd U.S. Rock Mechanics/Geomechanics Symposium, Seattle, WA, USA, 17–20 June 2018.

19. Bagheri, M.; Shariatipour, S.M.; Ganjian, E. Parametric study on the integrity of wellbores in CO_2 storage sites. *Constr. Build. Mater.* **2021**, *268*, 121060. [CrossRef]
20. Liu, K.; Gao, D.; Taleghani, A.D. Analysis on integrity of cement sheath in the vertical section of wells during hydraulic fracturing. *J. Pet. Sci. Eng.* **2018**, *168*, 370–379. [CrossRef]
21. Guo, S.; Bu, Y.; Yang, X.; Wang, C.; Guo, B.; Sun, B. Effect of casing internal pressure on integrity of cement ring in marine shallow formation based on XFEM. *Eng. Fail. Anal.* **2020**, *108*, 104258. [CrossRef]
22. Han, X.; Feng, F.; Zhang, J. Study on the whole life cycle integrity of cement interface in heavy oil thermal recovery well under circulating high temperature condition. *Energy* **2023**, *278*, 127873. [CrossRef]
23. Chu, W.; Shen, J.; Yang, Y.; Li, Y.; Gao, D. Calculation of micro-annulus size in casing-cement sheath-formation system under continuous internal casing pressure change. *Pet. Explor. Dev.* **2015**, *42*, 414–421. [CrossRef]
24. Bu, Y.; Tian, L.; Guo, B.; Wang, C.; Sun, B. Experiment and simulation on the integrity of cement ring interface in deep water shallow formation. *J. Pet. Sci. Eng.* **2020**, *190*, 107127. [CrossRef]
25. Liu, X.; Birchwood, R.; Hooyman, P.J. A new analytical solution for wellbore creep in soft sediments and salt. In Proceedings of the 45th U.S. Rock Mechanics/Geomechanics Symposium, San Francisco, CA, USA, 26–29 June 2011; American Rock Mechanics Association: Alexandria, VA, USA, 2011.
26. Orlic, B.; Buijze, L. Numerical modeling of wellbore closure by the creep of rock salt caprocks. In Proceedings of the 48th U.S. Rock Mechanics/Geomechanics Symposium, Minneapolis, MN, USA, 1–4 June 2014; American Rock Mechanics Association: Alexandria, VA, USA, 2014.
27. Orlic, B.; Thienen-Visser, V.; Schreppers, G. Numerical estimation of structural integrity of salt cavern wells. In Proceedings of the 50th US Rock Mechanics/Geomechanics Symposium, Houston, TX, USA, 26–29 June 2016; American Rock Mechanics Association: Alexandria, VA, USA, 2016.
28. Melo, D.; Fountoura, S.A.B.; Inoue, N.; Anjos, J.L.R. Finite Element Analysis of Casing-in-Casing Integrity due to Annulus Pressurization by means of Salt Creep. In Proceedings of the 49th US Rock Mechanics/Geomechanics Symposium, San Francisco, CA, USA, 28 June–1 July 2015; American Rock Mechanics Association: Alexandria, VA, USA, 2015.
29. Yu, B.; Hardy, M.P.; Zhao, H. Laboratory Testing of Casing-Cement Interface and Multi-Scale Modeling of Casing Integrity Within Salt. In Proceedings of the 51st US Rock Mechanics/Geomechanics Symposium, San Francisco, CA, USA, 25–28 June 2017; American Rock Mechanics Association: Alexandria, VA, USA, 2017.
30. Velilla, J.; Fountoura, S.A.B.; Inoue, N.; Anjos, J.L.R. Numerical modelling of casing integrity in salt layers including the effects of dissolution and creep. In Proceedings of the 49th US Rock Mechanics/Geomechanics Symposium, San Francisco, CA, USA, 28 June–1 July 2015; American Rock Mechanics Association: Alexandria, VA, USA, 2015.
31. Gholami, R.; Aadnoy, B.; Fakhari, N. A thermo-poroelastic analytical approach to evaluate cement sheath integrity in deep vertical wells. *J. Pet. Sci. Eng.* **2016**, *147*, 536–546. [CrossRef]
32. Jenab, A.; Taheri, A.K.; Jenab, K. The Use of ANN to Predict the Hot Deformation Behavior of AA7075 at Low Strain Rates. *J. Mater. Eng. Perform.* **2013**, *22*, 903–910. [CrossRef]
33. Erisov, Y.; Surudin, S.; Grechnikov, F. Hot Deformation Behavior of Al-Cu-Li-Mg-Zn-Zr-Sc Alloy in As-Cast and Hot-Rolled Condition. *Mater. Sci. Forum* **2018**, *920*, 244–249. [CrossRef]
34. Maia, C.A.; Poiate, E., Jr.; Falcao, J.L.; Coelho, L.F.M. Triaxial Creep Tests in Salt Applied in Drilling Through Thick Salt Layers in Campos Basin-Brazil. In Proceedings of the SPE/IADC Drilling Conference, Amsterdam, The Netherlands, 23–25 February 2005.
35. Poiate, E., Jr.; Costa, A.M.; Falcao, J.L. Well Design for Drilling Through Thick Evaporite Layers in Santos Basin—Brazil. In Proceedings of the IADC/SPE Drilling Conference, Miami, FL, USA, 21–23 February 2006.
36. da Costa, A.; Amaral, C.; Poiate, E.; Pereira, A.; Martha, L.; Gattass, M.; Roehl, D. Underground storage of natural gas and CO_2 in salt caverns in deep and ultra-deep water offshore Brazil. In Proceedings of the 12th ISRM Congress, Beijing, China, 16–21 October 2011; pp. 1659–1664. [CrossRef]
37. Costa, A.M.; Roehl, D.; da Costa, A.M.; Amaral, C.d.S.; Poiate, E., Jr. Rock Mechanics and Computer Modeling Applied in the Design of Underground Salt Caverns Opened by Solution Mining for Brine Production and Natural Gas Storage. In Proceedings of the 13th ISRM International Congress of Rock Mechanics, Montreal, QC, Canada, 10–13 May 2015.
38. Da Costa, A.M.; Costa, P.V.M.; Miranda, A.C.O.; Goulart, M.B.R.; Bergsten, A.; Meneghini, J.R.; Nishimoto, K.; Assi, G.R.S.; Sampaio, C.M.; Ebecken, N.F.F.; et al. Well Design for the Construction and Operation of an Experimental Salt Cavern Built Offshore in Ultra-Deep Water for CCS in Brasil. In Proceedings of the 54th U.S. Rock Mechanics/Geomechanics Symposium, Physical Event Cancelled, 28 June–1 July 2020.
39. Fan, M.; Liu, G.; Li, J.; Guo, L.; He, M.; Li, B. Study on Micro-gap Generation Mechanism of Cement Ring Interface Under Thermal Alternating Pressure. *Sci. Technol. Eng.* **2016**, *16*, 72–76.

energies

Article

A Coupled Poro-Elastic Fluid Flow Simulator for Naturally Fractured Reservoirs

Reda Abdel Azim [1], Saad Alatefi [2,*] and Ahmad Alkouh [2]

1 Petroleum Engineering Department, American University of Kurdistan, Sumel 42003, Iraq; reda.abdulrasoul@auk.edu.krd
2 Department of Petroleum Engineering Technology, College of Technological Studies, PAAET, P.O. Box 42325, Kuwait City 70654, Kuwait; ab.alkouh@paaet.edu.kw
* Correspondence: se.alatefi@paaet.edu.kw

Abstract: Naturally fractured reservoirs are characterized by their complex nature due to the existence of natural fractures and fissures within the rock formations. These fractures can significantly impact the flow of fluids within the reservoir, making it difficult to predict and manage production. Therefore, efficient production from such reservoirs requires a deep understanding of the flow behavior via the integration of various geological, geophysical, and engineering data. Additionally, advanced simulation models can be used to predict reservoir behavior under different production scenarios and aid in decision making and effective management. Accordingly, this study presents a robust mathematical two-phase fluid flow model (FRACSIM) for the simulation of the flow behavior of naturally fractured reservoirs in a 3D space. The mathematical model is based on the finite element technique and implemented using the FORTRAN language within a poro-elastic framework. Fractures are represented by triangle elements, while tetrahedral elements represent the matrix. To optimize computational time, short to medium-length fractures adopt the permeability tensor approach, while large fractures are discretized explicitly. The governing equations for poro-elasticity are discretized in both space and time using a standard Galerkin-based finite element approach. The stability of the saturation equation solution is ensured via the application of the Galerkin discretization method. The 3D fracture model has been verified against Eclipse 100, a commercial software, via a well-test case study of a fractured basement reservoir to ensure its effectiveness. Additionally, the FRACSIM software successfully simulated a laboratory glass bead drainage test for two intersected fractures and accurately captured the flow pattern and cumulative production results. Furthermore, a sensitivity study of water injection using an inverted five-spot technique was tested on FRACSIM to assess the productivity of drilled wells in complex fractured reservoirs. The results indicate that FRACSIM can accurately predict flow behavior and subsequently be utilized to evaluate production performance in naturally fractured reservoirs.

Keywords: naturally fractured reservoirs; poro-elastic environment; finite element technique

Citation: Abdel Azim, R.; Alatefi, S.; Alkouh, A. A Coupled Poro-Elastic Fluid Flow Simulator for Naturally Fractured Reservoirs. *Energies* **2023**, *16*, 6476. https://doi.org/10.3390/en16186476

Academic Editors: Pål Østebø Andersen, Haifeng Zhao and Yang Xia

Received: 16 July 2023
Revised: 19 August 2023
Accepted: 1 September 2023
Published: 7 September 2023

1. Introduction

Complexity is the rule rather than the exception in naturally fractured reservoirs, which are defined by the presence of natural fractures and fissures within the underlying formations. These fractures can significantly impact the flow of fluids within the reservoir, making it difficult to predict and manage production. Therefore, a thorough comprehension of the flow behavior is required for efficient production from such reservoirs, which can be achieved via the integration of multiple geological, geophysical, and engineering data.

The numerical modeling of naturally fractured reservoirs poses a real challenge to reservoir modelers due to the complex geological nature and the associated uncertainties in the characterization of existing fractures (i.e., the fracture attributes such as fracture network density, orientation, and mechanical properties). The presence of natural fractures plays an important role in the ultimate recovery and has a great effect on the reservoir performance; however, it is computationally expensive (if not technically infeasible) to take into account all fractures or fracture networks in standard reservoir simulators. Therefore, many studies have been published in the literature to propose different techniques on how to represent the natural fractures in reservoir simulators [1–20]. These studies can be divided into three groups based on how they account for the presence of natural fractures: 1—the dual continuum approach, 2—the discrete fracture network, and 3—the hybrid approach.

In the dual continuum approach, the matrix and fractures are two separate continua, where the matrix is considered the primary medium. The matrix–fracture media interact with each other via transfer functions that control their cross-flow and fluid exchange [21–24]. The main drawback of this approach is that it assumes an extremely simple geometrical representation of matrix and fracture.

In the discrete fracture approach, each fracture and matrix are discretized as 3D tetrahedral elements for matrix and triangle elements in a 2D space for fractures. The mathematical flow equation via matrix and fracture is modeled using various numerical approaches, such as finite element, finite volume methods, mixed finite element, and boundary element methods.

The discrete fracture approach has many advantages, such as the explicit representation of individual fractures on fluid flow and the approach not being constrained by grid fracture geometries; moreover, the fracture model is easily adjustable, and the fluid exchange between the matrix and fracture depends on the fracture geometry, relative permeability, and capillary pressure functions. Nevertheless, the main disadvantage of this approach is the complexities associated with the implementation process and the computational cost (i.e., the discretization of fractures in the created mesh requires local refinement and will most likely increase the computational time). To overcome such limitations, a hybrid approach was adopted by many researchers [25–29]. In this approach, the concept of effective permeability tensor is adopted, where the fractures are partitioned into short and long fractures depending on the cut-off value determined via the reservoir modeler. Each block with short fractures is replaced with a homogenous grid block with an equivalent 3D permeability tensor (i.e., the short fractures are considered local spatial heterogeneities within the matrix block). On the other hand, the long fractures are explicitly discretized and coupled with the 3D permeability tensor matrix.

The main objective of this study is to provide a simulation tool for discrete fracture media using the finite element method (FEM) for simulating two-phase fluid flow in a coupled poro-elastic framework. The simulation of the discrete fractured system contains the interaction of two domains: the first domain is the porous medium, and the second one is the fractures. In this study, the concept of effective permeability tensor to calculate the 3D equivalent permeability of matrix and fractures for small to medium fractures is used to save computational time by imposing the fracture elements into the matrix elements, whereas large fractures are discretized explicitly. In this model, fractures are discretized using triangle elements, while the matrix is represented by tetrahedral elements. The poro-elastic governing equations are discretized in space and time using a standard Galerkin-based finite element approach. The solution of the saturation equation in the finite element approach is stabilized using the Galerkin discretization method. The structure of this study is as follows: the derivation of the proposed coupled poro-elastic mathematical model is presented in Section 2. Then, the validation of the in-house simulator against (1) an exact analytical solution of a 2D poro-elasticity problem (Kirsch's Problem), (2) a 3D lab-scale drainage test of fractured micro-glass bead, and (3) history matching of real field dynamic data (well-test data) of a fractured basement reservoir are presented in Section 3 of this study. Finally, the conclusions of the study findings are presented in Section 4.

2. Derivation of Multiphase Flow Equations

2.1. Mass Conversation Equation

In porous media, the solid mass is represented as

$$M_s = (1-\phi)V\rho_s \tag{1}$$

ρ_s and ϕ are the density of solid constituents and sum of fracture–matrix porosity, respectively.

The conservation of solid mass in porous media can be written as

$$\frac{DM_s}{Dt} - \frac{D}{Dt}\int_v (1-\phi)V\rho_s = \int_v \left[\frac{\partial(1-\phi)\rho_s}{\partial t} + \frac{\partial(1-\phi)\rho_s u_s}{\partial x}\right]dv = 0 \tag{2}$$

Under continuum mechanism, Equation (2) is simplified to

$$\frac{\partial(1-\phi)\rho_s}{\partial t} + \frac{\partial(1-\phi)\rho_s u_s}{\partial x} = 0.0 \tag{3}$$

In the same manner, the fluid mass conversation (i.e., the two phases of mass: oil and water) inside the matrix–fracture systems can be presented as follows:

Two-phase fluid mass:

$$M_\Psi = (\phi)V\rho_\Psi \tag{4}$$

In terms of fluid constituents Equation (4) will be as follows:

$$\frac{DM_\psi}{Dt} = \frac{D}{Dt}\int_v \varphi S_\psi \rho_\psi = \int_v \left[\frac{\partial \varphi S_\psi \rho_\psi}{\partial t} + \frac{\partial \varphi S_\psi \rho_\psi u_\psi}{\partial x} - q_\psi\right]dv = 0 \tag{5}$$

In Equation (5), ρ, S, and q are fluid density, fluid saturation, and the fluid exchange rate between the matrix and fracture systems. Also, ψ refers to fluid phase (water or oil).

Assuming the existence of continuum state, Equation (5) can be written in terms of oil and water as follows:

$$\frac{\partial \varphi S_w \rho_w}{\partial t} + \frac{\partial \varphi S_w \rho_w U_w}{\partial x} - q_w = 0 \tag{6}$$

$$\frac{\partial \varphi S_o \rho_o}{\partial t} + \frac{\partial \varphi S_o \rho_o U_o}{\partial x} - q_o = 0 \tag{7}$$

where U is the intrinsic fluid velocity.

$$q_\psi = \frac{k_1 \rho_\psi}{\mu_\psi} k_{r\psi}(p_{1\psi} - p_{2\psi}) \tag{8}$$

where k_1 is the matrix permeability, $p_{1\psi}$ is the matrix pressure, and $p_{2\psi}$ is the fracture pressure.

The matrix and fracture Darcy's velocity in terms of oil and water is written as

$$\begin{aligned} u_w &= \varphi S_w(U_w - u_s) \\ u_o &= \varphi S_o(U_o - u_s) \end{aligned} \tag{9}$$

By re-arranging Equation (9), the two-phase intrinsic velocity is written as

$$\begin{aligned} U_w &= \tfrac{u_w}{\varphi S_w} + u_s \\ U_o &= \tfrac{u_o}{\varphi S_o} + u_s \end{aligned} \tag{10}$$

Using the intrinsic oil and water velocities in Equation (10) in Equation (7) gives

$$\frac{\partial}{\partial t}(\varphi S_w \rho_w) + \frac{\partial}{\partial x}(\varphi S_w \rho_w)\left(\frac{u_w}{\varphi S_w} + u_s\right) - q_w = 0.0 \tag{11}$$

$$\frac{\partial}{\partial t}(\varphi S_o \rho_o) + \frac{\partial}{\partial x}(\varphi S_o \rho_o)\left(\frac{u_o}{\varphi S_o} + u_s\right) - q_o = 0.0 \tag{12}$$

The expanded form of Equations (3) and (11) are as follows [30]:

$$\frac{\partial \rho_s}{\partial t} - \varphi\frac{\partial \rho_s}{\partial t} - \rho_s\frac{\partial \varphi}{\partial t} - \varphi\rho_s\frac{\partial u_s}{\partial x} - u_s\rho_s\frac{\partial \varphi}{\partial x} - u_s\varphi\frac{\partial \rho_s}{\partial x} + \rho_s\frac{\partial u_s}{\partial x} + u_s\frac{\partial \rho_s}{\partial x} = 0 \tag{13}$$

$$\begin{aligned}&\varphi S_w\tfrac{\partial \rho_w}{\partial t} + \rho_w\varphi\tfrac{\partial S_w}{\partial t} + S_w\rho_w\tfrac{\partial \varphi}{\partial t} + \varphi S_w\rho_w\tfrac{\partial u_s}{\partial x} + \varphi S_w u_s\tfrac{\partial \rho_w}{\partial t} + \varphi u_s\rho_w\tfrac{\partial S_w}{\partial x}\\&+u_s S_w\rho_w\tfrac{\partial \varphi}{\partial x} + u_w\tfrac{\partial \rho_w}{\partial x} + \rho_w\tfrac{\partial u_w}{\partial x} - q_w = 0\end{aligned} \tag{14}$$

The expanded form in Equations (13) and (14) can be re-arranged as

$$\left(\frac{\partial \rho_s}{\partial t} + u_s\frac{\partial \rho_s}{\partial x}\right) - \varphi\left(\frac{\partial \rho_s}{\partial t} + u_s\frac{\partial \rho_s}{\partial x}\right) - \rho_s\left(\frac{\partial \varphi}{\partial t} + u_s\frac{\partial \varphi}{\partial x}\right) + (1-\varphi)\rho_s\frac{\partial u_s}{\partial x} = 0 \tag{15}$$

$$\begin{aligned}&\varphi S_w\left(\tfrac{\partial \rho_w}{\partial t} + u_s\tfrac{\partial \rho_w}{\partial x}\right) + \rho_w\varphi\left(\tfrac{\partial S_w}{\partial t} + u_s\tfrac{\partial S_w}{\partial x}\right) + S_w\rho_w\left(\tfrac{\partial \varphi}{\partial t} + u_s\tfrac{\partial \varphi}{\partial x}\right) + \varphi S_w\rho_w\tfrac{\partial u_s}{\partial x} +\\&u_w\tfrac{\partial \rho_w}{\partial x} - q_w = 0\end{aligned} \tag{16}$$

Total derivative is defined as

$$\frac{D(*)}{Dt} = \frac{\partial(*)}{\partial t} + u\frac{\partial(*)}{\partial x} \tag{17}$$

Direct application of Equation (17) into Equations (15) and (16) gives

$$\left(\frac{D\rho_s}{Dt}\right) - \varphi\left(\frac{D\rho_s}{Dt}\right) - \rho_s\left(\frac{D\varphi}{Dt}\right) + (1-\varphi)\rho_s\frac{\partial u_s}{\partial x} = 0 \tag{18}$$

$$\varphi S_w\left(\frac{D\rho_w}{Dt}\right) + \rho_w\varphi\left(\frac{DS_w}{Dt}\right) + S_w\rho_w\left(\frac{D\varphi}{Dt}\right) + \varphi S_w\rho_w\frac{\partial u_s}{\partial x} + u_w\frac{\partial \rho_w}{\partial x} - q_w = 0 \tag{19}$$

Using Equation (18), the porosity change with time can be as follows:

$$\frac{D\varphi}{Dt} = \frac{1}{\rho_s}\left(\frac{D\rho_s}{\partial D}\right) - \frac{\varphi}{\rho_s}\left(\frac{D\rho_s}{Dt}\right) + (1-\varphi)\frac{\partial u_s}{\partial x} \tag{20}$$

Substituting Equation (20) into Equation (19) gives

$$\begin{aligned}&\varphi S_w\left(\tfrac{D\rho_w}{Dt}\right) + \rho_w\varphi\left(\tfrac{DS_w}{Dt}\right) + S_w\rho_w\left(\tfrac{1}{\rho_s}\left(\tfrac{D\rho_s}{\partial Dt}\right) - \tfrac{\varphi}{\rho_s}\left(\tfrac{D\rho_s}{Dt}\right) + (1-\varphi)\tfrac{\partial u_s}{\partial x}\right)\\&+\varphi S_w\rho_w\tfrac{\partial u_s}{\partial x} + u_w\tfrac{\partial \rho_w}{\partial x} - q_w = 0\end{aligned} \tag{21}$$

or

$$\varphi S_w\left(\frac{D\rho_w}{Dt}\right) + \rho_w\varphi\left(\frac{DS_w}{Dt}\right) + (1-\varphi)\frac{S_w\rho_w}{\rho_s}\frac{D\rho_s}{\partial Dt} + (1-\varphi)S_w\rho_w\frac{\partial u_s}{\partial x} + \rho_w\frac{\partial u_w}{\partial x} + u_w\frac{\partial \rho_w}{\partial x} - q_w = 0 \tag{22}$$

The rate of change in fluid and rock densities are related to the bulk modulus as follows:

$$\begin{cases}\frac{D\rho_w}{Dt} = \frac{\rho_w}{K_w}\frac{DP_w}{Dt}\\ \frac{D\rho_o}{Dt} = \frac{\rho_w}{K_o}\frac{DP_o}{Dt}\\ \frac{D\rho_s}{Dt} = \frac{\rho_s\left[\frac{(1-\varphi)}{K_s}\frac{DP}{Dt} + \frac{(1-\varphi)}{K_{ns}}\frac{DP}{Dt} - (1-\alpha)\frac{\partial u_s}{\partial x}\right]}{(1-\varphi)}\end{cases} \tag{23}$$

where K is the bulk modulus, and the subscripts (o, w, and s) refer to water, oil, and rock, respectively.

Direct substitution of Equation (23) into Equation (22) will give

$$\varphi S_w \frac{\rho_w}{K_w}\frac{DP_w}{Dt} + \rho_w \varphi\left(\frac{DS_w}{Dt}\right) + (1-\varphi)\frac{S_w \rho_w}{\rho_s}\left(\frac{\rho_s\left[\frac{(1-\varphi)}{K_s}\frac{DP}{Dt} + \frac{(1-\varphi)}{K_{ns}}\frac{DP}{Dt} - (1-\alpha)\frac{\partial u_s}{\partial x}\right]}{(1-\varphi)}\right)$$
$$+(1-\varphi)S_w \rho_w \frac{\partial u_s}{\partial x} + \rho_w \frac{\partial u_w}{\partial x} + u_w \frac{\partial \rho_w}{\partial x} - q_w = 0 \quad (24)$$

Equation (24) is the general form of the two-phase fluid flow equation; however, this equation can be simplified by applying the total derivative presented earlier in Equation (17), combining Equation (24) with Darcy's law, and neglecting solid velocity considering that it is small compared to other constituents; this will give a phase velocity as follows:

$$u_\psi = -\frac{k_\psi}{\mu_\psi} k_{r\psi}(P_\psi + \rho_\psi g h) \quad (25)$$

where ψ is the phase symbol (oil or water phase), and $k_{r\Psi}$ and ρ_Ψ are relative permeability and density of the phase, respectively.

Accordingly, the water flow equation is written as [30]

$$\varphi S_w \frac{\rho_w}{K_w}\frac{\partial P_w}{\partial t} + \rho_w \varphi\left(\frac{\partial S_w}{\partial t}\right) + S_w \rho_w \left(\left[\begin{array}{l} \frac{(1-\varphi)}{K_s}\left(S_o \frac{\partial P_o}{\partial t} + P_o \frac{\partial S_o}{\partial t} + S_w \frac{\partial P_w}{\partial t} + P_w \frac{\partial S_w}{\partial t}\right) \\ +\frac{(1-\varphi)}{K_{ns}}\left(S_o \frac{\partial P_o}{\partial t} + P_o \frac{\partial S_o}{\partial t} + S_w \frac{\partial P_w}{\partial t} + P_w \frac{\partial S_w}{\partial t}\right) \end{array}\right]\right)$$
$$-\rho_w \frac{k_w}{\mu_w} k_{rw} \frac{\partial(P_w + \rho_w g h)}{\partial x} - \rho_w \frac{k_w}{\mu_w} k_{rw}\left(P_{wm} - P_{wf}\right) + (1-\varphi) S_w \rho_w \frac{\partial u_s}{\partial x} - q_w = 0 \quad (26)$$

P_w is the water pressure, and the subscript terms (f and m) refer to the fracture and matrix systems, respectively.

Similarly, the oil phase fluid flow is as follows [30]:

$$\varphi S_o \frac{\rho_o}{K_o}\frac{\partial P_o}{\partial t} + \rho_o \varphi\left(\frac{\partial S_o}{\partial t}\right) + S_o \rho_o \left(\left[\begin{array}{l} \frac{(1-\varphi)}{K_s}\left(S_o \frac{\partial P_o}{\partial t} + P_o \frac{\partial S_o}{\partial t} + S_w \frac{\partial P_w}{\partial t} + P_w \frac{\partial S_w}{\partial t}\right) \\ +\frac{(1-\varphi)}{K_{ns}}\left(S_o \frac{\partial P_o}{\partial t} + P_o \frac{\partial S_o}{\partial t} + S_w \frac{\partial P_w}{\partial t} + P_w \frac{\partial S_w}{\partial t}\right) \end{array}\right]\right)$$
$$-\rho_o \frac{k_o}{\mu_o} k_{ro} \frac{\partial(P_o + \rho_o g h)}{\partial x} - \rho_o \frac{k_o}{\mu_o} k_{ro}\left(P_{om} - P_{of}\right) + (1-\varphi) S_w \rho_w \frac{\partial u_s}{\partial x} - q_o = 0 \quad (27)$$

2.2. Momentum Balance Equation

The total stresses σ_{ij} in porous media can be written as a function of effective stresses σ'_{ij} as follows:

$$\sigma_{ij} = \sigma'_{ij} - \alpha \delta_{ij} P \quad (28)$$

δ_{ij} and α are the Kronecker delta and the pore pressure ratio factor, respectively.

The effective stress can be written as a function of the elasticity matrix D_{ijkl} as follows:

$$\sigma'_{ij} = D_{ijkl}\varepsilon_{kl} \quad (29)$$

For a certain traction vector F applied on a solid body under total stress σ_{ij}, the equilibrium of solid motion will be as follows:

$$\sigma_{ij} + F = 0 \quad (30)$$

where F is the vector of tractions applied on the body.

The relationship between displacement and strain can be written as follows:

$$\varepsilon_{ij} = \frac{1}{2}(u_{ij} + u_{ji}) \quad (31)$$

2.3. Finite Element Discretization

This section gives a brief explanation of the finite element method used to simulate the coupled fluid flow and rock poro-elasticity equations. Moreover, the discretization technique of the problem domain into nodal system is also explained.

The four-node tetrahedral element presented in Figure 1 has the following shape functions in terms of local coordinate system (ξ, η, ζ) [30]:

$$\begin{aligned} N_1 &= 1 - \xi - \eta - \zeta \\ N_2 &= \xi \\ N_3 &= \eta \\ N_4 &= \zeta \end{aligned} \tag{32}$$

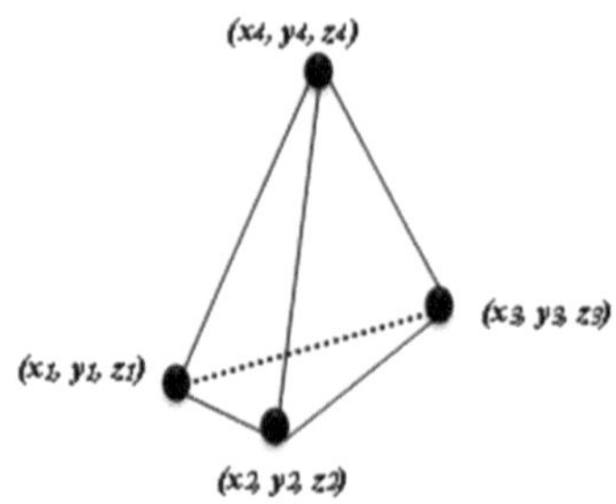

Figure 1. Linear tetrahedral element used in a finite element calculation.

Equation (33) can be used to transform the element geometry from the global coordinates (x, y, and z) to the local coordinates (ξ, η, ζ) as follows:

$$\begin{Bmatrix} \frac{\partial N_i}{\partial \xi} \\ \frac{\partial N_i}{\partial \eta} \\ \frac{\partial N_i}{\partial \zeta} \end{Bmatrix} = \begin{bmatrix} \frac{\partial x}{\partial \xi} & \frac{\partial y}{\partial \xi} & \frac{\partial z}{\partial \xi} \\ \frac{\partial x}{\partial \eta} & \frac{\partial y}{\partial \eta} & \frac{\partial z}{\partial \eta} \\ \frac{\partial x}{\partial \zeta} & \frac{\partial y}{\partial \zeta} & \frac{\partial z}{\partial \zeta} \end{bmatrix} \begin{Bmatrix} \frac{\partial N_i}{\partial x} \\ \frac{\partial N_i}{\partial y} \\ \frac{\partial N_i}{\partial z} \end{Bmatrix} = [J] \begin{Bmatrix} \frac{\partial N_i}{\partial x} \\ \frac{\partial N_i}{\partial y} \\ \frac{\partial N_i}{\partial z} \end{Bmatrix} \tag{33}$$

In Equation (33), the Jacobin matrix J has the following form:

$$[J] = \begin{bmatrix} \frac{\partial x}{\partial \xi} & \frac{\partial y}{\partial \xi} & \frac{\partial z}{\partial \xi} \\ \frac{\partial x}{\partial \eta} & \frac{\partial y}{\partial \eta} & \frac{\partial z}{\partial \eta} \\ \frac{\partial x}{\partial \zeta} & \frac{\partial y}{\partial \zeta} & \frac{\partial z}{\partial \zeta} \end{bmatrix} = \begin{bmatrix} \Sigma \frac{\partial N_i}{\partial \xi} x_i & \Sigma \frac{\partial N_i}{\partial \xi} y_i & \Sigma \frac{\partial N_i}{\partial \xi} z_i \\ \Sigma \frac{\partial N_i}{\partial \eta} x_i & \Sigma \frac{\partial N_i}{\partial \eta} y_i & \Sigma \frac{\partial N_i}{\partial \eta} z_i \\ \Sigma \frac{\partial N_i}{\partial \zeta} x_i & \Sigma \frac{\partial N_i}{\partial \zeta} x_i & \Sigma \frac{\partial N_i}{\partial \zeta} x_i \end{bmatrix} \tag{34}$$

To determine how the element's unknown variables vary, shape functions are utilized. In finite element space, interpolation functions are usually used to approximate these unknowns as follows:

$$u = N_u \overline{u} \tag{35}$$

$$P_w = N_p \overline{P}_w \tag{36}$$

$$P_o = N_p \overline{P}_o \tag{37}$$

$$\varepsilon = B\overline{u} \tag{38}$$

$\overline{u}$, $\overline{P}_w$, and $\overline{P}_o$ are the elements' unknowns, and N is the corresponding shape function.

The strain displacement matrix B is defined as

$$\left[B^T\right] = \begin{bmatrix} \frac{\partial N_i}{\partial x} & 0 & 0 & \frac{\partial N_i}{\partial y} & 0 & \frac{\partial N_i}{\partial z} \\ 0 & \frac{\partial N_i}{\partial y} & 0 & \frac{\partial N_i}{\partial x} & \frac{\partial N_i}{\partial z} & 0 \\ 0 & 0 & \frac{\partial N_i}{\partial z} & 0 & \frac{\partial N_i}{\partial y} & \frac{\partial N_i}{\partial x} \end{bmatrix} \tag{39}$$

The general equilibrium equation is defined as follows:

$$\int_v B^T \partial\sigma dv - \partial f = 0 \tag{40}$$

∂f and V are the boundary load vector and the volume of the element, respectively.

Combining Equations (39) and (40) and dividing by ∂t will give the following equation:

$$\int_v B^T DBdv \frac{\partial u}{\partial t} - \int_v B^T \alpha N dv \frac{\partial P}{\partial t} = \frac{\partial f}{\partial t} \tag{41}$$

The average pore pressure is defined as

$$P = S_o P_o + S_w P_w \tag{42}$$

Inserting the pressure term in Equation (42) into Equation (41) will give

$$\int_v B^T DBdv \frac{\partial u}{\partial t} - \int_v B^T \alpha N dv \left(S_o \frac{\partial P_o}{\partial t} + P_o \frac{\partial S_o}{\partial t} + S_w \frac{\partial P_w}{\partial t} + P_w \frac{\partial S_w}{\partial t} \right) = \frac{\partial f}{\partial t} \tag{43}$$

Adding the capillary pressure term to Equation (43) will give

$$\int_v B^T DBdv \frac{\partial u}{\partial t} - \int_v B^T \alpha N dv \left(\begin{array}{c} S_o \frac{\partial P_o}{\partial t} - P_o \frac{\partial S_w}{\partial P_c} \frac{\partial P_c}{\partial t} \\ + S_w \frac{\partial P_w}{\partial t} + P_w \frac{\partial S_w}{\partial P_c} \frac{\partial P_c}{\partial t} \end{array} \right) = \frac{\partial f}{\partial t} \tag{44}$$

The water flow equation inside the fracture system is discretized as follows:

$$\begin{array}{l} - \int_v \nabla N^T \rho_w \frac{k_w}{\mu_w} k_{rw} \nabla N (P_w + \rho_w g h) dv + \int_v N^T \varphi S_w \frac{\rho_w}{K_w} N \frac{\partial P_w}{\partial t} + \rho_w \varphi \left(\frac{\partial S_w}{\partial t} \right) \\ + \int_v N^T S_w \rho_w N \left(\left[\begin{array}{c} \frac{(1-\varphi)}{K_s} \left(S_w \frac{\partial P_w}{\partial t} + (1 - S_w) \frac{\partial P_o}{\partial t} - P_o \frac{\partial S_w}{\partial t} \frac{\partial P_c}{\partial t} + P_w \frac{\partial S_w}{\partial t} \right) \\ + \frac{(1-\varphi)}{K_{ns}} \left(S_w \frac{\partial P_w}{\partial t} + (1 - S_w) \frac{\partial P_o}{\partial t} - P_o \frac{\partial S_w}{\partial t} \frac{\partial P_c}{\partial t} + P_w \frac{\partial S_w}{\partial t} \right) \end{array} \right] \right) dv \\ - \int_v N^T \rho_w \frac{k_w}{\mu_w} k_{rw} N \left(P_{wm} - P_{wf} \right) + \int_v N^T (1 - \varphi) S_w \rho_w B \frac{\partial u_s}{\partial t} dv - Q_w = 0 \end{array} \tag{45}$$

Two-dimensional fluid flow inside discrete fracture is written as

$$\int\int_{\Omega} \mathrm{FEQ}\, \mathrm{d}\Omega = \int\int_{\Omega_m} \mathrm{FEQ}\, \mathrm{d}\Omega_m + b \times \int\int_{\Omega_f} \mathrm{FEQ}\, \mathrm{d}\overline{\Omega}_f \tag{46}$$

where Ω refers to the porous media domain, and the subscripts f and m refer to fracture and matrix domains.

2.4. Evaluation of Non-Linear Coefficients and Computational Procedure

In this work, the model's unknowns have been calculated at each time step using an iterative methodology. With this method, nonlinear coefficients like capillary pressure, relative permeability, and saturations can be evaluated. The primary unknowns affect these coefficients. The most current fluid pressure computations are used to update the coefficients at each iteration level. Using the updated calculations of capillary pressures, the new values of water saturation at matrix and fracture nodes are estimated from capillary pressure and saturation relationships. Therefore, in order to arrive at a stable solution of fluid pressures and displacement, a convergence criterion is applied to Equations (44) and (45) as follows:

$$\left\| \frac{{R_k}^{i+1} - R_k^i}{{R_k}^i} \right\| \leq \varepsilon \tag{47}$$

At each iteration number i, the convergence level of the nodal unknown R at the node number k should be less than or equal to a convergence limit (ε = 0.01).

3. Results and Discussion

The verification of the presented in-house model is important to ensure that the developed 3D coupled poro-elastic concept is implemented correctly into the in-house model. Therefore, in this section, the ability of the developed model (FRACSIM) to capture the complex flow behavior of naturally fractured reservoirs under a poro-elastic environment was validated against (1) an exact analytical solution of a 2D poro-elasticity problem (Kirsch's Problem), (2) a 3D lab-scale drainage test of fractured micro-glass bead, and (3) the history matching of real field dynamic data of a fractured basement reservoir. The discerption and analysis of each validation case will be presented next.

3.1. Validation Using Kirsch's Problem in Poroelasticity

The Kirsch model was originally developed to estimate the tangential and radial stresses around a hole in a homogenous infinite plate under unidirectional tension and assuming a plane-strain case (Kirsch 1898) [31]. For the case of a vertical borehole in a poro-elastic environment, Detournay and Cheng 1988 [32] presented an analytical solution of the pore pressure, the stress, and the displacement induced by drilling, producing, or pressurizing the borehole. In their study, they used the Laplace space transformation to derive the analytical solution, which then was transformed to the time domain using a numerical inversion technique. Herein, we present a verification of the poro-elastic in-house model (FRACSIM) against the analytical solution presented by Detournay and Cheng for a circular-shaped reservoir (see Figure 2).

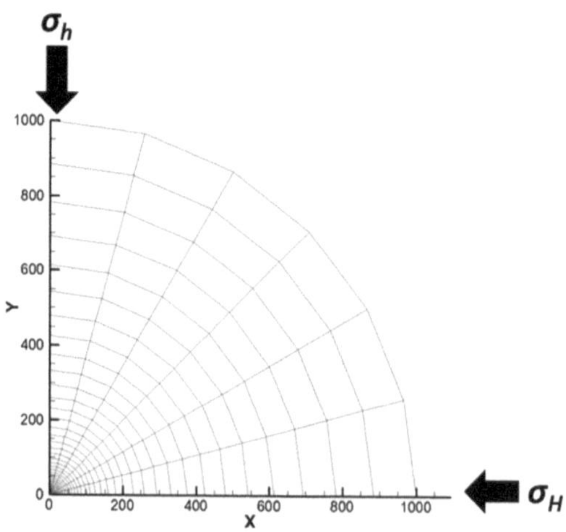

Figure 2. Two-dimensional circular reservoir used for validation of the poro-elastic in-house model (FRACSIM) with maximum horizontal stress σ_H and minimum horizontal stress σ_h.

The pore pressure, radial stress, tangential stress, radial displacement, and tangential displacement analytical solutions of a two-dimensional model of a circular-shaped reservoir with a maximum horizontal stress σ_H, a minimum horizontal stress σ_h, an initial pore pressure P_i, a wellbore pressure P_w, a wellbore radius r_w, and a drainage radius r_e are as follows [32]:

Pore pressure:

$$p(r,t) = P_i + (P_w - P_i)\, g(r,t) \tag{48}$$

Radial stress:

$$\sigma_{rr}(r,\theta) = \tfrac{\sigma_H+\sigma_h}{2}\left(1-\tfrac{r_w^2}{r^2}\right) + \tfrac{\sigma_H-\sigma_h}{2}\left(1+3\tfrac{r_w^4}{r^4}-4\tfrac{r_w^2}{r^2}\right)\cos(2\theta) + P_w\tfrac{r_w^2}{r^2} \\ +2\eta(P_w-P_i)\tfrac{r_w}{r}\, h(r,t) \tag{49}$$

Tangential stress:

$$\sigma_{\theta\theta}(r,\theta) = \tfrac{\sigma_H+\sigma_h}{2}\left(1-\tfrac{r_w^2}{r^2}\right) - \tfrac{\sigma_H-\sigma_h}{2}\left(1+3\tfrac{r_w^4}{r^4}\right)\cos(2\theta) \quad P_w\tfrac{r_w^2}{r^2} \\ -2\eta(P_w-P_i)\left(\tfrac{r_w}{r}h(r,t)+g(r,t)\right) \tag{50}$$

Radial displacement:

$$u_r(r,\theta) = \tfrac{r}{4G}(\sigma_H+\sigma_h)\left(1-2v+\tfrac{r_w^2}{r^2}\right) \\ +\tfrac{r}{4G}(\sigma_H-\sigma_h)\left(\tfrac{r_w^2}{r^2}\left(4-4v-\tfrac{r_w^2}{r^2}\right)+1\right)\cos(2\theta) - \tfrac{P_w}{2G}\tfrac{r_w^2}{r} \\ -\tfrac{\eta}{G}r_w(P_w-P_i)h(r,t) \tag{51}$$

Tangential displacement:

$$u_\theta(r,\theta) = -\frac{r}{4G}(\sigma_H-\sigma_h)\left(\frac{r_w^2}{r^2}\left(2-4v+\frac{r_w^2}{r^2}\right)+1\right)\sin(2\theta) \tag{52}$$

In Equations (48)–(52), G is Young's modulus, v is Poisson's ratio, η is the poro-elastic coefficient ($\eta = \alpha\frac{1-2v}{2(1-v)}$) with a physical range of variation of [0–0.5], θ is the angle relative to the direction of the maximum horizontal stress, and r is the radial distance from the borehole to the point of calculation for stresses and displacements. Moreover, the Laplace space transform of the functions $g\ (r, t)$ and $h\ (r, t)$ is as follows:

$$\tilde{g}(r,s) = \frac{K_0(\xi)}{sK_0(\beta)}\ and\ \tilde{h}(r,s) = \frac{1}{s}\left[\frac{K_1(\xi)}{\beta K_0(\beta)} - \frac{r_w}{r}\frac{K_1(\xi)}{\beta K_0(\beta)}\right] \tag{53}$$

where K_0 and K_1 are the first-order modified Bessel functions of the first and second kind.

The results of the developed poro-elastic in-house simulator were compared to the analytical solution presented in Equations (48)–(53) for a 2D circular reservoir with a wellbore radius of 0.1 m and a drainage radius of 1000 m. The reservoir rock and fluid properties used in this comparison are presented in Table 1.

Table 1. Parameters used in the verification of poro-elasticity solutions.

Parameter	Value
Young's modulus	40 GPa
Poisson's ratio	0.2
Porosity ϕ	0.1
Water compresibility C_w	1.0×10^{-4} psi^{-1}
Water viscosity μw	0.1 cp
Biot's coefficient	1.0
Maximum stress	5800 psi
Minimum stress	5500 psi
Initial reservoir pressure	5500 psi
Wellbore pressure	1000 psi
Formation permeability Kx	0.01 md
Formation permeability Ky	0.01 md
Wellbore radius r_w	0.1 m
Reservoir drainage radius r_e	1000 m

Figure 3 presents a comparison of the tangential stress of the analytical solution and the in-house simulator for different time steps (after one day, one month, and one year of production). It can be seen from this figure that the poro-elastic in-house model is in good agreement with the analytical solution. Furthermore, Figure 4A–C present the tangential displacement of the analytical solution and the in-house model after one day, one month, and one year of production; it is clear from this figure that the in-house code results are in close agreement with the analytical solution. It should be noted that the same level of agreement between the analytical solution and the in-house model was achieved for the radial stress and displacement for this validation exercise. Such results highlight the capability of the developed in-house model in capturing the stress redistribution and displacement due to production or pressurizing the borehole under the assumption that the rock behaves as a poro-elastic material with compressible constituents.

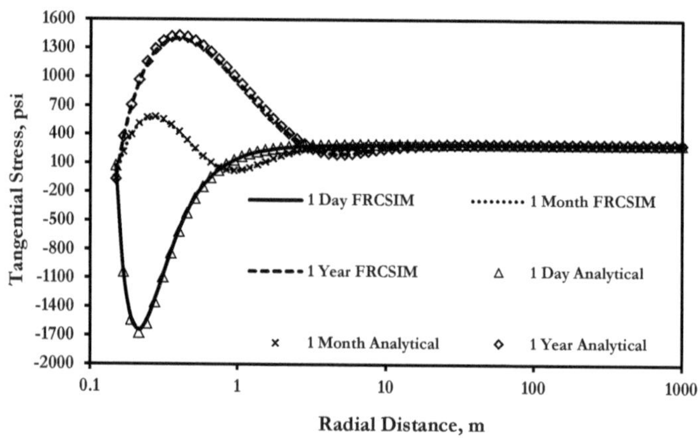

Figure 3. Comparison of tangential stress ($\sigma_{\theta\theta}$) of the analytical solution and the in-house simulator (FRACSIM) under σ_H = 5800 psi and σ_h = 5500 psi, P_r = 5500 psi, P_w = 1000 psi, k_x = 0.01 md, and k_y = 0.01 md.

3.2. *History Matching of Two-Phase Flow Real Data—Laboratory Scale*

The visualization experiments of the two-phase flow in fractured glass bead micro-models are a good tool to observe the flow behavior of fractured media and understand the interaction between matrix and fracture system as well as to investigate the effect of fracture properties on fluid flow. In this section, the in-house poro-elastic model was used to history match a drainage test result of a 200 mm × 100 mm × 2 mm multiple fractured glass bead model reported by Fahad et al., 2012 [33]. The drainage test was conducted on a homogeneous glass bead matrix with a mesh size range of 105 to 145 μm, while the fracture was made up of thin glass strips placed at the center of the model with different fracture orientations and densities. This drainage test studied the flow around two crossed fractures placed at the center of the glass bead matrix. Figure 5 presents a schematic of the experimental setup of the glass bead model used for the multiple crossed fracture drainage test. The matrix and fracture permeabilities are 3.4 and 10^4 Darcy, respectively. The experiment started with saturating the glass bead matrix with water, which has a viscosity of 1.002 cp and a density of 0.998 gm/cm^3. Then, a Soltrol-130 iso-paraffin oil, with a viscosity of 1.4 cp and a density of 0.75 gm/cm^3, was injected at the inlet of the glass bead matrix to displace the water at a constant injection rate of 2 cc/min, effluent volume produced from the displacement process are collected at the outlet of the glass bead model. Moreover, the displacement front was recorded using a digital camera with the help of a light source that was installed underneath the glass bead model to illumine the fractured glass bead matrix.

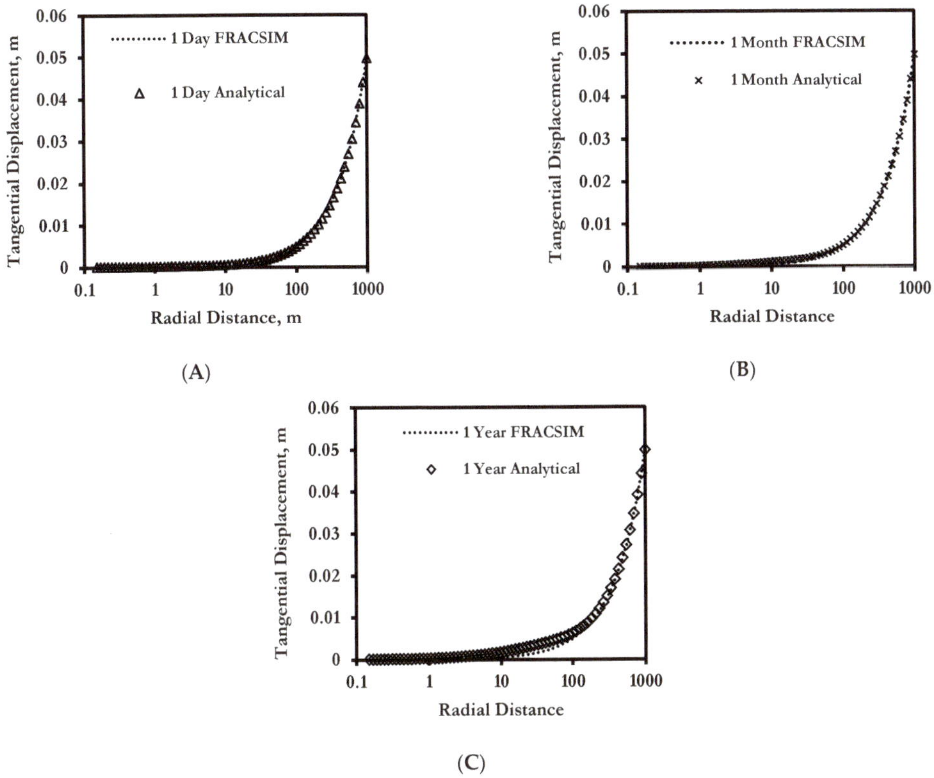

Figure 4. Comparison of tangential displacement (u_θ) of the analytical solution and the in-house simulator (FRACSIM) under $\sigma_H = 5800$ psi and $\sigma_h = 5500$ psi, $P_r = 5500$ psi, Pw = 1000 psi, $k_x = 0.01$ md, and $k_y = 0.01$ md (**A**) after one day of production, (**B**) after one month, and (**C**) after one year.

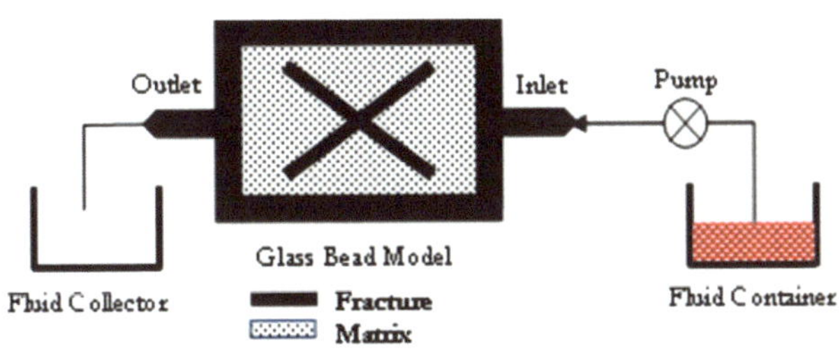

Figure 5. Schematic of the experimental setup of the glass bead model used for the multiple crossed fracture drainage test.

The developed in-house model was used to simulate the fractured glass bead drainage test. Figure 6A, top image, presents the generated in-house mesh to simulate the drainage experiment of two-phase flow on crossed-fractures at the center of a glass bead matrix. In this image, the tetrahedral elements of the mesh represent the matrix, while the fine triangle elements are used for meshing the two-crossed fractures. In addition, Figure 6A's middle and bottom images represent the drainage test visualization at 0.4 pore volume injected (0.4 PVI) and the simulated drainage test using the in-house poro-elastic model at the same pore volume injected, respectively. It can be noted from Figure 6A (middle and bottom images) that the in-house simulated drainage test image is in close agreement with the experimental drainage test image at the same pore volume inject (PVI = 0.4). Furthermore, Figure 6B presents a comparison of the measured produced water volume of the drainage experiment and that from the in-house model; it can be seen from this figure that the in-house model produces an excellent match to the measured water volume from the fractured glass bead experiment. Such results clearly highlight the capability of the developed model to capture the flow behavior of fractured media at the laboratory scale with a good level of accuracy. The validation of the in-house model with real field data will be presented in the next section.

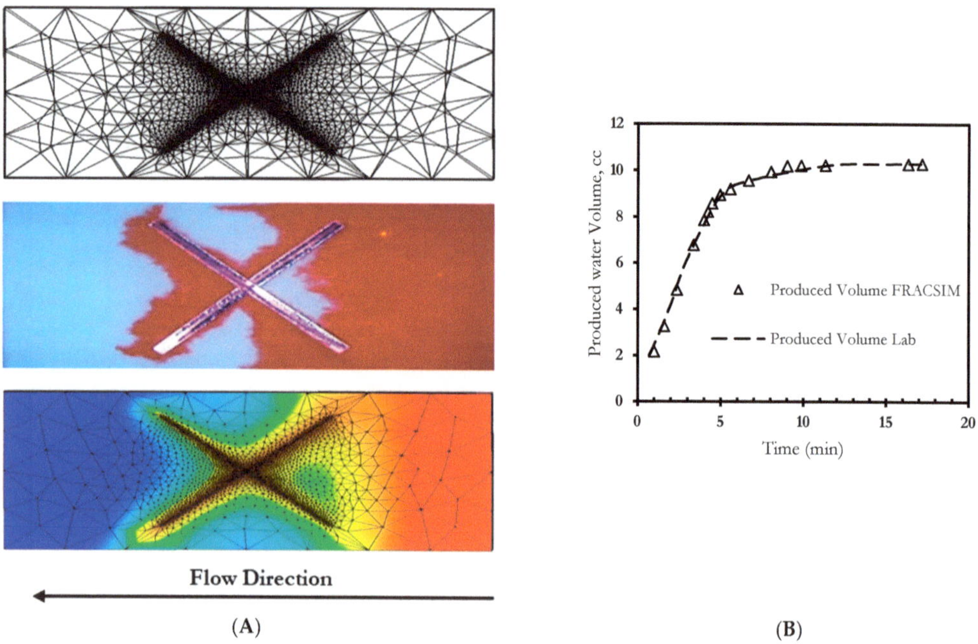

Figure 6. ((**A**)—Top) presents the generated in-house mesh to simulate the drainage experiment of two-phase flow on a crossed-fractures at the center of a glass bead matrix; ((**A**)—Middle) presents the drainage test visualization at 0.4 pore volume injected (0.4 PVI); (light blue is water, and dark red is Soltrol-130 oil); ((**A**)—Bottom) presents the simulated drainage test using the in-house poro-elastic model at 0.4 pore volume injected (0.4 PVI). (**B**) History matching the drainage test produced water volume using the in-house poro-elastic model.

3.3. *History Matching of Real Dynamic Data—Field Scale*

In this section, the developed in-house simulator is used to history match a well test data of a fractured granitic oil-bearing basement reservoir located in Southeast Asia using a hybrid discretization technique that couples the single continuum and the discrete fracture method in a poro-elastic environment. The hybrid approach consists of three main steps: 1—stochastic generation of 3D subsurface fracture network map based on field static data, 2—calculation of grid-based 3D permeability tensors accounting for short fractures, and 3—coupling of single continuum (grid-based permeability tensor) and explicit discrete long fractures. A brief description of the hybrid approach is given in the next paragraph.

In this study, the discrete fracture network map was generated via the statistical analysis of field data (static data) procedures developed by Doonechaly and Rahman (2012) [34]. This technique integrates different field data to determine a distribution range of fracture properties (fracture orientation, fracture density, and fractal dimension); the integrated field data include core analysis, conventional well logs, seismic attributes, and wellbore images.

The determined fracture density and fractal dimension distributions are used as an input to an object-based stochastic sequential Gaussian 3D simulator to generate different random realizations of fracture attributes. In this simulation, the fractures are treated as circular disk objects, as shown in Figure 7. Each fracture is defined based on its orientation (i.e., the dip and azimuth angles of the fracture), center point (i.e., fracture location), radius, and aperture. Moreover, the random realization of fractures continues until the total fracture intensity and fractal dimension of the studied area are met. At this point, the resulting model is the 3D subsurface fracture map of the studied reservoir. Afterward, the generated 3D subsurface map is divided into a number of grid blocks (N grid blocks) in order to calculate the grid-based permeability tensors, as shown in Figure 8A. For this purpose, a threshold fracture length (LF_{min}) is defined. Accordingly, fractures with a length less than the threshold fracture length that cut a certain block are used to calculate the permeability tensors of that block (i.e., fractures that have a length less than (LF_{min}) are considered as part of the matrix in the form of permeability tensor; in other words, the short fractures are considered local spatial heterogeneities within the matrix block), as shown in Figure 8B. Once the grid-based 3D permeability tensors are calculated, the matrix domain is coupled with the explicitly discretized long fractures (i.e., fractures with fracture length > LF_{min}). The matrix domain is discretized using four-node tetrahedral elements, while the triangular elements are used for the discretization of explicit long fractures, as shown in Figure 8C. The used grid blocks were built using an in-house developed mesh generator. It should be noted that the grid-based permeability tensor calculations, which include diagonal and off-diagonal terms, are performed using three-dimensional Darcy's law and continuity flow equations under pressure and flux periodic boundary conditions developed by Durlofsky (1991) [35]. Once the hybrid permeability tensor and discrete fracture mesh is ready, then it is used in the FRACSIM in-house poro-elastic model to simulate fluid flow around the fractured well and history match the dynamic production data (well test data).

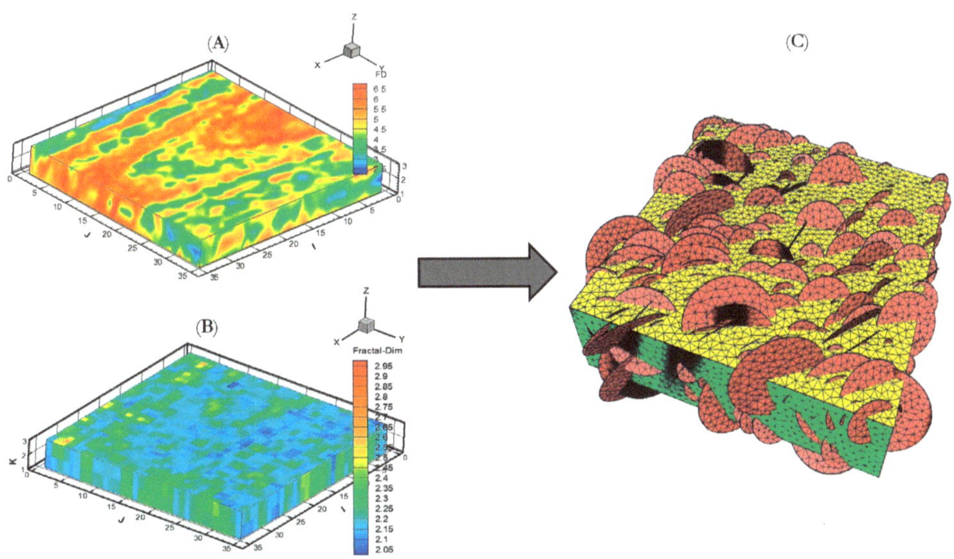

Figure 7. (**A**) Three-dimensional fracture density distribution; the fracture density varies between 2.5 m^{-1} and 6.5 m^{-1}. Red color indicates high fracture density. (**B**) Three-dimensional fractal dimension distribution; the fracture density varies between 2.5 m^{-1} and 6.5 m^{-1}. (**C**) Stochastic realization of discrete fracture map using object-based simulator; hence, fractures are represented as circular disk objects with their unique center point, dip, azimuth, and radius.

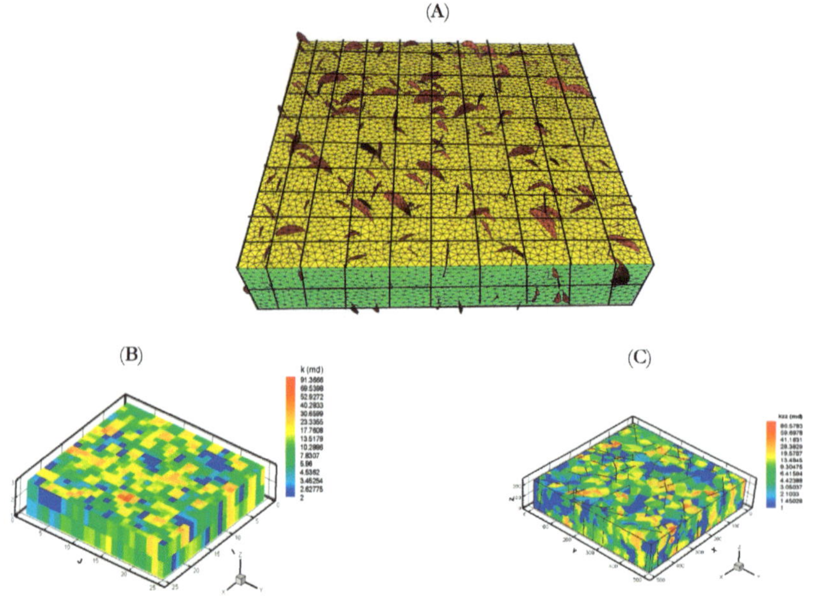

Figure 8. Demonstration of the hybrid approach: (**A**) division of the reservoir into a number of grid blocks, (**B**) conversion of short-to-medium fractures into 3D-grid-based permeability single continuum, and (**C**) coupling of the grid-based permeability tensors with discrete fracture network via an explicit discretization the long fractures.

A complete algorithm of the well-test history matching process is summarized as follows:

1. Generate the subsurface fracture realization using field data based on Doonechaly and Rahman (2012) approach [34].
2. Utilizing periodic boundary conditions Durlofsky (1991) [35], calculate the block-based permeability tensor of the single continuum taking into account the short fractures.
3. Couple the block-based permeability tensor with the discrete fracture network of long fractures using the in-house mesh generator (hybrid approach).
4. Start FRACSIM in-house model to simulate pressure build-up and draw-down cycles.
5. Compare the FRACSIM results with that of the measured well test data to estimate the error.
6. If the error from step 5 is less than the predefined error threshold, then stop and report the optimum fracture realization with the pressure change and pressure derivative results. Otherwise, go back to step 1 to modify the fracture attribute and generate a new fracture realization. The simulated well-test results are compared with that from the build-up and draw-down test data for each realization until the error is minimized.

Accordingly, the FRACSIM numerical model with the hybrid approach is used to simulate single-phase fluid flow via the subsurface fracture realization, which was generated using field data of a fractured granitic oil-bearing formation located in Southeast Asia. According to geological interpretation, the formation is highly fractured with an extremely low matrix porosity and permeability. The network of fractures formed the majority of storage spaces in the studied formation. A Drill Stem Test (DST) was conducted in this formation with controlled flow periods before shutting to better understand the reservoir extent and evaluate well deliverability. The reservoir and fluid input data of the simulation study are presented in Table 2. Different subsurface fracture realizations were generated using the available field data. A total of 3000 fractures were created, and the minimum fracture length threshold was set to 40 m. The grid-based permeability tensors were generated accounting for fractures with a length < 40 m. Long fractures with lengths > 40 m were discretized explicitly and coupled with the single continuum. The different fracture realizations were fed to the in-house FRACSIM model to simulate the main build-up and draw-down period of the DST test. Moreover, a commercial reservoir simulator (Eclipse100) was used to simulate the build-up test and draw-down periods. However, only the grid-based permeability tensor was passed to Eclipse100 since this simulator lacks the ability to explicitly discretize fracture network. Figure 9 presents the optimum coupled permeability tensor and explicit fracture network of the simulated area.

Table 2. Reservoir inputs data for a typical fractured basement reservoir.

Parameter	Value
Reservoir dimensions	500 m × 500 m × 250 m
Wellbore entry	partially penetration (90 m)
Matrix permeability	0.0095 mD
Matrix porosity	0.02
Initial water saturation	0.34
Fracture aperture	7.06×10^{-3} mm
Initial fracture intensity	0.15 m^{-1}
Fractal dimension (D)	1.25
Fracture permeability	100 D
Fracture porosity	0.1
Initial reservoir pressure	4200 psi
Horizontal stresses	4400 psi
Vertical stress	6000 psi
Fluid viscosity	1.38 cp
Fluid compressibility	10^{-5} psi^{-1}
Production time before shut in (t_p)	72 h
Production flow rate before shut in	5571 bbl/d

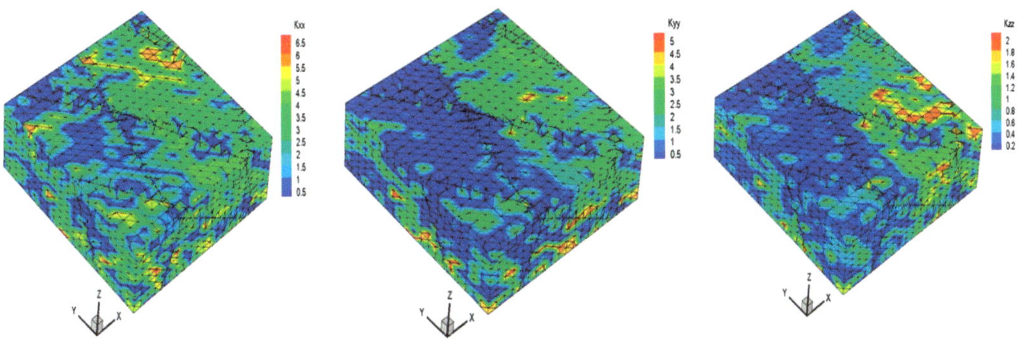

Figure 9. The optimum coupled permeability tensor and explicit fracture network of the simulated area.

Figure 10 presents the simulated shut-in pressure using FRACSIM and Eclipse100 compared to the measured shut-in pressure. It can be seen from this figure that the in-house simulator closely matches the measured shut-in pressure data. On the other hand, the Eclipse100 failed to provide a good match and underestimated the measured shut-in pressure. Similar to the in-house simulator, the simulated area in the Eclipse-100 model was divided into different grid blocks, and then the intersected fractures with each block have been accounted for in the form of permeability tensors for each grid block using the periodic boundary condition. However, the explicit discrete fracture approach represents fractures as a 2D triangle surface sandwiched between the 3D tetrahedral domain, giving it the flexibility to mesh any kind of fracture orientation and shape. Such fracture representation is not available in the ECLIPSE simulator, as the fractures were presented with the stack of grid blocks, which is the main reason that the Eclipse-100 model could not match the shut-in pressure as the effect of long fractures on pressure change is neglected. Furthermore, Figure 11 presents the simulated draw-down cycle of the DST using FRACSIM and Eclipse100 in comparison to the measured draw-down pressure. It can be noted from this that the FRACSIM produced a good match to the measured draw-down curve, while the Eclipse100 failed to match the measured draw-down pressure.

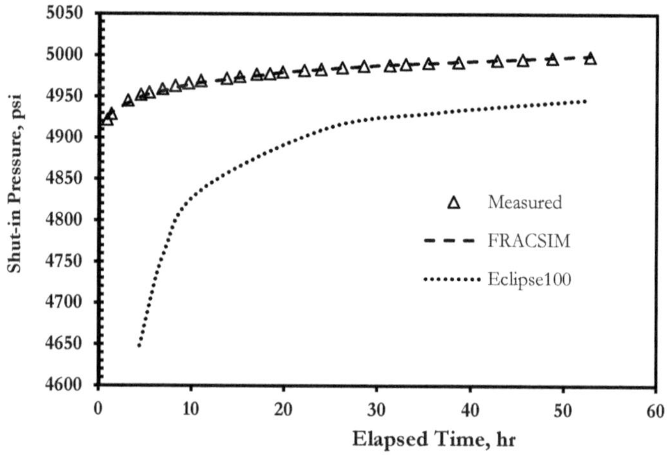

Figure 10. The simulated shut-in pressure of FRACSIM in-house model and Eclipse100 compared to the measured shut-in pressure.

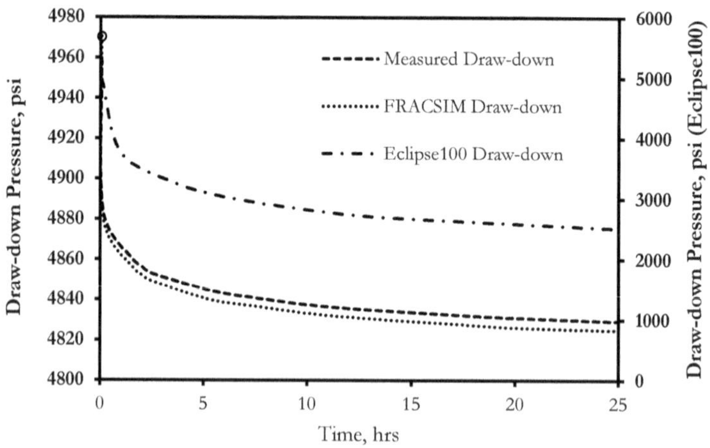

Figure 11. The simulated draw-down profile of FRACSIM in-house model and Eclipse100 compared to the measured draw-down pressure.

Figure 12A presents the pressure change and pressure derivative of the finite-difference-based reservoir simulator ECLIPSE-100 in comparison to the pressure change and pressure derivative. It can be noted from this figure that, again, the Eclipse100 model could not match the measured pressure change and pressure derivative data as a result of ignoring the effect of long fractures. Figure 12B presents the pressure change and pressure derivative results of the FRACSIM results compared to the measured data. It is clear from this figure that the in-house model results have a good match with the measured results. Furthermore, the FRACSIM in-house model and the measured data pressure derivative curves show similar flow regimes that include the spherical flow due to partial well penetration at early time (characterized by a negative slope in the derivative curve), followed by a short radial flow at middle time (characterized by flatness in the derivative curve with a zero slope) and a late linear flow due to the effect of fractures (characterized by a positive half-slop of the derivative curve). Such encouraging results highlight the good capability of the in-house poro-elastic simulator in simulating fractured reservoirs. Finally, the in-house single well model was run for one year under a natural depletion scenario in order to visualize the change in fluid velocity and reservoir pressure in the fracture network. Figure 13A–D present the velocity profile and the pressure profile of the single well model after one month and one year of production under a constant bottom hole pressure of 2300 psi. It can be noted from this figure that the permeability of the connected discrete fractures has a significant effect on fluid velocity and reservoir pressure diffusion, almost all connected fractures exhibit a high change in velocity, especially the fractures near the wellbore at the early production time (one month), as production continues to one year the velocity change is dominant in all fractures in the studied area. The same trend is present with the pressure distribution: after one year of production, the pressure change reaches the boundary of the studied area.

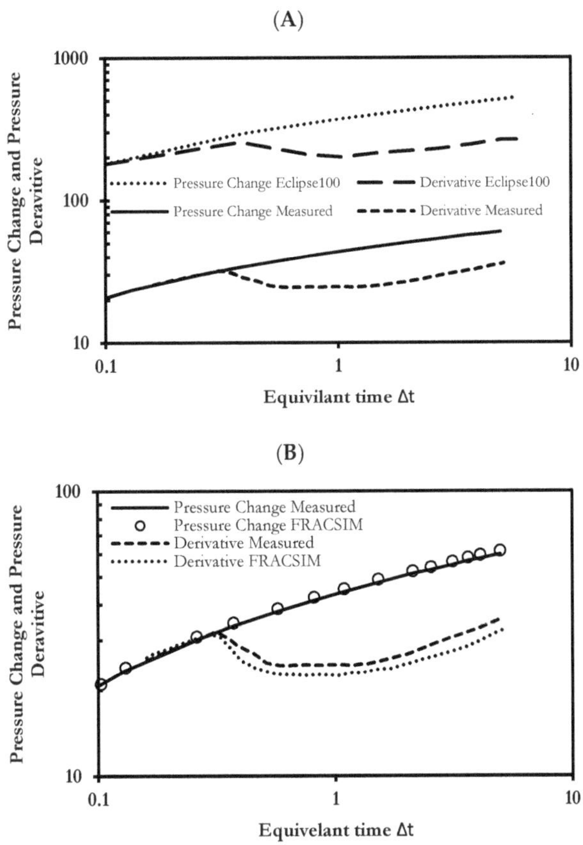

Figure 12. (**A**) The pressure change and pressure derivative of the finite-difference-based reservoir simulator ECLIPSE-100 and (**B**) the pressure change and pressure derivative of the FRACSIM in-house model compared to measured data.

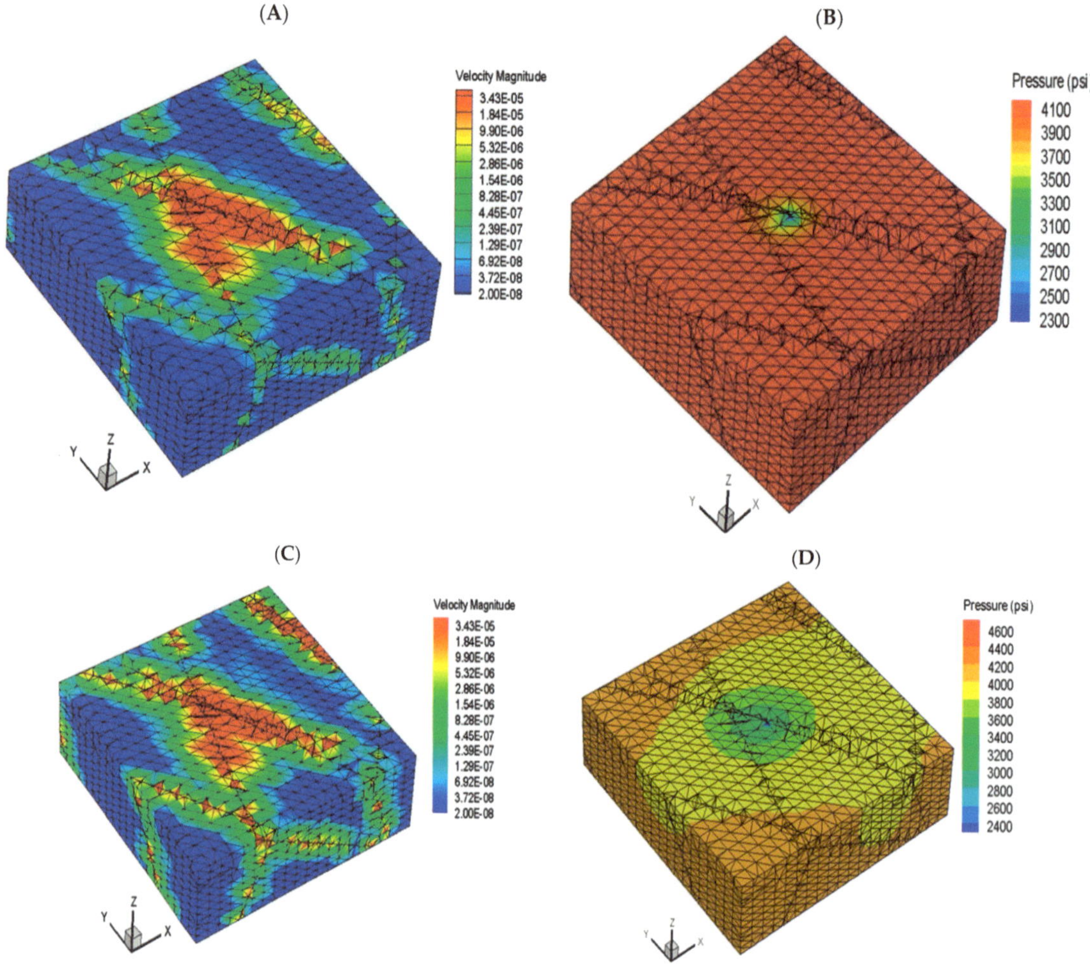

Figure 13. (**A**) The velocity profile and (**B**) the pressure profile of the single well model after one month of production; (**C**) the velocity profile and (**D**) the pressure profile of the single well model after one year of production. The well is located at the center of the block.

3.4. Sensitivity Study of Full Field Development Plan

In this section, the developed in-house simulator was utilized to investigate the performance of the reservoir (presented in Section 3.3) under two driving mechanisms: natural depletion and water flooding. The reservoir dimension of 4000 m × 5000 m × 250 m was used in the FRACSIM simulator with the fracture realization that was verified in the previous section. The first step was to investigate the reservoir performance under volumetric natural depletion since the reservoir does not have any aquifer support. The reservoir rock and fluid properties are similar to those presented in Table 2 (Section 3.3), and the relative permeability curves are shown in Figure 14A,B. Sixteen vertical producers were placed across the reservoir (see Figure 15). These producers were run under constant bottom hole pressure of 1000 psi, with a simulation period of 15 years. Figure 16 presents the oil production rate of the highest and lowest producers among the sixteen wells after 15 years of production. From this figure, it can be seen that the well production shows a typical production behavior of fractured reservoirs; that is, the wells start with a high production rate for a short period of time then production goes in a rapid decline. This is primarily driven by the fact that fluids initially flow via fractures, which led to a significant drop in wellbore pressure with a spike in fluid production. Then, the matrix begins to feed the fracture network with fluid. During this flow period, oil production at the wellbore drops to a steady low rate with a gradual decrease in wellbore pressure. The simulation results show that Well-P16 was the highest producer among the sixteen wells with a production rate of 13,146 bbl/day at the beginning of production; then, the production rate declined to 70 bbl/day after 6 years of production. Well-P2 was the lowest producer among all with an initial production of 4400 bbl/day; then, the production decreased to 20 bbl /day after 6 years of production. It should be noted that the flow rate profiles of all other wells were bounded between the highest and lowest producer rate profiles and followed the same trend as that seen in well-P6 and Well-P2 but with different ranges of oil rates. Moreover, the water production was negligible because the simulation was run with only connate water. The cumulative oil production ranges from 1.13×10^6 bbl at well-P2 to 2.98×10^6 bbl at well-P16. The total oil production for the studied reservoir is 3.27×10^7 bbl after a simulation of 15 years.

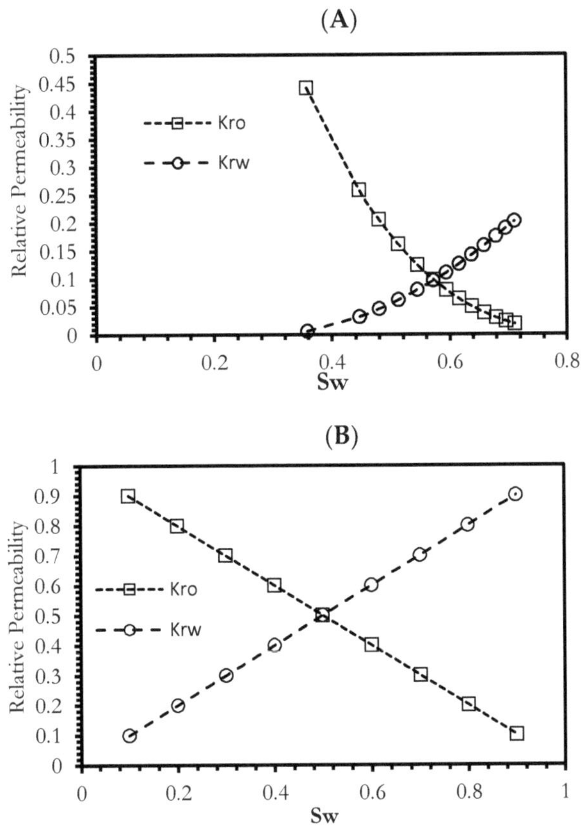

Figure 14. The reservoir relative permeability cures, (**A**) matrix relative permeability, and (**B**) fracture relative permeability.

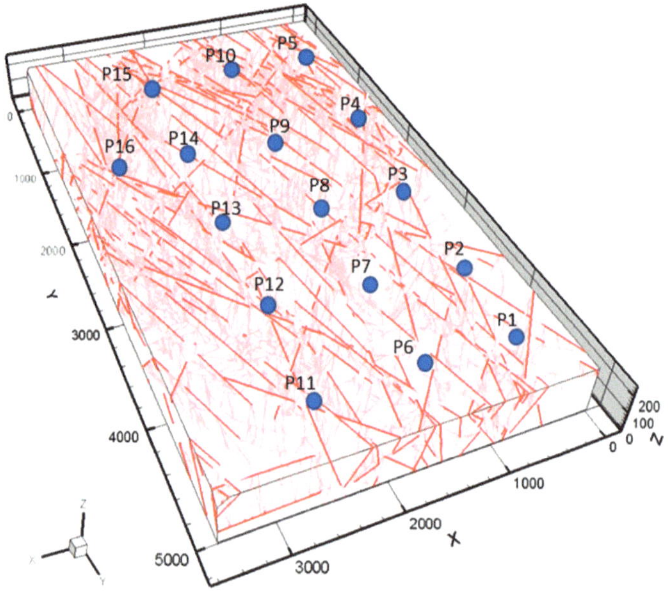

Figure 15. The discreet fracture map with the location of the sixteen producers represented by a blue circle.

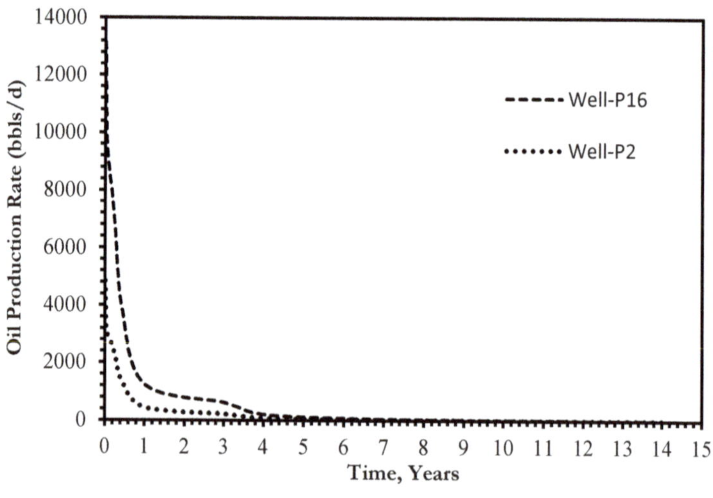

Figure 16. The oil production rate of the highest and lowest producers among the simulated sixteen wells.

In order to investigate the reservoir performance under water flooding, wells (P3, P7, P9, and P13) were converted into injection wells; these wells were chosen as they are located at the center of the studied area, as shown in Figure 17. The amount of water injected in each well was 6000 bbl/d. The reservoir thickness is 250 m; water was injected into the bottom region (the bottom 100 m of the reservoir thickness), while the oil was produced from the reservoir top section (the top 90 m of reservoir thickness). Thus, there is around a 60 m zone separating the producing and injection sections.

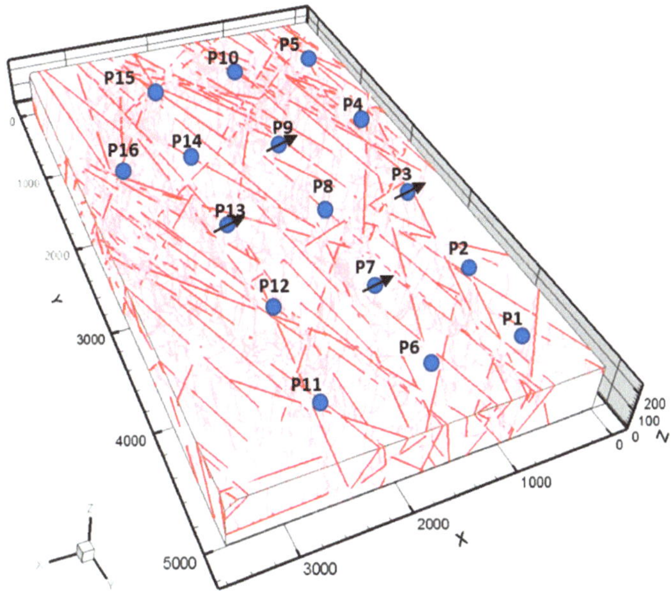

Figure 17. The discreet fracture map with the location of the twelve producers represented by a blue circle and the four injectors (P3, P7, P9, and P13) represented by a blue circle with slanted arrow.

Figures 18–20 present the oil production rate, the water cut, and the water production rates of the production wells, respectively. It can be seen from these three figures that Well-P8 has the highest oil and water production rate. This is due to the fact that Well-P8 location is bounded by four injectors, which, in turn, creates a high sweep efficiency area around Well-P8. Well-P5 has the lowest water production rate because the location of this well is far from the injection area. Furthermore, the water saturation map after one year is shown in Figure 21A, while the water saturation and pressure after five years of injection are shown in Figures 21B and 22 respectively. Figure 23 presents a cumulative production chart for each producer under the natural depletion and the water flooding mechanisms. Furthermore, Table 3 presents the percentage change in each producer's cumulative total oil production following the water flooding procedure in comparison to the cumulative total oil production due to natural depletion. It can be concluded from Figure 23 and Table 3 that all producers (except Well-P8) show a certain degree of improvement in the cumulative oil production due to water injection. For example, Wells P2, P4, P5, P6, P10, P12, and P14 show a promising percentage of improvement, reaching as high as 200% in the case of P2 and P4. The improvement in cumulative production is mainly attributed to the relative location of the well and the intensity of the fracture network intersecting that well. Wells that intersect high-intensity fractures will suffer from high water break though, such as Well-P15, which has the highest water cut among all producers. Furthermore, Well-P8 showed a reduction in total cumulative production compared to the natural depletion scenario due to the fact that it has the highest water production rate among all producers because it was bounded by four injectors. In summary, the presented results indicate that a comprehensive understanding of the geological characterization and fracture network structure via the integration of different resources including static and dynamic field data, will certainly help in achieving successful waterflooding in naturally fractured reservoirs.

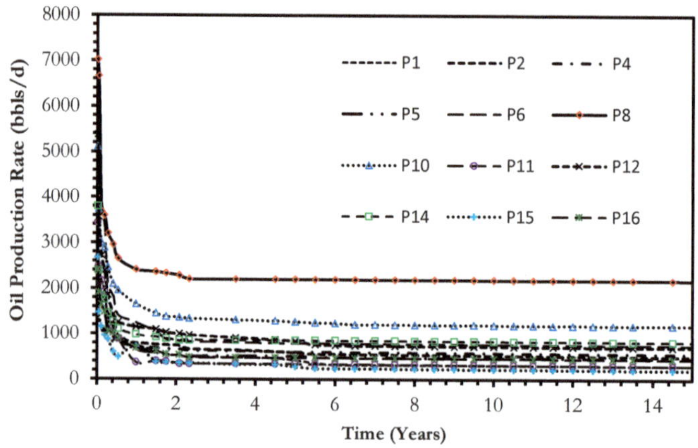

Figure 18. The oil production rate of the simulated twelve producers under water flooding.

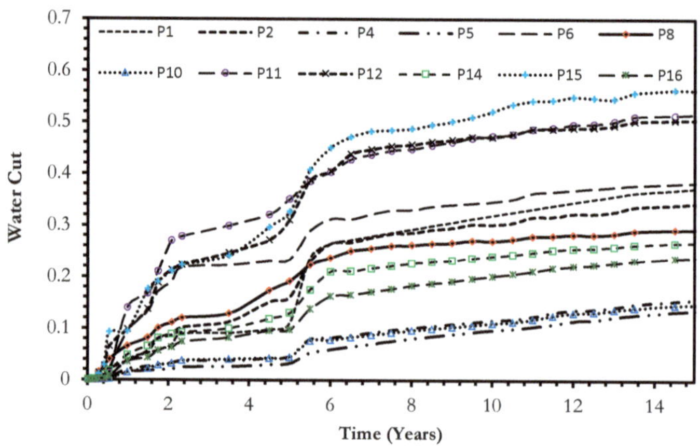

Figure 19. The water cut of the simulated twelve producers under water flooding.

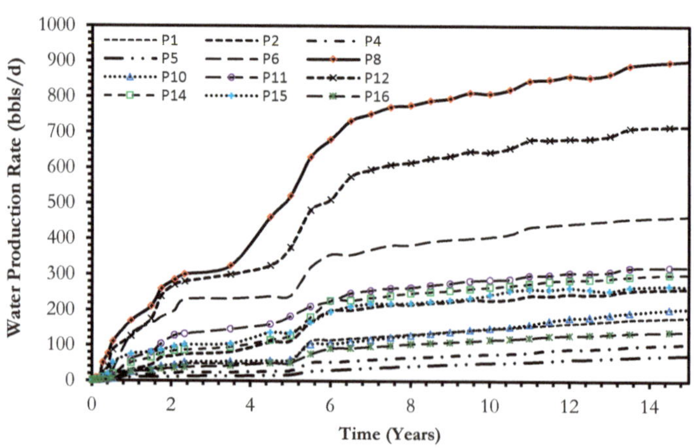

Figure 20. The water production rate of the simulated twelve producers under water flooding.

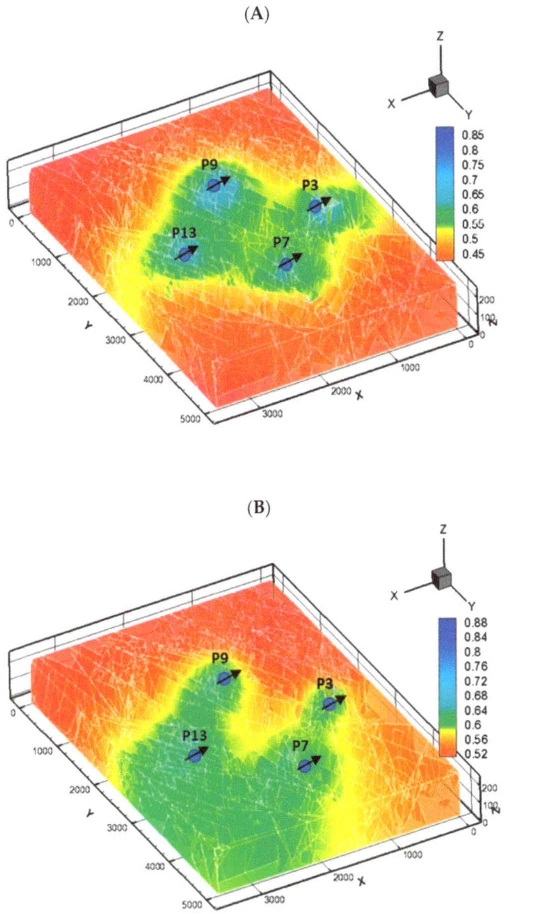

Figure 21. The water saturation map of around the injection wells (**A**) after one year of water injection and (**B**) after five years of water injection.

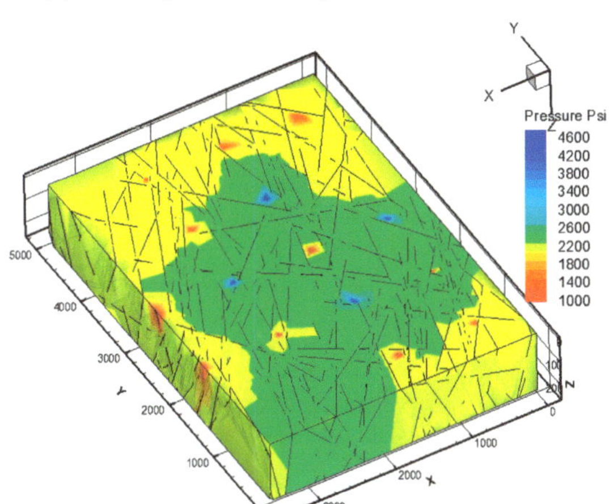

Figure 22. The reservoir pressure map around the injection wells after five years of water injection.

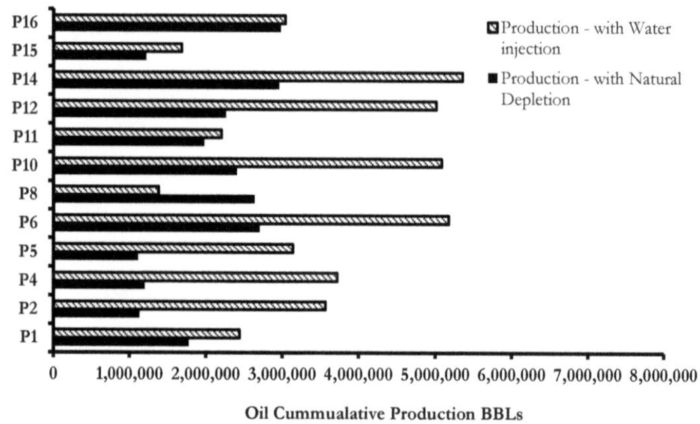

Figure 23. Total cumulative oil production bar chart for each producer under natural depletion and water flooding mechanisms.

Table 3. Comparison between cumulative oil production of each well before and after water injection.

Well #	Cumulative Oil Production—Natural Depletion	Cumulative Oil Production—Water Flooding	Difference %
P1	1.78×10^6	2.44×10^6	37.43
P2	1.13×10^6	3.57×10^6	214.89
P4	1.20×10^5	3.72×10^6	210.00
P5	1.11×10^6	3.14×10^6	182.23
P6	2.70×10^6	5.17×10^6	91.49
P8	2.63×10^6	1.38×10^6	−47.55
P10	2.40×10^6	5.08×10^6	111.28
P11	1.98×10^6	2.20×10^6	11.37
P12	2.26×10^6	5.01×10^6	121.77
P14	2.95×10^6	5.35×10^6	81.06
P15	1.21×10^6	1.68×10^6	38.42
P16	2.98×10^6	3.04×10^6	2.06
Total	2.41×10^7	4.18×10^7	73.51

3.5. Limitation and Future Work Recommendation

This presented study has a few limitations that should be considered. Firstly, it only simulates two-phase flow and does not account for other phases. Secondly, the long run time is observed for large mesh cases, particularly those with high-intensity fractures. In future research, efforts will be directed toward improving computational efficiency by employing parallel computing methods and incorporating three-phase flow equations into the model. Additionally, ongoing development aims to incorporate parameters such as thermal stresses in fractured reservoirs and the combined effects of thermal and chemical variations in fractured geothermal reservoirs.

4. Conclusions

This study presents a robust mathematical two-phase fluid flow model (FRACSIM) for the simulation of flow behavior of naturally fractured reservoirs in a 3D space. The mathematical model is based on the finite element technique and implemented using the FORTEAN language within a poro-elastic framework. Fractures are represented by triangle elements, while tetrahedral elements represent the matrix. To optimize computational time, short to medium-length fractures adopt the permeability tensor approach, while large fractures are discretized explicitly. The governing equations for poro-elasticity are discretized in both space and time using a standard Galerkin-based finite element approach.

The stability of the saturation equation solution is ensured via the application of the Galerkin discretization method. The 3D fracture model has been verified against Eclipse 100, a commercial software, via a well test case study of a fractured basement reservoir to ensure its effectiveness. Additionally, the FRACSIM successfully simulated a laboratory glass bead drainage test for two intersected fractures and accurately captured the flow pattern and cumulative production results. Furthermore, a sensitivity study of water injection using an inverted five-spot technique was tested on FRACSIM to assess the productivity of drilled wells in complex fractured reservoirs. The results indicate that FRACSIM software can accurately predict flow behavior and subsequently be utilized to evaluate production performance in naturally fractured reservoirs.

Author Contributions: Conceptualization, R.A.A. and S.A.; Methodology, R.A.A.; Software, R.A.A.; Validation, S.A.; Formal analysis, S.A. and A.A.; Investigation, R.A.A. and S.A.; Resources, A.A.; Writing—original draft, R.A.A.; Writing—review & editing, S.A.; Visualization, A.A. All authors have read and agreed to the published version of the manuscript.

Funding: This research received no external funding.

Data Availability Statement: The datasets used and/or analyzed during the current study are available from the corresponding author upon reasonable request.

Conflicts of Interest: The authors declare no conflict of interest.

References

1. Barenblatt, G.; Zheltov, I.; Kochina, I. Basic concepts in the theory of seepage of homogeneous liquids in fissured rocks [strata]. *J. Appl. Math. Mech.* **1960**, *24*, 1286–1303. [CrossRef]
2. Warren, J.; Root, P. The Behavior of Naturally Fractured Reservoirs. *Soc. Pet. Eng. J.* **1963**, *3*, 245–255. [CrossRef]
3. Kazemi, H. Pressure Transient Analysis of Naturally Fractured Reservoirs with Uniform Fracture Distribution. *Soc. Pet. Eng. J.* **1969**, *9*, 451–462. [CrossRef]
4. Snow, D.T. Anisotropie Permeability of Fractured Media. *Water Resour. Res.* **1969**, *5*, 1273–1289. [CrossRef]
5. Thomas, L.K.; Dixon, T.N.; Pierson, R.G. Fractured Reservoir Simulation. *Soc. Pet. Eng. J.* **1983**, *23*, 42–54. [CrossRef]
6. Noorishad, J.; Mehran, M. An Upstream Finite Element Method for Solution of Transient Transport Equation in Fractured Porous Media. *Water Resour. Res.* **1982**, *18*, 588–596. [CrossRef]
7. Baca, R.G.; Arnett, R.C.; Langford, D.W. Modelling fluid flow in fractured-porous rock masses by finite-element techniques. *Int. J. Numer. Methods Fluids* **1984**, *4*, 337–348. [CrossRef]
8. Pruess, K. A practical method for modeling fluid and heat flow in fractured porous media. *Soc. Pet. Eng. J.* **1985**, *25*, 14–26. [CrossRef]
9. Wu, Y.-S.; Pruess, K. A multiple-porosity method for simulation of naturally fractured petroleum reservoirs. *SPE Reserv. Eng.* **1988**, *3*, 327–336. [CrossRef]
10. Sudicky, E.A.; McLaren, R.A. The Laplace transform Galerkin technique for large-scale simulation of mass transport in discretely fractured porous formations. *Water Resour. Res.* **1992**, *28*, 499–514. [CrossRef]
11. Kim, J.G.; Deo, M.D. Finite Element, Discrete-Fracture Model for Multiphase Flow in Porous Media. *AIChE J.* **2000**, *46*, 1120–1130. [CrossRef]
12. Karimi-Fard, M.; Firoozabadi, A. Numerical Simulation of Water Injection in Fractured Media Using The Discrete-Fracture Model and The Galerkin Method. *SPEREE* **2003**, *6*, 117–126. [CrossRef]
13. Niessner, J.; Helmig, R.; Jakobs, H.; Roberts, J. Interface condition and linearization schemes in the newton iterations for two-phase flow in heterogeneous porous media. *Adv. Water Resour.* **2005**, *28*, 671–687. [CrossRef]
14. Hoteit, H.; Firoozabadi, A. Numerical Modeling of two Phase Flow in Heterogeneous Permeable media with Different capillarity Pressures. *Adv. Water Resour. J.* **2007**, *31*, 56–73. [CrossRef]
15. Wang, Q.; Chen, X.; Jha, A.N.; Rogers, H. Natural gas from shale formation–The evolution, evidences and challenges of shale gas revolution in United States. *Renew. Sustain. Energy Rev.* **2014**, *30*, 1–28. [CrossRef]
16. Wang, Q.; Li, R. Research status of shale gas: A review. *Renew. Sustain. Energy Rev.* **2017**, *74*, 715–720. [CrossRef]
17. Wang, Q.; Li, S.; Li, R.; Ma, M. Forecasting U.S. shale gas monthly production using a hybrid ARIMA and metabolic nonlinear grey model. *Energy* **2018**, *160*, 378–387. [CrossRef]
18. Wang, Q.; Li, S. Shale gas industry sustainability assessment based on WSR methodology and fuzzy matter-element extension model: The case study of China. *J. Clean. Prod.* **2019**, *226*, 336–348. [CrossRef]
19. Berre, I.; Doster, F.; Keilegavlen, E. Flow in fractured porous media: A review of conceptual models and discretization approaches. *Transp. Porous Media* **2019**, *130*, 215–236. [CrossRef]

20. Wong, D.L.Y.; Doster, F.; Geiger, S.; Francot, E.; Gouth, F. Fluid flow characterization framework for naturally fractured reservoirs using small-scale fully explicit models. *Transp. Porous Media* **2020**, *134*, 399–434. [CrossRef]
21. Kazemi, H.; Merrill, L., Jr.; Porterfield, K.; Zeman, P. Numerical simulation of water–oil flow in naturally fractured reservoirs. *Soc. Petrol. Eng. J.* **1976**, *16*, 317–326. [CrossRef]
22. Quandalle, P.; Sabathier, J.C. Typical features of a multipurpose reservoir simulator. *SPE Reserv. Eng.* **1989**, *4*, 475–480. [CrossRef]
23. Ramirez, B.; Kazemi, H.; Al-Kobaisi, M.; Ozkan, E.; Atan, S. A critical review for proper use of water/oil/gas transfer functions in dual-porosity naturally fractured reservoirs: Part I. *SPE Reserv. Eval. Eng.* **2009**, *12*, 200–210. [CrossRef]
24. Bourbiaux, B. Fractured reservoir simulation: A challenging and rewarding issue. *Oil Gas Sci. Technol.–Rev. De L'institut Français Du Pétrole* **2010**, *65*, 227–238. [CrossRef]
25. Lough, M.; Lee, S.; Kamath, J. An efficient boundary integral formulation for flow through fractured porous media. *J. Comput. Phys.* **1998**, *143*, 462–483. [CrossRef]
26. Park, Y.C.; Sung, W.M. Development of a FEM Reservoir Model Equipped with an Effective Permeability Tensor and its Application to Naturally Fractured Reservoirs. 2000. Available online: https://www.tandfonline.com/doi/abs/10.1080/00908310290086545 (accessed on 9 June 2023).
27. Gupta, A.; Penuela, G.; Avila, R. An Integrated Approach to the Determination of Permeability Tensors for Naturally Fractured Reservoirs. *J. Can. Pet. Technol.* **2001**, *40*. [CrossRef]
28. Teimoori, A.; Chen, Z.; Rahman, S.S.; Tran, T. Effective Permeability Calculation Using Boundary Element Method in Naturally Fractured Reservoirs. *Pet. Sci. Technol.* **2004**, *23*, 693–709. [CrossRef]
29. Hussain, S.T.; Rahman, S.S.; Azim, R.A.; Haryono, D.; Regenauer-Lieb, K. Multiphase fluid flow through fractured porous media supported by innovative laboratory and numerical methods for estimating relative permeability. *Energy Fuels Am. Chem. Soc. J.* **2021**, *35*, 17372–17388. [CrossRef]
30. Abdelazim, R. Integration of static and dynamic reservoir data to optimize the generation of subsurface fracture map. *J. Pet. Explor. Prod. Technol.* **2016**, *6*, 691–703. [CrossRef]
31. Kirsch, G. Theory of Elasticity and Application in Strength of Materials. *Z. Des Ver. Dtsch. Ingenieure* **1898**, *42*, 797–807.
32. Detournay, E.; Cheng, A.-D. Poroelastic response of a borehole in a non-hydrostatic stress field. *Int. J. Rock Mech. Min. Sci. Geomech. Abstr.* **1988**, *25*, 171–182. [CrossRef]
33. Fahad, M. Simulation of Fluid Flow and Estimation of Production from Naturally Fractured Reservoirs. Ph.D. Thesis, UNSW Sydney, Kensington, Australia, 2013.
34. Doonechaly, N.G.; Rahman, S. 3D hybrid tectono-stochastic modeling of naturally fractured reservoir: Application of finite element method and stochastic simulation technique. *Tectonophysics* **2012**, *541*, 43–56. [CrossRef]
35. Durlofsky, L.J. Numerical calculation of equivalent grid block permeability tensors for heterogeneous porous media. *Water Resour. Res.* **1991**, *27*, 699–708. [CrossRef]

MDPI AG
Grosspeteranlage 5
4052 Basel
Switzerland
Tel.: +41 61 683 77 34

Energies Editorial Office
E-mail: energies@mdpi.com
www.mdpi.com/journal/energies

www.ingramcontent.com/pod-product-compliance
Lightning Source LLC
Chambersburg PA
CBHW041402210926
51850CB00013B/3158